Y0-CAZ-089

THE GENE KNOCKOUT *FactsBook*

Other books in the FactsBook Series:

A. Neil Barclay, Albertus D. Beyers, Marian L. Birkeland, Marion H. Brown, Simon J. Davis, Chamorro Somoza and Alan F. Williams
The Leucocyte Antigen FactsBook, 1st edn

Robin Callard and Andy Gearing
The Cytokine FactsBook

Steve Watson and Steve Arkinstall
The G-Protein Linked Receptor FactsBook

Rod Pigott and Christine Power
The Adhesion Molecule FactsBook

Shirley Ayad, Ray Boot-Handford, Martin J. Humphries, Karl E. Kadler and C. Adrian Shuttleworth
The Extracellular Matrix FactsBook

Robin Hesketh
The Oncogene FactsBook, 1st edn

Grahame Hardie and Steven Hanks
The Protein Kinase FactsBook
The Protein Kinase FactsBook CD-Rom

Edward C. Conley
The Ion Channel FactsBook
I: Extracellular Ligand-Gated Channels

Edward C. Conley
The Ion Channel FactsBook
II: Intracellular Ligand-Gated Channels

Edward C. Conley and William J. Brammar
The Ion Channel FactsBook
IV: Voltage-Gated Channels

Kris Vaddi, Margaret Keller and Robert Newton
The Chemokine FactsBook

Marion E. Reid and Christine Lomas-Francis
The Blood Group Antigen FactsBook

A. Neil Barclay, Marion H. Brown, S.K. Alex Law, Andrew J. McKnight, Michael G. Tomlinson and P. Anton van der Merwe
The Leucocyte Antigen FactsBook, 2nd edn

Robin Hesketh
The Oncogene and Tumour Suppressor Gene FactsBook, 2nd edn

Jeffrey K. Griffith and Clare E. Sansom
The Transporter FactsBook

THE GENE KNOCKOUT *FactsBook*

Tak W. Mak
Editor-in-chief
Ontario Cancer Institute/Princess Margaret Hospital and Amgen Institute, Departments of Medical Biophysics and Immunology, University of Toronto, Toronto, Canada

Josef Penninger
Ontario Cancer Institute/Princess Margaret Hospital and Amgen Institute, Departments of Medical Biophysics and Immunology, University of Toronto, Canada

John Roder
Program in Development and Fetal Health, Samuel Lunenfeld Research Institute, Mount Sinai Hospital, Departments of Immunology and Molecular and Medical Genetics, University of Toronto, Canada

Janet Rossant
Program in Development and Fetal Health, Samuel Lunenfeld Research Institute, Mount Sinai Hospital, Departments of Molecular and Medical Genetics, and Obstetrics/Gynaecology, University of Toronto, Canada

Mary Saunders
Senior Scientist/Scientific Editor, Amgen Institute, Toronto, Canada

Academic Press
SAN DIEGO LONDON BOSTON NEW YORK
SYDNEY TOKYO TORONTO

This book is printed on acid-free paper

Academic Press
525 B Street, Suite 1900, San Diego, California 92101-4495, USA
http://www.apnet.com

Academic Press
24–28 Oval Road, London NW1 7DX, UK
http://www.hbuk.co.uk/ap/

ISBN 0-12-466044-4 (2 VOLUME SET)
ISBN 0-12-466045-2 (VOL. 1)
ISBN 0-12-466046-0 (VOL. 2)

A catalogue record for this book is available from the British Library

Typeset in Great Britain by Alden Group, Oxford.
Printed in Great Britain by WBC, Bridgend, Mid Glamorgan

98 99 00 01 02 03 WB 9 8 7 6 5 4 3 2 1

Contents

Section I INTRODUCTION

Section II THE KNOCKOUT GENES

Contents

Contents

Preface

The importance of mutant animals in furthering our understanding of genetic disease and the physiological functions of specific genes has been clear for over half a century. The ability to generate animal models by manipulating, in a predetermined manner, the germ line of animals such as rodents represents a significant advance towards these goals. Two major developments in the 1980s made it possible for biologists to genetically alter specific genes in the germ line of mice. The first was the development of methods for the culture *in vitro* of embryonic stem (ES) cells derived from mouse blastocysts. Culture conditions were discovered that maintained the propensity of these cells to give rise to fertile chimeras capable of transmitting the ES cell genome to their progeny. The second development was the devising of homologous recombination techniques which allowed the manipulation of the genome of a mammalian cell; in this case, the genome of a mouse ES cell. The combination of these two advances resulted in the creation of gene-targeted or "knockout" mutant mice in 1989. These techniques are now routinely employed in scores of laboratories, and the generation of these mutant animals has mushroomed into a veritable cottage industry. Academic laboratories as well as industrial partners have joined forces in using these animals to define the normal and patho-physiological roles of specific genes. Since the generation of the first gene-targeted mutant mice, close to one thousand different genes have been "knocked out", a number that has almost quadrupled since I was first approached to assemble a Gene Knockout Factsbook cataloging knockout mouse phenotypes. While I had initially thought that such a volume would be useful to a select group of biologists actively investigating various aspects of physiology, I now believe that such a handbook will benefit scientists engaged in many disciplines of basic and applied medical science.

The magnitude of the task of assembling the *Gene Knockout FactsBook* necessitated the combined efforts of a team of Editors, which included Josef Penninger, John Roder, Janet Rossant, Mary Saunders and myself. We have attempted to adhere to a standard format and level of content throughout this volume, although each Editor's individual stylistic flair will be apparent. In addition, the Editors soon realized that there could be no better source of information than the original creators of the mutants. Accordingly, we called for submissions by print and electronic media, and were gratifyingly deluged with responses from the scientific community. These submissions greatly eased our labors, and those individuals who contributed are listed in the Acknowledgements section of each entry in the *FactsBook*. We ask for your understanding when we note that space constraints forced us to make some editorial adjustments and to limit the number of references in some cases. In fact, the magnitude and scope of the literature describing these mutant mice has made it necessary to restrict the number of knockout phenotypes cataloged in these volumes.

In the *Gene Knockout FactsBook*, we provide a compendium of 600 knockout mouse mutants, published and some unpublished, which have proven particularly informative for understanding gene function. It is by no means a complete list of all knockouts in the literature to date, and exclusion of your favorite gene from this volume should not be construed to mean that it is not important or has not been knocked out. For updated listings of targeted mutations, you are referred to two databases on the Internet:

1. T-base: <http://www.bis.med.jhmi.edu/Dan/tbase/tbase.html>
2. Biomednet's Mouse Knockout and Mutation Database <http://biomednet.com/db/mkmd

For the gene symbols and accession numbers, we consulted databases maintained by MGD, Mouse Genome Informatics, The Jackson Laboratory (Bar Harbor, Maine, USA).

For the official gene symbols (as of 5 May 1998):
<http://www.informatics.jax.org/noforms.html>

For accession numbers (as of 5 May 1998):
<http://www.informatics.jax.org/accession.html>

A word about our editorial approach: The entries are listed alphabetically by the abbreviation of the "common name" of the protein affected by mutation of the gene. Synonyms for the common name are given as "Other names", followed by the official gene symbol for the mutated gene. The accession ID number as listed in the Jackson Laboratories Informatics database is also included. Physiologic compartments in which a major phenotypic effect was reported are listed under "Area of impact". Descriptions of the original knockout constructs are followed by the phenotype of the mouse, with differences in various mutant strains noted as appropriate. The first reference(s) in the References section are those for the original knockout constructs. In addition to these and other references that may be quoted within the text, background references that may be helpful to readers are included in the Reference lists of some entries.

Again, we gratefully acknowledge the help of the contributors, without whom this task would have been insurmountable. The enthusiastic response of the scientific community has reflected the high degree of interest in the compilation of these phenotypes under one cover. We appreciate the assistance of the Help Line of the Mouse Genome Database (MGD, Mouse Genome Informatics, The Jackson Laboratory, Bar Harbor, Maine, USA) in verifying the official gene symbols. We also thank Andrew Hessel for technical support, Jeremy Cook and Jeffrey Henderson for the preparation of some entries, and Judith Mapleson, Vera Danyliw and Susan Adlam for expert administrative assistance.

This has been an exhilarating project which has afforded us much appreciated contact with our fellow researchers around the globe. Our collective efforts cannot fail to help decipher the mysteries of the gene and its expression *in vivo*, and we are confident that our *FactsBook* with its collected wisdom will make a positive contribution in this regard.

There will undoubtedly be omissions and errors in this volume although we hope that they will be infrequent. We would greatly appreciate being informed of any inaccuracies by writing to the Editor, The Transporter FactsBook, Academic Press, 24–28 Oval Road, London NW1 7DX, UK, so that these can be rectified in future editions.

Tak. W. Mak

– on behalf of Josef Penninger, John Roder, Janet Rossant and Mary Saunders.

Left to right: John Roder, Josef Penninger, Mary Saunders and Tak Mak.

This work is dedicated to Shirley S-W Mak
and all victims of breast cancer.

IAP

Other names
Integrin-associated protein, CD47

Gene symbol
Itgp

Accession number
MGI: 96617

Area of impact
Immunity and inflammation

General description

IAP is a glycoprotein of the Ig superfamily which spans the cell membrane five times and has a short cytoplasmic tail. IAP associates with the β3 family of integrins, including αvβ3 (vitronectin). It is expressed on virtually all cells, including erythrocytes. It is also a receptor for the thrombospondin C-terminal domain.

KO strain construction

The neomycin-resistance gene was inserted into exon 2 of the *Itgp* gene, which encodes the immunoglobulin variable domain and the signal peptide cleavage site. The HSV-*tk* gene was used for negative selection against random integration. The vector was electroporated into AB2.1 ES cells, and chimeric mice were backcrossed onto the C57BL/6 background. Homozygous inactivation of the IAP gene was confirmed by Southern blot analysis and by flow cytometry.

Phenotype

IAP$^{-/-}$ mice were viable and exhibited normal gross appearance, histology and blood counts. No increase in spontaneous infection was observed. The mice exhibited decreased resistance against *E. coli* peritonitis, characterized by an early defect in PMN accumulation at the site of the infection. *In vitro,* IAP$^{-/-}$ PMNs were deficient in β3 integrin-dependent ligand binding, activation of oxidative burst, and Fc-receptor-mediated phagocytosis. β2 integrin signaling was normal. In contrast to normal neutrophils, neutrophils from IAP-deficient mice were not activated by Arg-Gly-Asp-containing extracellular matrix proteins.

Comments

IAP-deficient mice exhibit a defect in host defense, probably largely due to defects in both PMN migration to the site of infection, and PMN activation of effector responses at the site of infection. The phenotype of these mice suggests a more potent role for PMN–extracellular matrix interactions in host defense against virulent pathogens than was previously realized.

Acknowledgements
Frederik Lindberg
Washington University School of Medicine, St. Louis, MO, USA

References

[1] Lindberg, F.P. et al. (1996) Science 274, 795–798.
[2] Lindberg, F.P. et al. (1996) J. Cell Biol. 134, 1313–1322.
[3] Gao, A.G. et al. (1996) J. Cell Biol. 135, 533–544.
[4] Lindberg, F.P. et al. (1994) J. Biol. Chem. 269, 1567–1570.
[5] Lindberg, F.P. et al. (1993) J. Cell Biol. 123, 485–496.

ICAM-1

Other names
Intercellular adhesion molecule 1, CD54

Gene symbol
Icam1

Accession number
MGI: 96392

Area of impact
Immunity and inflammation, adhesion

General description

ICAM-1 is an Ig superfamily member expressed widely on hematopoietic and non-hematopoietic cells. It is the ligand for Mac-1 and LFA-1 (CD18/CD11a,b) on leukocytes. ICAM-1 expression is highly upregulated during inflammation on endothelial cells. Its expression on resting leukocytes is weak, but is increased after stimulation by inflammatory cytokines. ICAM-1 is required for the transendothelial migration of neutrophils and lymphocytes, and has been shown *in vitro* to provide costimulatory signals critical for proper T and B lymphocyte activation. The mouse ICAM-1 gene consists of seven exons: exon 1 encodes the 5′ untranslated region and signal peptide; exons 2–6 encode the five Ig-like domains; and exon 7 encodes the transmembrane and cytoplasmic domains

KO strain 1 construction[1]

The targeting vector was prepared by cloning a 5.5 kb segment containing exons 4–7 of the mouse ICAM-1 gene into pBluescript II KS(−). A neomycin-resistance cassette containing a truncated version of the promoter for RNA polymerase II and the BGH polyA signal was inserted into exon 5. The targeting vector was electroporated into the AB1 ES cell line and positive clones were injected into C57BL/6J blastocysts. Chimeric males were mated with C57BL/6J females and heterozygous offspring were intercrossed to generate ICAM-1$^{-/-}$ mutants. The null mutation was confirmed by RT-PCR of lung RNA, by immunofluorescent staining of lung tissue, and by flow cytometry of stimulated splenic B cells.

KO strain 2 construction[2]

A replacement vector was made in which a neomycin-resistance cassette under the control of the PGK promoter was inserted in exon 4 of the mouse ICAM-1 gene (originally derived from the AKR mouse strain). The targeting construct also contained the HSV-*tk* gene to allow for negative selection. The targeting vecor was electroporated into J1 ES cells and a positive clone was expanded and microinjected into C57BL/6 blastocysts to obtain chimeras. Chimeric males were bred to (C57BL/6 × DBA/2) F1 females and heterozygotes were intercrossed to generate homozygous mutants. The null mutation was confirmed by flow cytometry and immunofluorescence of mitogen-stimulated thymocytes and splenocytes, and by histochemical staining of lung sections.

Phenotype

ICAM-1-deficient mice were born at the expected Mendelian frequency and showed no gross anomalies in either development or fertility. However, mutant mice showed elevated numbers of neutrophils and lymphocytes, as well as diminished allogeneic T cell responses and decreased contact hypersensitivity to 2,4-dinitrofluorobenzene (2,4-DNFB). Neutrophil emigration in response to peritonitis induced by thioglycollate injection was reduced but not abolished, suggesting the existence of an ICAM-1-independent transendothelial migration mechanism. ICAM-1 mutant mice were resistant to the lethal effects of high doses of endotoxin, a finding that correlated with a significant decrease in neutrophil accumulation in the liver. However, levels of inflammatory cytokines (including TNFα and IL-1) produced by ICAM-$1^{-/-}$ macrophages or other leukocytes were normal, suggesting that this protection was related instead to a decrease in leukocyte–endothelial interactions. ICAM-$1^{-/-}$ mice sensitized with D-galactosamine were also resistant to the lethal effect of low doses of exotoxin (SEB), but not to low doses of endotoxin (LPS).

Comments

The phenotype of the ICAM-$1^{-/-}$ mutant is complementary to that of TNF-R1-deficient mice. The axis CD14/IL-1α/ICAM-1 seems to be required for facilitating neutrophil-mediated injury during high-dose LPS shock. ICAM-1 plays an important role in contact hypersensitivity, and is necessary for MLR stimulation, consistent with its proposed role as a critical costimulatory molecule for T cell function. The ICAM-$1^{-/-}$ mice will be important in assessing the effects of anti-ICAM-1 therapies used in various chronic inflammatory disease models.

Acknowledgements
Jose Gutierrez-Ramos
Millennium Pharma., Cambridge, MA, USA

References

1 Sligh Jr., J.E. et al. (1993) Proc. Natl Acad. Sci. USA 90, 8529–8533.
2 Xu, H. et al. (1994). J. Exp. Med. 180, 95–109.
3 Dustin M. et al. (1986) J. Immunol. 137, 245–254.
4 Siu, G. et al. (1989) J. Immunol. 143, 3813–3820.
5 Gutierrez-Ramos, J. and Bluethmann, H. (1997) Immunol. Today 18, 329–334.

ICAM-2

Other names
CD102

Gene symbol
Icam2

Accession number
MGI: 96394

Area of impact
Immunity and inflammation, adhesion

General description

ICAM-2 is an adhesion receptor of the Ig superfamily that binds to the integrin LFA-1 on leukocytes. ICAM-2 is expressed at high levels in all endothelia and at low levels in specific subsets of hematopoietic cells. ICAM-2 has been shown to be the only ICAM expressed on platelets and stromal cells. It is constitutively expressed and is not induced in response to LPS or cytokines.

KO strain construction

The targeting vector was designed such that a *neo*-resistance cassette was inserted in exon 2 of the *Icam2* gene. Exon 2 encodes the first Ig domain. The targeting construct also contained the tk gene to allow for negative selection. The targeting vector was electroporated into J1 ES cells. Several ES clones that underwent homologous recombination were expanded and microinjected into C57BL/6 blastocysts to make chimeric mice.

Phenotype

ICAM-$2^{-/-}$ mice showed decreased numbers of megakaryocyte progenitors but no differences in platelet numbers. There was impaired accumulation of eosinophils in the lung in response to allergens. Irradiation-chimera experiments showed that this phenotype was due to the lack of ICAM-2 in the endothelia. (The lack of ICAM-2 on hematopoietic cells did not affect this phenotype.) Mutant mice showed no differences in the recirculation of lymphocytes to the lymph nodes, or in basal or inflammatory situations. No differences in NK cell responses were observed.

Comments

Study of the ICAM-$2^{-/-}$ mouse should increase our understanding of the differential migration of eosinophils to sites of allergic inflammation.

Acknowledgements
Jose Gutierrez-Ramos
Millennium Pharma, Cambridge, MA, USA

References

1 The ICAM-2$^{-/-}$ mouse is not yet published.
2 Dustin, M.L. et al. (1988) J. Cell Biol. 107, 321–331.
3 Staunton, D.E. et al. (1989) Nature 339, 61–64.
4 de Fougerolles, A.R. et al. (1991) J. Exp. Med. 174, 253–267.

Other names
Interleukin-1 beta convertase, interleukin 1β (IL-1β)-converting enzyme, caspase-1, EC 3.4.22.36

Gene symbol
Casp1

Accession number
MGI: 96544

Area of impact
Apoptosis

General description

ICE processes the 31 kDa IL-1β precursor (pIL-1β) to its active 17.5 kDa form. It may also be involved in generating the active form of IL-1γ. ICE is the first member of a newly described family of cysteine proteases, termed caspases, that require aspartic acid in the P1 position of the substrate cleavage site.

KO strain 1 construction[1]

The targeting vector pICERV1 was designed to delete part of exon 6 (from the internal *Hin*dIII restriction site) and exon 7. This genomic material was replaced with a neomycin-expression cassette in the anti-sense direction. In the process of genomic recombination, the replacement vector was inserted, creating a partial gene duplication. Insertion of the replacement vector resulted in one copy of the gene that was deficient at its 3′ end (missing part of exon 6 and exons 7–10), and a second copy of the gene that was missing the promoter region and exons 1 and 2. No subsequent intrachromosomal recombination events to generate a wild-type gene were observed. Northern and Western blot analyses indicated that the ICE gene was not expressed in the cells and tissues of these mice.

The ICE knockout was generated using AB2.1 ES cells (129SvEv) and was investigated on the C57BL/6 and B10.RIII mouse backgrounds.

KO strain 2 construction[2]

The replacement-type targeting vector disrupted exon 6 of the ICE gene (from 129Sr) by the introduction of PGK*neo*. The targeting vector was electroporated into D3 Es cells. Positive clones were injected into C57BL/6 blastocysts and chimeras were crossed to C57BL/6.

Phenotype

ICE-deficient mice were viable and fertile. They produced normal litter sizes and their lifespan under SPF conditions was comparable to that of wild-type littermates. ICE$^{-/-}$ mice developed normally and the majority of apoptotic processes were unaffected, indicating that ICE is not required for programmed cell death. CTL-mediated cell killing, as well as T cell and macrophage apoptosis, occurred normally in $ICE^{-/-}$ mice. There was a modest effect on Fas ligand-induced apoptosis in these mice. In cells from wild-type mice, this process is inhibited by *crmA*, and to some extent by a tetrapeptide ICE inhibitor.

Cells from $ICE^{-/-}$ mice did not process pIL-1β via an ICE-dependent mechanism and did not secrete IL-1β. These mice lacked the 45 kDa ICE proenzyme. Cells from $ICE^{-/-}$ mice contained normal intracellular levels of pIL-1β when activated. Levels of mature IL-1β were reduced by more than 95%. In heterozygous mice, the level of IL-1β was 50% of that found in wild-type mice, indicating that mature IL-1β production is determined by the amount of ICE. However, a small amount of biologically active mature IL-1β production was still generated in $ICE^{-/-}$ mice. The production of this material was not inhibited by a potent ICE inhibitor and represented less than 1% of the mature IL-1β produced by wild-type mice.

IL-1α levels in LPS-challenged $ICE^{-/-}$ mice were also reduced by 50–90%. This observation suggests a role for ICE in the release of both IL-1α and IL-1β. In stark contrast to IL-1β-deficient mice, $ICE^{-/-}$ mice developed normal fever response in response to LPS or turpentine, but were resistant to lethal LPS-challenge. $ICE^{-/-}$ mice also developed collagen-induced arthritis (CIA) comparable to that of wild-type littermates. The development of CIA in $ICE^{-/-}$ mice could be blocked by the administration of neutralizing antibodies to murine IL-1β.

Comments

ICE is not required for programmed cell death.

Acknowledgements
John S. Mudgett and Michael J. Tocci
Merck Research Laboratories, Rahway, NJ, USA

References

[1] Smith, D.J. et al. (1997) J. Immunol. 158, 163–170.
[2] Li, P. et al. (1995) Cell 80, 401–411.
[3] Visco, D.J. et al. (1996) Orthopedic Research Association Annual Meeting.
[4] Tocci, M.J. (1997) in: Vitamins & Hormones, vol. 53 (G. Littwack, ed.). New York: Academic Press, pp. 27–63.

Other names
Interferon-γ consensus sequence binding protein

Gene symbol
Icsbp

Accession number
MGI: 96395

Area of impact
Immunity and inflammation, signaling

General description

ICSBP is a transcription factor which belongs to the interferon (IFN) regulatory factor family (IRF). Members of this family bind to IFN-stimulated responsive elements (ISRE) and are primarily involved in the regulation of type I IFN genes and genes affected by these growth factors. ICSBP is restricted to cells of the immune system. It is thought that ICSBP is a negative regulator of INFγ-stimulated genes.

KO strain construction

The targeting vector was constructed using genomic sequences from a mouse BALB/c library. The *Icsbp* gene was mutated by the insertion of the PGK*neo* cassette into the second exon, which encompasses the major part of the DNA-binding domain. The PGK*tk* cassette was appended 3′ of the homologous ICSBP sequences. The targeting vector was electroporated into E14.1 ES cells and positive clones were injected into C57BL/6 blastocysts. Chimeras were crossed to both C57BL/6 and 129Ola mice. The null mutation was confirmed by Northern analysis of splenic mRNA and immunoblotting of spleen cell extracts. A single mRNA species of less than normal size and disrupted reading frame was identified, but no ICSBP protein was detected. Heterozygotes showed a gene dosage effect on ICSBP protein levels.

Phenotype

Mice with a null mutation of the *Icsbp* gene showed defective ISRE-binding activity with reduced amounts of IRF2 protein. Development and selection of T cells appeared to be normal. Levels of IFNγ produced in spleen cells in response to ConA or LPS stimulation were reduced by 3- and 100-fold, respectively. The mutant mice were more susceptible to certain virus infections. Although they were able to mount a normal humoral response against vesicular stomatitis virus, they were defective in their ability to mount normal cytotoxic T cell responses against vaccinia virus or lymphocytic choriomeningitis virus, and often died as a result of these infections. Mutant mice also developed with 100% penetrance a disease with significant similarities to human chronic myelogenous leukemia (CML). The mice showed enhanced counts of mainly mature granulocytes, and monocytic and B lymphoid cells in spleen, lymph

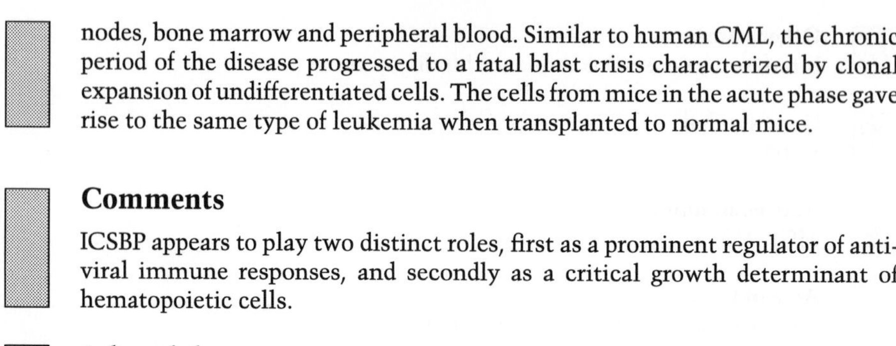

nodes, bone marrow and peripheral blood. Similar to human CML, the chronic period of the disease progressed to a fatal blast crisis characterized by clonal expansion of undifferentiated cells. The cells from mice in the acute phase gave rise to the same type of leukemia when transplanted to normal mice.

Comments

ICSBP appears to play two distinct roles, first as a prominent regulator of antiviral immune responses, and secondly as a critical growth determinant of hematopoietic cells.

Acknowledgements
Ivan Horak
Department of Molecular Genetics, Institute of Molecular Pharmacology, Berlin, Germany

References

1 Nelson, N. et al. (1996) J. Immunol. 156, 3711–3720.
2 Holtschke, T. et al. (1996) Cell 87, 307–317.
3 Fehr, T. et al. (1997) J. Exp. Med. 180, 921–931.

Other names
Interferon gamma, type 2 interferon

Gene symbol
Ifng

Accession number
MGI: 107656

Area of impact
Immunity and inflammation, cytokines

General description

IFN-γ is a cytokine secreted primarily by activated T cells and NK cells. It possesses inherent antiviral activity and has pleiotropic immunomodulatory effects on a broad range of immune system cells. IFN-γ plays a key role in the activation of macrophages and primes these cells for nitric oxide production. IFN-γ also induces the expression of MHC class II molecules on the macrophage cell surface. NK cells exposed to IFN-γ *in vivo* or *in vitro* show enhanced cytolytic activity. IFN-γ has also been implicated in the regulation of the function and proliferation of activated T cells, and has been shown to exert antiproliferative effects on some T cells and B cells. IFN-γ has been established as a crucial mediator of endotoxic shock.

KO strain construction

The targeting vector was designed to insert a 2 kb neomycin-resistance cassette into exon 2 of the IFN-γ gene, introducing a translation stop codon after the first 30 amino acids of the protein. The neomycin-resistance cassette was flanked on the 5′ side by 5 kb and on the 3′ side by 937 bp of genomic IFN-γ sequences. Exons 1 and 3 were included in the targeting vector. Two copies of the HSV-*tk* gene flanked the IFN-γ sequences on either side. The targeting vector was transfected into AB-1 ES cells and positive clones were injected into C57BL/6 blastocysts to obtain chimeras. Heterozygous progeny were intercrossed to generate homozygous IFN-$\gamma^{-/-}$ mutant mice.

To confirm inactivation of the IFN-γ gene, supernatants of mutant splenocytes cultured with ConA were assayed by ELISA: no IFN-γ protein was detected. No detectable activity was found in an IFN-γ-specific bioassay.

Phenotype

Mutant mice were born at the expected frequency and were grossly normal, healthy and fertile when kept under pathogen-free conditions. No gross or histological anomalies of any organs, including lymphoid, were observed. No alterations to spleen or thymus in terms of cell number or lymphocyte subpopulations were found in IFN-$\gamma^{-/-}$ mice. However, macrophages from BCG-infected mutant mice produced very little nitric oxide in response to LPS, a defect that could be reversed by treatment with IFN-γ. Production of reactive oxygen intermediates by macrophages of BCG-infected mutant mice was decreased but not abolished. MHC class II expression was reduced on the surfaces of IFN-$\gamma^{-/-}$ macrophages. Mutant mice infected with a sublethal dose of BCG experienced significantly increased mortality. IFN-$\gamma^{-/-}$ splenocytes treated with ConA proliferated at the same rate as wild-type cells early in the response, but showed increased proliferation late in the response and a higher number of viable cells. This effect could be reversed by the addition of IFN-γ at the start of culture. Examination of the development and function of allogeneic CTLs in mutant mice showed that IFN-$\gamma^{-/-}$ splenocytes used as responders continued to proliferate after wild type cells had ceased to do so. Proliferation of mutant cells was arrested by the addition of IFN-γ. Mutant cells used as effectors showed enhanced cytolytic activity against target cells compared to wild-type effectors. IFN-$\gamma^{-/-}$ NK cells had a significantly decreased resting NK activity at all effector-to-target ratios. However, treatment with the type I IFN inducer poly[I : C] resulted in wild-type levels of NK cell activity.

TNF failed completely to stimulate nitric oxide release in IFN-$\gamma^{-/-}$ macrophages. IFN-$\gamma^{-/-}$ mice infected with BCG and treated with LPS showed resistance to endotoxic shock lethal to wild-type mice. This resistance correlated with dramatically reduced production of TNFα in mutant mice[2].

Wild-type BALB/c mice are resistant to the induction of experimental allergic encephalomyelitis (EAE). IFN-$\gamma^{-/-}$ BALB/c mice were found to be strikingly susceptible to the induction of EAE, showing dramatic increases in CD3γ chain and TNFα mRNA levels in the spinal cord, and perivascular infiltration in the CNS[3].

Comments

IFN-γ is not essential for the development of the immune system but is required for the normal function of several types of immune system cells. Priming for the respiratory burst is not entirely dependent on IFN-γ. IFN-γ is not necessary for the generation of cytolytic effector cells but may suppress the response to allogeneic cells. Type I IFNs can replace IFN-γ for NK cell activation. TNF cannot replace IFNγ for macrophage activation. IFN-γ confers resistance to EAE in BALB/c mice.

References

1 Dalton, D.K. et al. (1993) Science 259, 1739–1742.
2 Kamijo, R. et al. (1994) Hokkaido J. Med. Sci. 69, 1332–1338.
3 Krakowski, M. and Owens, T. (1996) Eur. J. Immunol. 26, 1641–1646.

Other names
IFN gamma receptor, type 2 IFN receptor

Gene symbol
Ifngr

Accession number
MGI: 107655

Area of impact
Immunity and inflammation, cytokines

General description

IFN-γ is a pleiotropic cytokine with immunomodulatory and intrinsic antiviral activity. It stimulates macrophages and NK cells, and enhances the expression of MHC class I and II molecules. *In vitro*, $CD4^+$ T cells treated with IFN-γ differentiate into Th1 cells, and IFN-γ has been found to regulate Th1 cell-mediated immune responses required for the elimination of intracellular pathogens, including viruses. IFN-γ has also been implicated as a critical mediator of endotoxic shock. The IFN-γ receptor, which is expressed almost ubiquitously, is a 90 kDa heterodimer composed of α and β subunits. The high affinity α chain is required for signaling and mediates the specificity of IFN-γ binding.

KO strain construction

The targeting vector was designed such that the IFN-γR gene (derived from an EMBL/3 genomic library of C57BL/6 mouse brain DNA) was disrupted by the insertion of a neomycin-resistance cassette into exon 5. Exon 5 encodes the extracellular, membrane-proximal portion of the receptor, while exon 6 encodes the transmembrane region. The vector contained 10 kb of the IFN-γR gene comprising exons 5–7. Any protein translated from the intact extracellular portion of the receptor was not expected to be functional, not only because it would lack the normal cytoplasmic domain but also because the insertion was upstream of a cysteine residue considered critical for ligand binding. The targeting vector was introduced into AB-1 ES cells. Chimeric males were mated with C57BL/6 or 129SvEv females to obtain heterozygous (129SvEv × C57BL/6)F1 or 129SvEv F1 progeny. These were intercrossed to obtain IFN-$\gamma R^{-/-}$ mutant mice.

The null mutation was confirmed by the lack of binding of ^{125}I-labeled IFN-γ to mutant splenocytes, and by an assay for antiviral activity. Mutant macrophages failed to respond to IFN-γ treatment in an assay for nitrite release, but were able to respond normally to IFN-α/β.

Phenotype

Mutant mice were born at the expected frequency, were viable and showed no gross anomalies for up to 12 months. The immune system developed normally, as judged by analysis of surface markers on mutant thymocytes and splenocytes. NK cell activity in response to IFN-α/β was normal, but undetectable in response to IFNγ. A dose of *Listeria monocytogenes* that was sublethal for the wild type proved lethal for most IFN-γR$^{-/-}$ mice, and mutants also showed increased susceptibility to vaccinia virus infection. However, normal cytotoxic and T helper responses to vaccinia and lymphocytic choriomeningitis virus (LCMV) were observed *in vitro*. Total serum IgG2a concentrations were decreased in mutant mice and specific IgG2a responses were impaired.

Surprisingly, IFN-γR$^{-/-}$ mice were able to generate a Th1-type profile of antiviral cytokines in response to treatment with live attenuated pseudorabies virus. However, these mice had impaired generation of antiviral antibodies and succumbed more readily to the infection[2]. IFN-γR$^{-/-}$ mice were also more resistant to LPS-induced endotoxic shock than wild-type mice. Weight loss, lymphopenia and thrombocytopenia were all reduced, and serum TNF levels were reduced 10-fold in LPS- or LPS/D-GalN-treated mutant mice. The expression of TNF receptors was normal. Bone marrow and splenic macrophages exhibited a 4–6-fold reduction in LPS-binding and a similar decrease in CD14[3].

Comments

IFN-γR is not required for normal embryonic development, the development of the immune system or the generation of a specific immune response. Signaling through IFN-γR is necessary for the generation of a normal IgG2a response, but not for switching to this isotype. A functional IFN-γR is crucial for early host defense against certain infectious agents. Susceptibility to endotoxic shock is influenced by signaling through IFN-γR.

References

1 Huang, S. et al. (1993) Science 259, 1742–1745.
2 Schijns, V.E.C.J. et al. (1994) J. Immunol. 153, 2029–2037.
3 Car, B.D. et al. (1994) J. Exp. Med. 179, 1437–1444.

IFNAR1

Other names
Type I interferon receptor (component 1), AR 1, IFN receptor α chain; IFN-α/βR

Gene symbol
Ifnar

Accession number
MGI: 107658

Area of impact
Immunity and inflammation, cytokines

General description

The *Ifnar* gene encodes a component of the receptor for type I interferons (IFNs) which include (in humans) multiple IFN-α subtypes, IFN-β and IFN-ω. These IFNs have pleiotropic and broadly overlapping effects on target cells which may render them resistant to viral infection and may inhibit their proliferation. These IFNs also activate most effector cells of the immune system. IFNAR1 is necessary for the transduction of signals in response to type I IFNs, although its precise role in ligand binding is unclear. IFNAR1 acts together with another receptor component, designated IFNAR2, in ligand binding and signal transduction.

KO strain 1 construction[1]

The targeting construct was made using a 5.7 kb genomic DNA fragment which included exons 3–6 of the murine *Ifnar* gene isolated from a 129SvJ murine genomic library. The gene structure was similar to that of the human *IFNAR1* gene. The neomycin-resistance cassette from pMC1*neo*polyA was ligated into a unique *Sna*BI site (nucleotide 632 of the IFNAR1 cDNA) in exon 5, and an HSV-*tk* cassette was ligated to the 5′ end of intron 2. D3 ES cells were electroporated with the targeting construct and two targeted clones were isolated. Targeted clones were microinjected into BALB/c blastocysts to generate germ line chimeras, which were bred to BALB/c females to generate heterozygous mice. These were mated to produce homozygous IFNAR1$^{-/-}$ mutant mice which were housed in a conventional animal facility throughout.

The absence of expression of IFNAR1 was confirmed in homozygous mutant mice and IFNAR1$^{-/-}$ cell lines by Northern blot and RT-PCR analyses.

KO strain 2 construction[2]

The targeting vector was of the replacement type, in which a fragment of pMC1*neo*bpolyA was introduced into exon III of the IFN-α/βR (*Ifnar*) gene isolated from a 129SvEv ES cell library. An HSV-*tk* cassette was included 3′ of the *Ifnar* sequences. The targeting vector was electroporated into D3 ES cells. The null mutation was confirmed by Northern analysis of total RNA from mutant primary embryonic fibroblasts.

Phenotype

Crosses of IFNAR1$^{+/-}$ mice produced offspring of the three expected genotypes at the expected Mendelian frequency of 1:2:1. Both male and female IFNAR1$^{-/-}$ mice reproduced at the normal frequency and were morphologically normal at the macroscopic and microscopic levels. Studies of IFN signaling in primary embryo fibroblasts (PEFs) and bone marrow-derived macrophages (BMMs) showed that there was no response to IFNs α or β as judged by induction of the IFN-responsive enzyme 2′-5′ oligoadenylate synthetase (OAS), by activation of IFN-stimulated gene factor 3 (ISGF3) or gamma-activated factor (GAF) as determined by gel shifts, or by assay of antiviral or antiproliferative activities. By all of these criteria, IFNγ-mediated signaling was unaffected by the absence of IFNAR1. These studies also showed that the constitutive level of 2′-5′ OAS activity found in IFNAR1$^{+/+}$ cells was reduced in IFNAR1$^{+/-}$ cells and absent in IFNAR1$^{-/-}$ cells, consistent with the existence of a basal level of type I IFN production and the activation of response gene(s) through the cell surface receptor.

Abnormalities of hematopoietic cells were detected in IFNAR1$^{-/-}$ mice. Elevated levels (2–3-fold) of myeloid lineage cells were detected in peripheral blood and bone marrow aspirates by staining with Mac-1 and Gr-1 antibodies; increased expression of both markers was also evident in bone marrow aspirates. There were no anomalies in the major lymphocyte subsets, although cell surface IgM was slightly decreased in mutant mice. BMM derived from IFNAR1$^{-/-}$ mice showed abnormal responses to *in vitro* incubation with M-CSF, LPS and TNF. Indeed, these results demonstrated that the production of type I IFNs by BMMs was responsible for the majority of the inhibition of proliferation induced by LPS and TNF.

IFNAR1$^{-/-}$ mice showed dramatic susceptibility to acute viral infections. Neonatal mice infected with Semliki Forest virus normally survive for up to 96 hours, but IFNAR1$^{-/-}$ mice died at about 24 hours after inoculation. Viral titers examined in organs of mice killed 22 hours post-inoculation showed that there was no detectable virus in IFNAR1$^{+/+}$ mice, indicating that the type I IFN system is a major acute antiviral defense. However, in the IFNAR1$^{-/-}$ mice, significant viral replication was detected in all organs studied. Similar results were obtained with encephalomyocarditis virus, vaccinia virus and lymphocytic choriomeningitis virus (LCMV). In contrast, IFNAR1$^{-/-}$ mice were as resistant to *Listeria monocytogenes* as wild-type mice, indicating an intact IFN-γ defense system in mutant mice. No increase in the rate of spontaneous tumor incidence was observed in mutant mice of up to 6 months of age.

Comments

IFNAR1-mediated signals are not required for normal development or reproduction. IFN-α/β and IFN-γ exert their anti-viral activities through separate, non-redundant pathways. Type I IFNs play an important role in regulating hemopoietic cell proliferation, differentiation and response. This may have important implications for the control of infection and septic shock. Type I IFNs may participate in the response to other cytokines such as TNFα. IFNAR1$^{-/-}$ mice will be useful for the study of the control of viral replication.

Acknowledgements
Paul Herzog
Institute of Reproduction & Development, Monash University, Clayton, Victoria, Australia

References

[1] Hwang, S.Y. et al. (1995) Proc. Natl Acad. Sci. USA 92, 11284–11288.
[2] Müller, U. et al. (1994) Science 264, 1918–1921.
[3] Hamilton, J.A. et al. (1996) J. Immunol. 156, 2533–2557.
[4] Steinhoff, U. et al. (1995) J. Virol. 69, 2153–2158.
[5] Kalinke, U. et al. (1996) Eur. J. Immunol. 26, 2801–2806.
[6] Rousseau, V. et al. (1995) J. Interferon Cytokine Res. 15, 785–789.

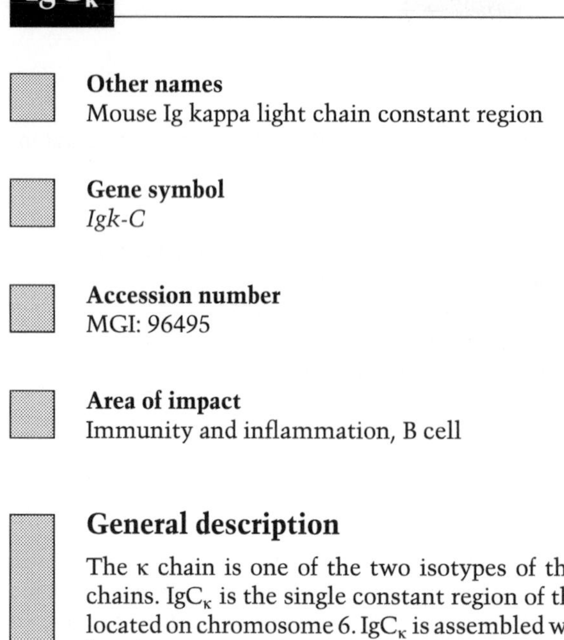

Other names
Mouse Ig kappa light chain constant region

Gene symbol
Igk-C

Accession number
MGI: 96495

Area of impact
Immunity and inflammation, B cell

General description

The κ chain is one of the two isotypes of the mouse immunoglobulin light chains. IgC$_\kappa$ is the single constant region of the mouse κ chain locus which is located on chromosome 6. IgC$_\kappa$ is assembled with a rearranged κ variable region (V$_\kappa$J$_\kappa$) by RNA splicing to generate an immunoglobulin κ chain molecule. Rearrangement and expression of the κ chain is initiated at the pre-B cell stage in the mouse bone marrow. In normal mice, the majority of B cells (95%) express the kappa light chain, whereas the rest express the λ light chain.

KO strain 1 construction[1]

The 5′ and 3′ homologous sequences in the IgC$_\kappa$ replacement targeting vector were a 3.96 kb *Sph*I/*Bsu*36I and a 1.1 kb *Bgl*II/*Bam*HI 129 strain-derived genomic fragment, respectively. Successful targeting of the mouse κ chain locus resulted in the replacement of the 0.54 kb *Bsu*36I/*Bgl*II region carrying the *Igk-C* gene with a 1.1 kb pMC1*neo*poly1A expression cassette. The targeting vector was introduced into the 129Sv strain-derived E14.1 ES cell line. The genetic background of homozygous IgC$_\kappa^{-/-}$ mutant mice was 129Sv/C57BL6J.

KO strain 2 construction[2]

The targeting vector contained 5 kb of genomic DNA, including J$_{\kappa 1-5}$ and the κ intronic enhancer (iEκ). The IgC$_\kappa$ exon and its 3′ untranslated region (containing the polyadenylation site) were replaced by a neomycin-resistance gene under the control of the HSV-*tk* promoter and the polyoma enhancer. The HSV-tk gene was situated at the 5′ end of the genomic sequences. The targeting vector was electroporated into B6III ES cells (derived from C57BL/6) and positive clones (designated CκT) were injected into CB.20 blastocysts. Chimeric males were bred to C57BL/6 females and heterozygous progeny were intercrossed to generate IgC$_\kappa^{-/-}$ mice which were entirely of C57BL/6 origin.

The null mutation was confirmed by flow cytometry of mutant spleen cells. No κ-producing cells were detected in the periphery of IgC$_\kappa^{-/-}$ mice.

Phenotype

$IgC_\kappa^{-/-}$ mice were born at the expected Mendelian frequency but were deficient in κ chain expression. The B cell population, which numbered about 50% of that found in wild-type mice, expressed the λ light chain exclusively. The number of λ-expressing B cells in homozygous and heterozygous $IgC_\kappa^{-/-}$ mice was about 7- to 10- and 2-fold higher, respectively, than that detected in wild-type mice. $V_\kappa J_\kappa$ rearrangements still occurred in $IgC_\kappa^{-/-}$ mice, although the frequency of RS (recognition sequence) rearrangement was lower than in the wild type.

Comments

Gene rearrangement in the Igλ locus can occur in the absence of gene rearrangement at Igκ. B cells expressing the λ chain can appear in surprisingly large numbers in both the bone marrow and the periphery without prior $V_\kappa J_\kappa$ or RS rearrangement.

Mice homozygous for the IgC_κ deletion are useful in evaluating the ability of the mouse λ V genes to compensate for the absence of the much larger and more diverse κ gene repertoire. The mice can also be used to assess the ability of mouse and xenogeneic unrearranged Ig κ chain constructs to rescue κ chain expression and to elucidate the regulation of κ/λ light chain rearrangement and expression. Mouse strains derived from breeding $IgC_\kappa^{-/-}$ mice with heavy chain-inactivated mice were also useful in assessing the ability of DNA fragments cloned from the human heavy and κ chain loci to reconsititute B cell development and to induce production of fully human antibodies.

Acknowledgements
Aya Jakobovits
Abgenix, Inc., Fremont, CA, USA

References

[1] Green, L.L. et al. (1994) Nature Genet. 7, 13–21.
[2] Zou, Y-R. et al. (1993) EMBO J. 12, 811–820.
[3] Mendez, M.J. et al. (1997) Nature Genet. 15, 146–156.

Other names
Immunoglobulin D

Gene symbol
Igh-5

Accession number
MGI: 96447

Area of impact
Immunity and inflammation

General description

IgD is an antigen receptor co-expressed with IgM on mature, resting B lymphocytes. IgD expression starts when B cells leave the bone marrow and stops when B cells are activated in the course of an immune response. In the mouse, IgD is found only as a membrane-bound receptor (mIgD). The function of IgD has remained elusive, although evidence has been brought forward that mIgD plays a role in protecting the B cell from tolerance induction.

KO strain 1 construction[1]

The targeting construct spanned a region from the third constant domain exon (Cδ3) to the second membrane exon (δM2). A translational stop codon and a neomycin-resistance gene were introduced into Cδ3. The HSV-*tk* gene was added to the 5′ short arm. The targeting vector was electroporated into D3 ES cells (129Sv) and positive clones were injected into C57BL/6 blastocysts to generate chimeras. Offspring of chimeras included $IgD^{-/-}$ mutant mice. The null mutation was confirmed by flow cytometric analysis of mutant spleen cells and Northern blotting of spleen cell RNA.

KO strain 2 construction[2]

The targeting vector was constructed using site-directed mutagenesis to convert the fourth codon of exon Cδ1 into a stop codon and to introduce a *Sal*I site. The neomycin-resistance gene derived from pMC1*neo*polyA was used to replace a 500 bp *Sal*I/*Pvu*II fragment containing most of Cδ1 and some 3′ flanking intron sequences. Frameshift and stop codons were introduced into Cδ1 and Cδ3 to inactivate the Ig domains, precluding the possibility of a truncated δ_H chain competing with μ_H for L chains. Cδ2 is a pseudogene. The HSV-*tk* gene was attached to the 3′ end of the downstream homology region. The targeting vector was introduced into E14.1 ES cells (129Ola-derived) and positive clones were injected into C57BL/6 blastocysts. Male chimeras were bred with C57BL/6 females and heterozygous offspring were intercrossed to generate $IgD^{-/-}$ mice.

The null mutation was confirmed by Northern blotting of splenic $polyA^+$ RNA. No transcripts capable of translation into a functional IgD or an aberrant IgM–IgD hybrid molecule were detected. Flow cytometric analysis of B cells from a variety of lymphoid tissues in mutant mice showed that no surface IgD was present.

Phenotype

IgD$^{-/-}$ mice were born at the expected Mendelian frequency and showed normal B cell development. IgD$^{-/-}$ B cells in peripheral lymphoid organs had a 2–3-fold higher density of membrane-bound IgM, and a somewhat lower expression of CD45, giving the cells an immature phenotype. CD5^{+} B cell distribution and surface expression levels of CD22 and CD23 were normal. Frequency of peripheral T cells and CD4^{+}/CD8^{+} ratios were normal. When the development and maturation of B cells with a wild-type allele and a targeted allele were compared in the bone marrow of F1 animals, the development of cells expressing the targeted allele and their pool size were normal. In contrast, the number of B cells in the periphery expressing the targeted allele (IgD negative) was half the number of the cells expressing the wild-type allele (IgD positive). This reduction in the number of peripheral IgD$^{-/-}$ B cells appeared to be predominantly caused by a loss of long-lived B cells. The reduction in the number of peripheral B cells was most evident when IgD$^{-/-}$ mice were kept under conventional, non-specific-pathogen-free housing conditions, implying that IgD$^{-/-}$ B cells fail to enter the long-lived pool when they are in competition with IgD-bearing cells, and when they are under increased pathogenic stress. BrdU incorporation by IgD$^{-/-}$ B cells was normal in KO strain 2 mice. In mice heterozygous for the KO strain 2 mutation, $\mu^{+}\delta^{+}$ B cells were preferentially recruited over $\mu^{+}\delta^{-}$ B cells in a T cell-dependent immune response.

Serum immunoglobulin levels were normal in IgD$^{-/-}$ mice with the exception of IgG2b, which was lower, and IgA, which was higher. Immune responses to T cell-independent antigens were normal, but responses to T-dependent antigens were 2–5-fold lower. Germinal center formation was normal. Affinity maturation of serum antibodies was delayed in KO strain 2 mutant mice.

Comments

The phenotype of IgD$^{-/-}$ mice is compatible with the notion that mIgD functions as a survival factor for the B cell, protecting the cell from tolerance induction and prolonging its lifespan. IgD may also or alternatively function as an antigen receptor promoting optimal recruitment of B cells during antigen-driven immune responses.

Acknowledgements
Marinus Lamers
Max-Planck-Institute für Immunobiologie, Freiburg, Germany

References

1. Nitschke, L. et al. (1993) Proc. Natl Acad. Sci. USA 90, 1887–1891.
2. Roes, J. and Rajewsky, K. (1993) J. Exp. Med. 177, 45–55.
3. Roes, J. and Rajewsky, K. (1991) Int. Immunol. 3, 1367–1371.

Other names
Immunoglobulin E

Gene symbol
Igh-7

Accession number
MGI: 96449

Area of impact
Immunity and inflammation, B cell

General description

Naive B lymphocytes use membrane-bound IgM (mIgM) and IgD as antigen receptors, but after antigen contact, they can switch and use IgG, IgA or IgE. The nature and effects of the signals generated by mIgs other than IgM and IgD are mostly unknown, but may control affinity maturation, memory induction and differentiation into plasma cells. IgE contributes least in terms of concentration to the serum immunoglobulins. Its specific function is not understood, but the IgE-triggered release of mast cell mediators in response to antigen is a well-known cause of allergic reactions.

KO strain 1 construction[1]

The replacement-type targeting vector contained 3.2 kb of genomic DNA (from 129Sv) 5′ of the Cε gene and 10 kb of 3′ homology. IgE constant region exons 1–4 were deleted and replaced with the PGK*neo* cassette. The HSV-*tk* gene was included 5′ of the Cε sequences. The targeting vector was electroporated into J1 ES cells and positive clones were injected into blastocysts. Chimeras were bred with Black Swiss mice and heterozygous progeny were crossed to generate homozygous $IgE^{-/-}$ mutants. The null mutation was confirmed by Northern analysis of RNA from IL-4/LPS-stimulated splenocytes. Mutant splenocytes failed to secrete detectable IgE into culture supernatants. Levels of IgG, IgA and IgM were normal in sera of mutant mice.

KO strain 2 construction[2]

The targeting vector included 7 kb of genomic DNA which encompassed the membrane exons M1 and M2 of the IgE gene. A base exchange was introduced into the third codon of exon M2 (TGG to TGA, resulting in W to stop), followed by a frameshift. The neomycin-resistance gene and the HSV-*tk* gene, flanked by Cre-recombinase recognition (*loxP*) sequences, were introduced 3′ of exon M2. A third *loxP* site was introduced 5′ of exon M1. Targeting vectors were introduced into BALB/c ES cells. After the primary targeting event, the ES cells were "treated" with Cre recombinase, leading to a small and a large deletion: the ΔM1M2 line lacks the transmembrane and cytoplasmic domains of IgE, while the KVKΔtail line can only express a cytoplasmic tail of three amino acid residues (KVK) identical to the cytoplasmic domain of mIgM and mIgD.

The gene disruptions were confirmed by Southern analysis and by measurement of serum IgE. Serum IgE was reduced by 94–98% in ΔM1M2 mice, and by 50% in KVKΔtail mice. Other Igs were present at normal levels in sera of both types of mutants.

Phenotype

KO strain 1 $IgE^{-/-}$ mice which were sensitized became anaphylactic upon subsequent antigen challenge. Tachycardia and changes to pulmonary function similar to those recorded for wild-type mice were observed. Mutant animals suffered less airway obstruction than did wild-type mice. Mutant mice exhibited greatly elevated plasma histamine and vascular leak, and died within minutes.

Mutant mice of both the ΔM1M2 and KVKΔ tail KO strain 2 lines appeared normal except for defects in the IgE immune response. In the ΔM1M2 line, specific IgE was barely detectable after immunization with T cell-dependent antigens. Only after a secondary immunization with the helminth *Nippostrongylus brasiliensis* was a measurable IgE response found. In the KVKΔtail line, specific IgE antibody responses to both T cell-dependent antigens and parasites were reduced by 40–80%, without a clear secondary response. The reduction in IgE levels was not due to defective switch induction, but to a reduction in the number of IgE-producing cells. Therefore, the membrane domain of IgE is indispensable for *in vivo* IgE secretion, and the cytoplasmic tail influences both the quality and quantity of the IgE response.

Comments

Although IgE is central to many allergic reactions, anaphylaxis can occur independently of its presence. Such IgE-independent hypersensitivity requires a functional immune system but does not depend on complement activation.

Signals generated via mIg may be critical at all times, not only for the maturation of antigen-specific cells, but also for their expansion. Alternatively, it may be necessary that antigen be presented to T helper cells during the entire antibody response, and that only the antigen receptor is capable of effective antigen capture for presentation.

Acknowledgements
Marinus Lamers
Max-Planck-Institute for Immunobiology, Freiburg, Germany

References

1 Oettgen, H.C. et al. (1994) Nature 370, 367–370.
2 Achatz, G. et al. (1997) Science 276, 409–411.

Other names
Insulin-like growth factor 1

Gene symbol
Igf1

Accession number
MGI: 96432

Area of impact
Hormone, metabolism

General description

Insulin-like growth factors (IGF-I and IGF-II) are related to insulin and can mediate growth and differentiation. IGF-I is a single-chain polypeptide of 70 amino acids long that affects glucose metabolism, organ homeostasis, immune responses, neurological systems and linear growth. IGFs are also involved in organ and soft tissue regeneration, skeletal muscle innervation, erythropoiesis, and the formation and breakdown of lipids. The effects of IGF-I are mainly mediated via IGF1R, though IGF-I can also bind to the insulin receptor at high concentrations.

KO strain 1 construction[1]

A neomycin-resistance cassette was inserted into exon 3 of the murine IGF-I gene corresponding to amino acid 15 of the mature protein. CMV-*tk* was placed at the 5′ end of the construct. Targeted AB.1 ES cells were injected into C57BL/6 blastocysts.

KO strain 2 construction[2]

A portion of exon 4 was replaced by a neomycin-resistance cassette which abolished the IGF-I functional domains. HSV-*tk* was placed at the 5′ end. CCE ES cells were targeted and injected into C57BL/6 and MF1 blastocysts.

Phenotype

IGF-I$^{+/-}$ mice were fertile, healthy, and displayed normal histology of organs, although they were 10–20% smaller than those of their IFG-1$^{+/+}$ littermates. In particular, muscle, bone, and organ mass were reduced in IGF-I$^{+/-}$ mice. IGF-I$^{-/-}$ mice were severely growth-deficient (60% of the normal birthweight), died perinatally (>95%) and exhibited underdeveloped muscle tissues and lungs. However, some IGF-I$^{-/-}$ mice reached adulthood, depending on their genetic background. Post-natal growth of surviving mice was significantly reduced in IGF-I$^{-/-}$ mice[3]. Adult IGF-I$^{-/-}$ mice were infertile and showed reduced bone ossification, but otherwise behaved normally and exhibited normal proportions.

References
1 Powell-Braxton, L. et al. (1993) Genes Dev. 7, 2609–2617.
2 Liu, J.-P. et al. (1993) Cell 75, 59–72.
3 Baker, J. et al. (1993) Cell 75, 73–82.

Other names
Insulin-like growth factor 2

Gene symbol
Igf2

Accession number
MGI: 96434

Area of impact
Hormone, metabolism

General description

Insulin-like growth factors (IGF-I and IGF-II) are produced in many tissues and can act as hormones and paracrine/autocrine factors. IGFs mediate growth and differentiation during development of multiple organ systems. IGF-II can bind to IGF1R, the insulin receptor, and IGF2R. IGF2R is identical to the cation-independent mannose-6-phosphate receptor (CI-MPR, IGF2/M6P), has low affinity for IGF-1, does not bind insulin, and mediates the turnover of excess IGF-II. The main receptor for the growth-promoting effects of IGF-II appears to be the insulin receptor.

KO strain construction

A neomycin-resistance cassette replaced 0.25 kb of the first coding exon. HSV-*tk* was placed at the 3′ end of the construct. CCE ES cells were electroporated and targeted clones injected into CD1, MF1 and C57BL/6 blastocysts.

Phenotype

IGF-II$^{-/-}$ mice showed severe growth retardation of both fetus and placenta during the last two-thirds of embryogenesis. These mice had normal skin and neuronal histologies and normal bone ossification. IGF-II$^{-/-}$ mice were viable and fertile although dwarfed. After birth, the growth curves of IGF-II$^{-/-}$ mice paralleled the growth of wild-type mice, implying that IGF-II acts mainly to promote growth during embryogenesis. IGF-II$^{-/-}$IGF1R$^{-/-}$ double mutant mice (30% of normal body weight) displayed a more severe phenotype than IGF-II$^{-/-}$ (60% of normal body weight) or IGF1R$^{-/-}$ (45% of normal body weight) single mutant mice[2,3] and invariably died after birth due to respiratory failure. Similarly, IGF-I$^{-/-}$IGF-II$^{-/-}$ mice (30% of normal birthweight) displayed a more severe growth retardation than single IGF-I$^{-/-}$ (60% of normal birthweight) and IGF-II$^{-/-}$ (60% of normal birthweight) mice and these double mutant mice died shortly after birth due to respiratory failure[2,3].

Comments

IGF-II is subjected to imprinting; that is, the paternal IGF-II is expressed and the maternal allele is silent in most tissues. Thus, IGF-II$^{+/-}$ mice that inherited the mutant allele from the father are indistinguishable from IGF-II$^{-/-}$ mice[4].

References

[1] De Chiara, T.M. et al. (1990) Nature 345, 78–80.
[2] Liu, J.-P. et al. (1993) Cell 75, 59–72.
[3] Baker, J. et al. (1993) Cell 75, 73–82.
[4] De Chiara, T.M. et al. (1991) Cell 64, 849–859.

IGF1R

Other names
IGF-I receptor

Gene symbol
Igf1r

Accession number
MGI: 96433

Area of impact
Hormone, metabolism

General description

IGF1R is a transmembrane disulfide-linked tetraheteromeric ($\alpha2\beta2$) glycoprotein receptor that contains an extracellular ligand-binding and a cytoplasmic tyrosine kinase domain. IGF1R binds IGF-I with higher (15–20 times) affinity than IGF-II. It is widely expressed in embryonic tissues but expression declines post-natally.

KO strain construction

Parts of exon 3 were deleted by insertion of a neomycin-resistance cassette into the mouse *Igf1r* gene. HSV-*tk* was placed at the 3′ end. CCE ES cells were targeted and injected into C57BL/6 and MF1 blastocysts.

Phenotype

IGF1R$^{+/-}$ mice are fertile, healthy, and do not display any apparent phenotype as compared to IFG-I$^{+/+}$ littermates. IGF1R$^{-/-}$ mice were severely growth deficient (45% of the normal birthweight), and died invariably at birth due to respiratory failure. IGF1R$^{-/-}$ mice exhibited general organ hypoplasia, muscle hypoplasia, delayed bone ossification, and changes in the CNS (reduced numbers of O1^{+} oligodendrocyte precursors) and skin (lower number of hair follicles and thin stratum spinosum).

Comments

IGF-I$^{-/-}$IGF1R$^{-/-}$ double mutant mice do not differ in phenotype from IGF1R$^{-/-}$ single mutant mice, indicating that IGF1R is the principal receptor mediating the effects of IGF-I.

References
[1] Liu, J.-P. et al. (1993) Cell 75, 59–72.
[2] Baker, J. et al. (1993) Cell 75, 73–82.

Other names
Cation-independent mannose-6-phosphate receptor, CI-MPR, IGF2/M6P, MPR300, Igf2/Mpr

Gene symbol
Igf2r

Accession number
MGI: 96435

Area of impact
Hormone, metabolism

General description

Insulin-like growth factors (IGF-I and IGF-II) mediate growth and differentiation during development of multiple organ systems. IGF-II can bind to IGF1R, the insulin receptor and IGF2R. IGF2R is identical to the cation-independent mannose-6-phosphate receptor (CI-MPR, IGF2/M6P), has low affinity for IGF-I, does not bind insulin, and mediates the turnover of excess IGF-II. The main receptors for the growth-promoting effects of IGF-2 appear to be the insulin receptor and IGF1R. IGF2R is a single-chain polypeptide of 300 kDa that has no tyrosine kinase activity. The main function of IGF2R is to bind to the phosphomannosyl recognition marker of lysosomal hydrolases to target molecules to lysosomes. IGF2R is imprinted and expressed from the maternal allele.

KO strain 1 construction[1]

A 0.33 kb 5 flanking region fragment and 38 codons of exon 1 were replaced with a neomycin-resistance cassette. HSV-*tk* was placed at the 3′ end of the construct. Targeted CCE-32 ES cells were injected into C57BL/6 blastocysts.

KO strain 2 construction[2]

A 96 bp fragment of exon 1 was replaced by a *lacZ* reporter gene and an RSV-neomycin-resistance cassette. HSV-*tk* was placed at the 3′ end. Targeted D3 ES cells were injected into C57BL/6 blastocysts.

Phenotype

IGF2R$^{-/-}$ mice had increased serum and tissue levels of IGF-II and displayed overgrowth (135% of normal birthweight). All organs were increased in size. These mice had kinky tails, post-axial polydactyly, enhanced bone ossification, split sternum, shortened facial long axis, heart defects, and edemas. Most mice died after birth. Depending on the genetic background, some IGF2R$^{-/-}$ mice survived and even reproduced. These mice also missorted mannose-6-phosphate-tagged proteins.

Comments

The phenotype of IGF2R$^{-/-}$ mice is rescued by introduction of an IGF-II$^{-/-}$ or IGF1R-null mutation, but triple IGF2R$^{-/-}$IGF1R$^{-/-}$IGF-II$^{-/-}$ mice are non-viable dwarfs (30% of normal size). Moreover, IGF-II$^{-/-}$IGF2R$^{-/-}$ and CD-MPR$^{-/-}$ (encoding the cation-dependent mannose-6-phosphate receptor) triple mutants survive at a low frequency within the first weeks after birth, implying that mannose-6-phosphate-regulated lysosomal molecule trafficking is essential for survival[1].

References

[1] Ludwig, T. et al. (1996) Dev. Biol. 177, 517–535.
[2] Wang, Z.-Q. et al. (1994) Nature 372, 464–467.
[3] Lau, M.M.H. et al. (1994) Genes Dev. 8, 2953–2963.

IGFBP-2

Other names
IGF-binding protein-2, BP-2

Gene symbol
Igfbp2

Accession number
MGI: 96437

Area of impact
Hormone, metabolism

General description

Insulin-like growth factors (IGF-I and IGF-II) are produced in many tissues and can act as hormones and paracrine/autocrine factors. IGFs mediate proliferation and differentiation during development of multiple organ systems. They circulate through the plasma in association with IGF-binding proteins (IGFBPs). Six IGFBPs exist which are encoded by six different genes. IGFBPs form binary complexes with other IGFBPs or trimeric complexes with IGFBP3 and an 88 kDa subunit. Circulatory IGFBPs may inhibit IGF actions, whereas tissue IGFBPs may either inhibit or enhance IGF actions. IGFBPs can modulate binding of IGFs to their receptors *in vitro*. All IGFBPs are expressed during embryogenesis. In adults, IGFBPs display tissue-specific expression. The expression level of IGFBP-2 parallels that of IGF-II in fetal development.

KO strain construction

The targeting vector was a replacement type construct in which *neo* replaced almost all of exon 3. The mutation was expected to abolish IGF binding to any fusion or truncated protein that could be made *in vivo*. The HSV-*tk* gene was included at the 3′ end. ES cells (129SvEv) were electroporated with the vector and C57BL/6 blastocysts were microinjected with targeted ES cells.

Phenotype

IGFBP-2$^{-/-}$ mice were fertile and viable. However, they had spleens of decreased size (30% reduction compared to normal) and increased levels of other circulating IGFBPs in the serum.

References

1. Wood, T.L. et al. (1993) Growth Regul. 3, 5–8.
2. Pintar, J.E. et al. (1996) Horm. Res. 45, 172–177.

IgH intronic enhancer

Other names
Immunoglobulin heavy chain intron enhancer, iEμ, Eμ

Gene symbol
Igh (intronic enhancer)

Accession number
MGI: 96442

Area of impact
Immunity and inflammation, B cell

General description

IgH intronic enhancer (iEμ) is located in the heavy chain locus in the intron between the J_H gene segments and exon 1 of the Cμ constant region. It is important for driving IgH transcription during B cell maturation, and this transcription is thought to be necessary for the initiation of VDJ recombination at the IgH locus.

KO strain 1 construction[1]

An insertion-type vector was used for the targeted deletion of the IgH intronic enhancer (iEμ) via the "Hit and Run" procedure. This resulted in a mutated IgH locus in which the entire 1 kb *Xba*I fragment encompassing the iEμ region was deleted and replaced by a short diagnostic oligonucleotide. This approach made it possible to analyze the effect of the removal of iEμ without introducing heterologous transcriptional enhancer elements in the *neo* or HSV-*tk* cassettes. The targeting vector was electroporated into D3 ES cells (129Sv) and a positive clone was injected into C57BL/6 blastocysts to produce chimeric animals. Spleen cells were prepared from the chimeras and the mutant ES cell (129Sv)-derived $sIgD^{a+}$ B cells were isolated by cell sorting utilizing magnetic and FACS techniques. The rearrangement status of the genomic IgH locus was subsequently analyzed by PCR.

KO strain 2 construction[2]

The targeting vector was of the replacement type, in which the entire 1 kb *Xba*I-*Xba*I fragment containing the core Eμ (enhancer) sequence and the flanking MARs (matrix-associated regions) was replaced by PGK*neo*. The HSV-*tk* selection cassette was added to the 3′ end of the targeting vector. The targeting construct was electroporated into CCE ES cells. Positive clones were injected into C57BL/6J blastocysts to generate chimeras. $E\mu^{-/-}$ pre-B cell lines were established from chimeric mice by Abelson murine leukemia virus (A-MuLV) transformation.

Phenotype

VDJ recombination in the B cells was impaired but not blocked by targeted deletion of iEμ. Quantitative PCR analyses demonstrated that about 15–30% of the mutated loci in mature B cells were unrearranged, in striking contrast to the wild-type alleles (<2% unrearranged). The remainder of the mutated loci underwent D-to-J (65–80%) as well as V-to-DJ rearrangements, albeit the latter less frequently (3–6%).

Comments

The "Hit and Run" procedure was used to delete iEμ in KO strain 1 to ensure that no selection marker gene was left in the targeted locus. This was of particular importance, since it has been shown that the presence of a *neo* gene cassette can affect the mutant phenotype[2]. The intronic enhancer is important for normal IgH VDJ recombination, but rearrangement can occur in its absence, albeit at reduced frequency. These data also argue for the presence of other *cis* elements contributing to the initiation of VDJ recombination.

Acknowledgements
Fred Sablitzky
Max-Delbrück-Laboratorium, Cologne, Germany

References

[1] Serwe, M. and Sablitzky, F. (1993) EMBO J. 12, 2321–2327.
[2] Chen, J. et al. (1993) EMBO J. 12, 4635–4645.
[3] Hasty, P. et al. (1991) Nature 350, 243–246.
[4] Valancius, V. and Smithies, O. (1991) Mol. Cell. Biol. 11, 1402–1408.

IgH joining region plus intronic enhancer

Other names

Joining region and intronic enhancer of the Ig heavy chain locus, J_H–iEμ region

Area of impact

Immunity and inflammation, B cell

General description

Isotype switching is a gene rearrangement process that occurs in B cells following antigen activation. The IgH locus DNA is looped out such that the $V_HD_HJ_H$ variable region is brought into proximity to either the γ, α or ε constant region genes, deleting the μ and δ constant region genes. Transcription then results in the production by members of the B cell clone of Igs of differing isotypes. Isotype switching results from "switch recombination" between defined "switch region" sequences (S regions) positioned upstream of each of the constant region genes except Cδ. The alignment of two S regions on the same chromosome permits the looping out and deletion of the intervening DNA. Isotype switching is influenced by cytokines, particularly by IL-4 which promotes switching from μ to γ1 and ε. It is thought that IL-4 stimulates the transcription of germ line (sterile) transcripts from the Cγ1 or Cε genes which are necessary to make the S region of Cγ or Cε accessible to elements of the switch recombination mechanism.

KO strain construction

The targeting vector used the Cre-*loxP* recombination system to create a deletion that included the sequences of the joining region (J_H) and the intronic enhancer (iEμ) of the Ig heavy chain. The mutation was designated as J_HT. Two *loxP* sites were introduced to bracket a 1.3 kb genomic fragment containing the J_H-iEμ locus. HSV-*tk* and neomycin-resistance expression cassettes were incorporated 3′ of the Ig genomic sequences, just 5′ of the second *loxP* site. The targeting vector was electroporated into E14-1 ES cells. Positive clones were transfected with supercoiled Cre-encoding plasmid DNA by electroporation to cause deletion of the J_H-iEμ region. ES cells carrying the J_HT mutation were injected into C57BL/6 blastocysts and male chimeric progeny were mated to C57BL/6 females to obtain heterozygotes. Heterozygotes were intercrossed to obtain J_HT mice.

To confirm the functional deletion of the J_H-iEμ region (which is necessary for functional Ig molecule assembly), the B cell compartment of mutant mice was examined by flow cytometry. The J_HT mutation was expected to completely abolish functional B cell generation. Examination of bone marrow cells of homozygous J_HT mice showed that B cell development was indeed blocked at the pre-B cell stage. Mice heterozygous for the J_HT mutation had only B cells that expressed IgM^b from the wild-type allele.

Phenotype

Homozygous mutant mice were born at the expected frequency and had no overt phenotype. B cell development was completely blocked at the pre-B cell stage and no functional B cells were present in mutant mice. Analysis of isotype switch recombination in heterozygous splenic B cells (containing one-wild type and one inactivated J_HT chromosome) activated by LPS plus IL-4 treatment showed that switch recombination at the Sμ region on the J_HT chromosome was strongly suppressed (but not abolished). In contrast, switch recombination at the Sγ1 region on the J_HT chromosome in response to LPS plus IL-4 was carried out with normal efficiency. S regions of downstream constant region genes did not appear to be involved in switch recombination at Sγ1. No Cμ transcripts from the J_HT chromosome were detected in heterozygous B cells. In B cells from mutant mice, DNA methylation was increased at the H1 site close to the Sμ region of the J_HT chromosome.

Comments

The J_HT mutant mouse furnishes an ideal animal model of complete functional B cell deficiency. The $V_HD_HJ_H$-iEμ region of the IgH locus is important for efficient switch recombination at the Sμ region. Sγ1 switch recombination is independent of switching at Sμ and is not influenced by the Eμ intronic enhancer. The $V_HD_HJ_H$-iEμ region may be involved in control of the methylation of the Sμ region in B cells. Demethylation may be necessary to confer accessibility to the Sμ region for switch recombination.

Reference

[1] Gu, H. et al. (1993) Cell 73, 1155–1164.

Other names
Mouse immunoglobulin heavy chain joining region

Gene symbol
Igh-J

Accession number
MGI: 96461

Area of impact
Immunity and inflammation, B cell

General description

J_H is one of the three multigene DNA segments (V_H, D_H and J_H) on the mouse immunoglobulin (Ig) heavy chain locus on chromosome 12. Heavy chain variable regions ($V_HD_HJ_H$) are formed by sequential recombination of one of the four mouse J_H segments with one of the D_H segments. Functional recombination and expression of Ig heavy chain genes precedes light chain rearrangement, antibody production and B cell differentiation.

KO strain 1 construction[1]

The 5′ and 3′ homologous sequences in the J_H replacement targeting vector pmHΔJ were BALB/c-derived genomic fragments of 2.68 kb *Eco*RI/*Xho*I and 1.12 kb *Nae*I/*Eco*RI respectively. Successful targeting of the mouse heavy chain locus resulted in the replacement of the 2.29 kb *Xho*I/*Nae*I region containing the DQ52 and the four J_H segments with a 1.15 kb pMC1*neo*polyA expression cassette. The targeting vector was electroporated into 129Sv-derived E14TG2a ES cells. Positive clones were injected into C57BL/6 blastocysts and chimeric mice were generated. The genetic background of homozygous $J_H^{-/-}$ mutant mice was therefore 129/C57BL6J.

KO strain 2 construction[2]

The targeting vector was of the replacement type and resulted in a 2.3 kb deletion which included the four J_H segments and the DQ52 segment. These were replaced by the PGK*neo* expression cassette. The PGK*tk* cassette was included 5′ of the genomic sequences for counterselection. The targeting vector was electroporated into AB-1 ES cells and positive clones were injected into C57BL/6 blastocysts to generate chimeras. Chimeras were crossed with C57BL/6J females and heterozygous progeny were intercrossed to obtain $J_H^{-/-}$ homozygotes. The null mutation was confirmed by Southern analysis.

Phenotype

The deletion of the J_H region resulted in the disruption of mouse heavy chain rearrangement. Mice homozygous for the J_H deletion were deficient in mouse antibody production and were completely devoid of mature B cells in the bone marrow and in the periphery. Precursor B cells in the bone marrow of homozygous mutant mice accumulated at the pro-B/Pre-BI ($B220^-$, μ^-, c-kit^+, $CD25^-$) stage or in fraction B ($B220^{lo}$, $CD43^+$, HSA^{med}, BP-1^-), indicating an arrest of B cell development at the point at which heavy chain rearrangement is initiated. However, these precursor B cells were able to assemble κ light chains at a low level in the absence of μ_H chains. T cell development and function were not affected by the J_H deletion.

Comments

The phenotype of the $J_H^{-/-}$ mice proves that heavy chain gene rearrangement is a prerequisite for antibody production and B cell development, but not for κ light chain rearrangment *in vivo*.

Homozygous $J_H^{-/-}$ mice, being B cell-deficient, are useful for the investigation of the role of B cells in the induction of immune responses upon infection and immunization, and in the development of autoimmune diseases. In addition, the lack of heavy chain rearrangement provides a null background for assessing the ability of mouse and xenogeneic unrearranged Ig heavy chain constructs to reconstitute B cell development and antibody production. Homozygous $J_H^{-/-}$ mutant mice, when bred with mice deficient in mouse κ light chain expression, were suitable for assessing the ability of DNA fragments cloned from the human heavy and κ chain loci to reproduce a human antibody response in mice, and to produce fully human antibodies[3,4].

Acknowledgements
Aya Jakobovits
Abgenix, Inc., Fremont, CA, USA

References

1 Jakobovits, A. et al. (1993) Proc. Natl Acad. Sci. USA 90, 2251–2555.
2 Chen, J. et al. (1993) Int. Immunol. 5, 647–656.
3 Green, L.L. et al. (1994) Nature Genet. 7, 13–21.
4 Mendez, M.J. et al. (1997) Nature Genet. 15, 146-156.

Igκ intronic enhancer

Other names
Intronic enhancer in the kappa light chain locus, iEκ

Gene symbol
Igk (intronic enhancer)

Accession number
MGI: 96494

Area of impact
Immunity and inflammation, B cell

General description

A mature plasma cell secretes immunoglobulins containing either κ or λ light chains but not both, a phenomenon called haplotype or κ-λ exclusion. In κ-producing B cells, the λ chain genes are usually found in germ line configuration, while the κ genes in λ-producing B cells are either deleted or non-productively rearranged. The κ locus contains two enhancers: the Igκ 3′ enhancer, located 10 kb downstream of the C_κ exon; and the Igκ intronic enhancer (iEκ), located between J_κ and C_κ. The intronic enhancer is thought to be sufficient for Igκ gene transcription in mature B cells.

KO strain construction

The targeting vector was designed such that positions 69–504 of iEκ were replaced with a neomycin-resistance gene controlled by the rabbit β-globin promoter plus an enhancer element containing four tandem repeats of the GT-IIC + GT-I motif. This sequence is active in ES cells but apparently not in lymphoid tissues. Elements missing from the mutated iEκ were the octamer site, the NFκB binding site, κE1, κE2, κE3, and the kappa silencer. The HSV-*tk* gene was included at the 3′ end of the genomic sequences. The targeting vector was electroporated into D3 ES cells and positive clones (designated iEκT) were injected into C57BL/6 blasocysts. Chimeric offspring were crossed with C57BL/6 females to obtain heterozygous mice, which were intercrossed to obtain homozygous iEκT mutant mice.

Northern blot analysis of mutant spleens showed mRNAs much larger than wild-type C_κ transcripts that hybridized to both C_κ and *neo* probes, as expected. Lambda chain transcripts were present in increased abundance.

Phenotype

iEκT mutant mice were born at the expected ratio and had a similar phenotype to that of $IgC_\kappa^{-/-}$ mice (mice lacking the κ chain constant region) in some respects. iEκT mice produced about half the wild-type number of B cells in bone marrow, characterized by a 7-fold increase in those cells expressing λ chains. In mice heterozygous for the iEκT mutation, $V_\kappa J_\kappa$ rearrangement occurred only in the normal allele, and no $V_\kappa J_\kappa$ rearrangement occurred in homozygous iEκT mice. There was little, if any, RS rearrangement in iEκT mice. This stands in contrast to the case of $IgC_{\kappa}^{-/-}$ mice, which showed somewhat decreased RS rearrangement. In homozygous iEκT mice, splenic B cells were only slightly reduced in number. The surface Ig on B cells was exclusively the λ chain, indicating that κ rearrangement is not required for λ gene rearrangement. Mice homozygous for another mutated allele in which *neo* was inserted into the 3′ end of the intronic κ enhancer showed impaired, but not abolished, $V_{\kappa}J_{\kappa}$ recombination.

Comments

iEκ is essential for $V_{\kappa}J_{\kappa}$ recombination but Igλ rearrangement can occur in the absence of prior $V_{\kappa}J_{\kappa}$ and RS rearrangements.

References

[1] Takeda, S. et al. (1993) EMBO J. 12, 2329–2336.

[2] Zou, Y.-R. et al. (1993) EMBO J. 12, 811–820.

IgM transmembrane exon

Other names
Immunoglobulin μ gene (membrane), mIgM, surface IgM, sIgM

Area of impact
Immunity and inflammation, B cell

General description

Immunoglobulin μ heavy chains are expressed either as a secreted form in association with light chain (IgM), or as a membrane-bound form, also associated with light chain (sIgM). The transmembrane and intracytoplasmic portions of sIgM are encoded by two exons, and expression of sIgM vs. secreted IgM is controlled by differential RNA processing. In addition to IgM and sIgM, the μ heavy chain is also expressed on the cell surface of developing B cells in association with surrogate light chain, which is encoded by the λ5 and $V_{pre\text{-}B}$ genes. The surrogate L chain/μ H chain pre-B cell receptor complex is thought to be important in promoting B cell differentiation through the pre-B cell stage to mature, naive B cells.

KO strain construction

A 9 kb genomic DNA fragment originating from BALB/c mouse liver, containing the membrane exons of the μ heavy chain gene and the δ constant region gene, were used to create the targeting vector. The neomycin-resistance gene was inserted near the 5′ end of the first membrane exon, resulting in a stop codon being generated at the third codon of the exon just upstream of the *neo* insertion site. The HSV-*tk* gene was placed at the 3′ end of the construct for negative selection. D3 ES cells were electroporated with the targeting vector, and positive clones were injected into C57BL/6 blastocysts to generate chimeric animals. The successful targeting of the IgM μ chain transmembrane exon (the mutation was designated as μMT) was confirmed by Southern blot analysis and by flow cytometry. The original mutant mouse strain had a mixed background of 129Sv × C57BL/6, but this mouse has since been backcrossed to C57BL/6 mice and is available from the Jackson Laboratory.

Phenotype

B cell development in μMT/μMT mice was arrested at the large pre-B cell stage, resulting in a severe decrease in the number of small pre-B cells, as well as a complete lack of B cells. B cells were also absent in the periphery. The developmental defect appeared to stem from the absence of cell proliferation at the end of the large pre-B cell stage. Some limited levels of light chain rearrangement (20-fold less than wild-type mice) did take place prior to this developmental block, probably due to rare, early light chain rearrangement in some large pre-B cells. It was shown in heterozygotes that allelic exclusion of the μ heavy chain locus was abrogated in the absence of membrane-bound μ heavy chain. Membrane-bound Dμ protein was also demonstrated, produced from a heavy chain allele with D_HJ_H rearranged in reading frame II, signaling the arrest of pro-B cell development.

Comments

The membrane-bound μ heavy chain plays an important role in the progression of developing B cells through the pre-B cell stage, including the efficient initiation of light chain gene rearrangement. It is also necessary for allelic exclusion of the second heavy chain allele. The B cell-deficient mice described above have also been used by several researchers to study the requirement for B cells and/or antibodies in effective immunity to various pathogens or antigens.

Acknowledgements
Daisuke Kitamura
Research Institute for Biological Sciences, Science University of Tokyo, Noda City, Chiba, Japan

References

[1] Kitamura, D. et al. (1991) Nature 350, 423–426.
[2] Ehlich, A. et al. (1993) Cell 72, 695–704.
[3] Kitamura, D. et al. (1992) Nature 356, 154–156.
[4] Gu, H. et al. (1991) Cell 65, 47–54.
[5] Beutner, U. et al. (1994) J. Exp. Med. 179, 1457–1466.
[6] Brundler, M.-A. et al. (1996) Eur. J. Immunol. 26, 2257–2262.

Other names
Invariant chain

Gene symbol
Ii

Accession number
MGI: 96534

Area of impact
Immunity and inflammation, T cell

General description

Ii is involved in antigen presentation by class II molecules of the major histocompatibility complex. It is a highly conserved glycosylated type II membrane protein encoded by a gene unlinked to the MHC class II locus. Two evolutionarily conserved isoforms of Ii exist: the p31 and p41 proteins, which are derived from alternatively spliced mRNAs. After synthesis in the rER, MHC class II α/β heterodimers are associated with pre-existing invariant chain multimers, targeting the class II molecules from the ER to antigen-processing compartments of the endocytic route. The invariant chain acts as a chaperone and blocks the peptide-binding groove of MHC molecules during this transport. The Ii and MHC $\alpha\beta$ heterodimers form a nonamer structure before exiting from the ER. The $(\alpha\beta\text{Ii})_3$ complex is transported through the Golgi to an acidic compartment, where Ii is proteolytically cleaved and the MHC class II heterodimers are released for transport to the cell surface. The invariant chain is essential for normal MHC class II function *in vivo*.

KO strain 1 construction[1]

The targeting vector was designed such that a neomycin-resistance cassette replaced a 1.5 kb genomic *Nde*I fragment that included the first two protein-coding domains (exons 2 and 3) of the eight-exon Ii gene. No *lox* sites or HSV-*tk* selection cassette were included in the construct. Since exons 2 and 3 encode the transmembrane region, and any alternative splicing joining exon 1 to exon 4 would result in an mRNA containing a frameshift-preventing translation, the Ii gene was expected to be completely inactivated. The targeting vector was electroporated into the D3 ES cell line (129 background). Positive clones were injected into blastocysts to generate chimeras. Heterozygous offspring of chimeras were intercrossed to obtain $\text{Ii}^{-/-}$ mice, which subsequently have been backcrossed onto several inbred backgrounds, primarily C57BL/6 and B10.Br. Both are available at repositories.

The null mutation was confirmed by cytofluorometric analysis of permeabilized cells; immunohistological analysis of thymic, lymph node and spleen sections; and, immunoprecipitation of $[^{35}\text{S}]$methionine-labeled proteins. No form of Ii was detected in $\text{Ii}^{-/-}$ mice.

KO strain 2 construction[2]

The targeting vector was designed such that homologous recombination would result in the deletion of 3.8 kb of the eight-exon Ii gene and the replacement of the first intron and 11 nucleotides of exon 2 by the MC1*neo*polyA neomycin-resistance cassette. The HSV-*tk* gene was included at the 5′ end of the construct. The targeting vector was electroporated into CCE ES cells and positive clones were injected into MF1 or C57BL/6J blastocysts to generate chimeras. Male chimeras were mated with C57BL/6J females and heterozygous progeny were intercrossed to obtain $\text{Ii}^{-/-}$ mutants.

The null mutation was confirmed by cytoplasmic staining of permeabilized mutant spleen cells; biosynthetic labeling and immunoprecipitation; and RNAase protection assays.

KO strain 3 construction[3]

The targeting vector contained a 5′ fragment of 2.1 kb encoding the promoter and exon 1 of the Ii gene, and a 3′ fragment of 3.0 kb encoding exons 2 and 3. An internal 400 bp *Stu*I fragment was deleted from the 5′ fragment and replaced with the neomycin-resistance cassette. Two copies of the HSV-*tk* gene were included 5′ of the Ii sequences. The targeting vector was electroporated into D3 ES cells and positive clones were injected into C57BL/6 blastocysts. Heterozygous progeny of chimeric mice were intercrossed to obtain homozygous $\text{Ii}^{-/-}$ mutant mice.

The null mutation was confirmed by immunofluorescence confocal microscopy of LPS-treated splenocytes. No Ii expression was detected either on the plasma membrane or in any intracellular compartments in mutant mice.

Phenotype

$\text{Ii}^{-/-}$ mice were normally healthy and fertile, and showed no gross anomalies of either anatomy or development. However, mutant spleen cells showed poor assembly of MHC class II heterodimers, aberrant transport and low cell surface levels of class II with improper biochemical maturation. MHC class II molecules in mutants lacked a typical compact conformation and behaved in peptide-binding experiments as if empty or occupied by a peptide that could easily be displaced. Poor antigen presentation (for most antigens) was observed, as well as deficient negative selection of $CD4^+$ T cells, although this was found to vary with T cell specificity. Numbers of $CD4^+CD8^-$ T cells were significantly reduced in the thymus and periphery. No phenotype was observed in $\text{Ii}^{+/-}$ heterozygotes.

Double label confocal and electron microscopy were used to analyze cells of KO strain 3 $\text{Ii}^{-/-}$ mice to determine the precise site of MHC α and β chain accumulation. Although the α and β chains were capable of significant assembly and folding in the absence of Ii, they were mistakenly identified as "misfolded proteins" and relegated to subcompartments of the ER/Golgi system with the putative function of collecting those proteins which are incorrectly folded. Class II molecules in $\text{Ii}^{+/-}$ mice were not terminally glycosylated in the absence of Ii, consistent with a lack of passage through the Golgi[3].

Isoform knockout mutants

The p41 isoform of Ii differs from the p31 isoform in the presence of the p41-specific exon 6b, which encodes a 64 amino acid cysteine-rich thyroglobulin-like segment. This region lies immediately adjacent to the C-terminal residues involved in homotrimer formation. "Hit-and-run" gene targeting was used to generate Ii p31$^{-/-}$[4] and Ii p41$^{-/-}$[5] mice. Mice expressing only the p31 isoform had spleen cells that were indistinguishable from the wild type with respect to MHC class II assembly, transport, peptide acquisition, surface expression, and the ability to present intact protein antigens. Normal numbers of thymic and peripheral CD4^{+} T cells were observed, and CD4^{+} T-dependent proliferative responses to soluble antigen were normal.

Unlike Ii p41$^{-/-}$ mice, Ii p31$^{-/-}$ mice exhibited a small population of MHC class II molecules which had reduced mobility on SDS-PAGE and appeared to be incompletely processed. However, class II surface expression, peptide occupancy, CD4^{+} T cell maturation and proliferative responses to intact protein antigens were normal. Spleen cells of both isoform mutants showed similar dose–response curves in antigen-presentation assays, indicating redundancy of Ii function.

Comments

Ii is crucial for efficient transport and cell surface display of MHC class II. While it is not required for the assembly of MHC class II molecules, Ii is necessary for normal MHC class II conformation and evasion of the quality control mechanisms in the ER. The p31 and p41 isoforms of Ii have largely redundant functions. A complete lack of Ii severely impairs antigen presentation and decreases negative selection of CD4^{+} T cells.

Acknowledgements
Christophe Benoist
Institut de Génétique et de Biologie Moléculaire et Cellulaire, CNRS INSERM
Université Louis Pasteur, Strasbourg, France

References

[1] Viville, S. et al. (1993) Cell 72, 635–648.
[2] Bikoff, E.K. et al. (1993) J. Exp. Med. 177, 1699–1712.
[3] Elliot, E.A. et al. (1994) J. Exp. Med. 179, 681–694.
[4] Takaesu, N.T. et al. (1995) Immunity 3, 385–396.
[5] Takaesu, N.T. et al. (1997) J. Immunol. 158, 187–199.
[6] Tourne, S. et al. (1995) Eur. J. Immunol. 25, 1851–1856.
[7] Bodmer, H. et al. (1994) Science 263, 1284–1286.

Other names
NF-κB1α, nuclear factor of kappa light chain gene enhancer in B cells inhibitor alpha, inhibitor of NF-κB transcription factor, MAD-3

Gene symbol
Nfkbia

Accession number
MGI: 104741

Area of impact
Immunity and inflammation, signaling, neurology

General description

IκBα is a member of an inhibitory family of proteins (the IκB family) that bind to NF-κB transcription factors. This interaction is responsible for both the inhibition of DNA binding and cytoplasmic retention of NF-κB proteins. Treatment of cells with activators of NF-κB, such as TNF, leads to the phosphorylation and degradation of IκB proteins, allowing free NF-κB transcription factors to translocate to the nucleus. In the nucleus, NF-κB proteins can activate various genes but have their primary effects on those genes involved in immune functions and inflammation. The IκBα gene is also an NF-κB target gene, which may be the basis for a potential autoregulatory loop that allows for only transient activation of NF-κB.

KO strain 1 construction[1]

The IκBα locus (derived from a 129Sv mouse liver genomic library) was targeted in a manner that allowed replacement of the coding region with the bacterial *lacZ* gene. Disruption was such that the expression of the bacterial *lacZ* gene could be regulated by the IκBα promoter. (This enabled the determination of NF-κB activation in cells following treatment with various agents and the study of the developmental regulation of these proteins.) The PGK*tk* selection cassette was included at the 3′ end of the targeting construct. The targeting vector was electroporated into J1 ES cells and positive clones were injected into C57BL/6 blastocysts. Chimeras were bred to C57BL/6 females and heterozygotes interbred to obtain IκBα$^{-/-}$ mutants. The null mutation was confirmed by Southern blot and by Western blotting of mutant thymocytes and splenocytes.

KO strain 2 construction[2]

The targeting vector was constructed such that most of the IκBα-coding sequence (encoding amino acids 31–317) was replaced by the PGK*neo*bpolyA expression cassette. The HSV-*tk* expression cassette was inserted 3′ to the IκBα gene. The targeting vector was electroporated into W9.5 ES cells (129Sv strain) and positive clones were injected into C57BL/6 blastocysts, followed by implantation into (C57BL/6 × CBA) F1 hybrid foster mothers. Chimeras

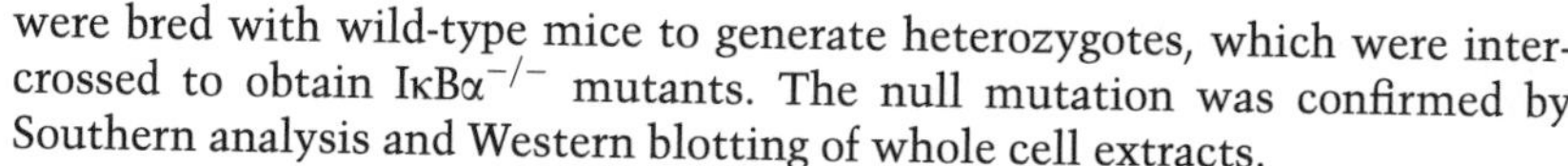

were bred with wild-type mice to generate heterozygotes, which were intercrossed to obtain IκBα$^{-/-}$ mutants. The null mutation was confirmed by Southern analysis and Western blotting of whole cell extracts.

Phenotype

Although normal at birth, IκBα$^{-/-}$ mice exhibited severe runting, skin defects and extensive granulopoiesis post-natally, typically dying by 8 days. Spleens of 5–9-day-old pups showed an increased percentage of monocytes/macrophages. Hematopoietic tissues from these mice displayed high levels of nuclear NF-κB as well as elevated levels of mRNAs of NF-κB-responsive genes, including G-CSF. However, fibroblasts showed normal levels of nuclear NF-κB proteins. The underlying basis for post-natal lethality in IκBα$^{-/-}$ mice is still unclear.

Stimulation of IκBα$^{-/-}$ embryonic fibroblasts with TNF or LPS led to normal NF-κB activation, suggesting that other IκB proteins present in IκBα$^{-/-}$ cells can allow NF-κB activation. However, nuclear localization of NF-κB was prolonged in IκBα$^{-/-}$ fibroblasts, indicating a defect in the post-activation repression of NF-κB-binding activity in the nucleus.

The phenotype of the IκBα$^{-/-}$ mouse is consistent with hyperactivity of NF-κB. The post-natal lethality in these mice appears to be a consequence of aberrant NF-κB regulation, since double knockout mouse mutants lacking both IκBα and p50 (a member of the NF-κB family) resulted in survival till weaning. Double knockout mouse mutants lacking both IκBα and c-Rel (also a member of the NF-κB family) resulted in the complete rescue of lethality.

Comments

IκBα is not required for signal-dependent activation of NF-κB in fibroblasts but is required for post-induction repression of NF-κB in these cells. IκBα and related NF-κB inhibitors, IκBβ and IκBε, may have different physiological roles which appear to be tissue-specific.

Acknowledgements
Amer Beg
Columbia University, New York, NY, USA

References

1 Beg, A. et al. (1995) Genes Dev. 9, 2736–2746.
2 Klement, J.F. et al. (1996) Mol. Cell. Biol. 16, 2341–2349.
3 Baeuerle, P.A. and Henkel, T. (1994) Annu. Rev. Immunol. 12, 141–179.

Gene symbol
Ikaros

Accession number
MGI: 96535

Area of impact
Immunity and inflammation, transcription factors

General description

The Ikaros proteins are a family of conserved zinc-finger DNA-binding transcription factors expressed primarily in the fetal and adult hematopoietic systems. Ikaros is expressed during embryogenesis only in sites of hematopoiesis, and in the adult in early B cells, T cell progenitors and mature T cells. The five known isoforms of Ikaros (Ik-1, Ik-2, Ik-3, Ik-4 and Ik-5) are all derived via alternative splicing of the single *Ikaros* gene. All Ikaros isoforms share a common C-terminal domain but vary in the number of N-terminal zinc-finger domains. At least three N-terminal zinc fingers are required for an Ikaros protein to bind DNA specifically and with high affinity. In addition, the C-terminal domain is essential for dimerization and high-affinity DNA binding. Ikaros proteins with fewer than three N-terminal zinc fingers can act in a dominant-negative fashion, interfering with the DNA binding of other Ikaros isoforms. High-affinity binding sites for Ikaros have been identified in the regulatory regions of many lymphocyte-specific genes.

KO strain 1 construction[1]

The targeting vector was designed to remove zinc fingers 1, 2 and 3 in the N-terminal high affinity DNA-binding domain, abolishing the ability of Ik-1–4 to bind to their recognition sites. An 8.5 kb genomic fragment (from a 129Sv library) containing part of exon 3 and all of exon 4 was replaced with a neomycin-resistance cassette. The C-terminal dimerization domain was left intact by this strategy. The targeting vector was electroporated into J1 ES cells and positive clones were injected into C57BL/6 or BALB/c blastocysts. Chimeras were bred to mice of the genetic background of the blastocyst and heterozygotes were intercrossed to obtain Ikaros $N^{-/-}$ mice.

RT-PCR analysis of splenic RNA in mutant mice showed that mRNAs lacking exons 3 and 4 were produced. Low amounts of protein corresponding in size to an mRNA containing exons 1, 2, 5, 6 and 7 were detected in Western blots. Ikaros-binding activity determined in mutant nuclear extracts using a gel retardation assay showed none of the four DNA-binding complexes present in extracts from control mice.

Phenotype of Ikaros $N^{-/-}$ mice

Ikaros $N^{-/-}$ mice were born at the expected Mendelian frequency but were smaller than normal at one week of age and failed to thrive. The majority of mutant mice died between weeks 1 and 3, apparently of opportunistic infections. Livers were often necrotic and numerous bacterial species were present in the intestinal tract. A rudimentary thymus was found but no definitive T cell precursors were present. All lymph nodes and Peyer's patches were absent, as were lymphocyte follicles in the gastrointestinal tract. Dendritic epidermal $\gamma\delta$ T cells were also missing. Bone marrow of mutant mice contained 10-fold fewer cells, and the $CD45^{+}$ population of B cells was absent. Of T lineage cells, only a small population of Thy-1^{lo} cells was found in mutant bone marrow. The percentage of erythrocyte precursors in mutant marrow was nearly twice that of myeloid precursors at 2 weeks of age. In contrast to the thymus, the spleen in Ikaros $N^{-/-}$ mice was enlarged by 1.5–3-fold and heavily populated with cells of erythroid and myeloid origin. The morphology of the spleen was disorganized. Mature T cells were absent but a small population of Thy-1^{lo} cells was again present. Pro-B cells were also missing from the spleen, as were NK cells. No evidence of NK cell activity was found in NK functional assays of mutant spleen cells.

In mice heterozygous for this allele, mutant Ikaros proteins showed a dominant interference effect on wild-type isoforms, resulting in the development of lymphomas and leukemias.

KO strain 2 construction[2]

Because proteins transcribed from the Ikaros $N^{-/-}$ allele still have intact the C-terminal domain required for dimerization and interaction with other proteins, and could exert dominant-negative effects over functional Ikaros isoforms, a mutant allele was designed in which the C-terminal zinc-finger dimerization domain was deleted. The targeting vector was constructed to remove the last translated exon (exon 7) which contains domains involved in transcription activation, dimerization and protein interactions. A 1.35 kb genomic fragment (derived from 129Sv) containing exon 7 and including its 5′ splice acceptor site was replaced with a neomycin-expression cassette. Proteins derived from such an allele should be transcriptionally inactive and inert with respect to dominant-negative interactions. The targeting vector was electroporated into J1 ES cells and homozygous Ikaros $C^{-/-}$ mice generated as described above.

Northern blots of RNA from Ikaros $C^{-/-}$ thymocytes showed a short transcript lacking exon 7 sequences. Immunohistochemical analysis showed a lack of characteristic Ikaros nuclear staining. Western blots of thymocytes showed a short mutant protein present at 100-fold lower concentration than Ikaros in wild-type thymocytes. This mutated protein was unstable and rapidly degraded within cells, so that Ikaros $C^{-/-}$ cells were functionally null for any Ikaros protein.

Phenotype of Ikaros $C^{-/-}$ mice

Ikaros $C^{-/-}$ mice showed a less drastic phenotype than did Ikaros $N^{-/-}$ mice, in that $C^{-/-}$ mice lived for up to 4 months and males were fertile. Both fetal- and adult-derived B cells were absent in Ikaros $C^{-/-}$ mice, and fetal, but not postnatal, thymocyte differentiation was impaired. Definitive thymocytes were detected in the thymus at 3–5 days after birth. These clonal populations were able to expand to reach nearly normal numbers in the adult thymus. T cell progenitors in the Ikaros $C^{-/-}$ thymus were able to differentiate into αβ T cells but γδ T cells were drastically reduced in number. The differentiation of the thymic αβ T cells was abnormal and skewed towards $CD4^{+}CD8^{-}$ cells and their precursors. Thymocytes showed enhanced TCR-mediated proliferative responses, leading to the appearance of oligoclonal expansions. No γδ T cells were detected in adult spleen. Dendritic epidermal Vγ3 T cells and NK cells were missing, and the numbers of intestinal intra-epithelial γδ T cells (IELs) and $CD8^{-}$ IELs were decreased. In contrast, αβ IELs were present in normal numbers. Lymph nodes were absent in Ikaros $C^{-/-}$ mice and thymic APCs were severely reduced. Bone marrow and spleen contained normal to increased numbers of erythroid and myeloid cells.

Comments

Ikaros proteins are not required for the production of totipotent hematopoietic stem cells (HSC). Ikaros proteins are essential for the development or differentiation of fetal HSC into lymphoid lineages. However, there is partial redundancy for the development of adult HSCs into some lymphoid lineages.

References

1 Georgopoulos, K. et al. (1994) Cell 79, 143–156.
2 Wang, J.-H. et al. (1996) Immunity 5, 537–549.

Other names
Interleukin 1β

Gene symbol
Il1b

Accession number
MGI: 96543

Area of impact
Immunity and inflammation, cytokines

General description

Mammalian IL-1β is a pro-inflammatory cytokine that is produced particularly by mononuclear phagocytes, but also by numerous other cell types, in response to injury and infection. Interleukin-1 has been implicated in a broad spectrum of inflammatory, physiologic, hematopoietic and immunologic activities. It is believed to be a mediator of inflammation in both human and animal models and has been proposed as an attractive target for therapeutic intervention in the treatment of inflammatory diseases.

KO strain construction

A targeting vector was constructed in which the first six of the seven exons of the *IL1b* gene were deleted and replaced with the PGK*neo* sequence. The 5′ and 3′ homologies were 4.5 kb and 1.3 kb, respectively. The HSV-*tk* cassette was attached at the end of the 3′ homology for negative selection. The targeting vector was electroporated into AB2.1 ES cells, and chimeric mice were bred with 129SvEv inbred mice, and C57BL/6J/129SvEv hybrid mice. The targeted deletion of the *IL1b* gene was confirmed by Southern blot analysis.

Phenotype

The IL-1β-deficient mice developed normally and were apparently healthy and fertile. The lack of IL-1β expression was confirmed by Northern blot and ELISA analyses of LPS-stimulated mice. The mice responded normally in models of contact and delayed-type hypersensitivity or following bacterial endotoxin LPS-induced inflammation. The production of acute-phase proteins was unimpaired. The IL-1β-null mice showed equivalent resistance to *Listeria monocytogenes* compared to wild-type controls. In contrast, when challenged with turpentine, which causes localized inflammation and tissue injury, the IL-1β mutant mice exhibited an impaired acute-phase inflammatory response and were completely resistant to fever development and anorexia.

Comments

These results demonstrate a central role for IL-1β as a pyrogen and a mediator of the acute-phase response in a subset of inflammatory disease models, and support the notion that blocking the action of a single key cytokine can alter the course of specific immune and inflammatory responses.

Acknowledgements

Hui Zheng
Department of Genetics and Molecular Biology, Merck Research Laboratories, Rahway, NJ, USA.

References

[1] di Giovine, F.S. and Duff, G.W. (1990) Immunol. Today 11, 13–20.
[2] Dinarello, C.A. (1992) Int. J. Tissue React. 14, 65–75.
[3] Kluger, M.J. (1991) Physiol. Rev. 71, 93–127.
[4] Kozak, W. et al. (1995) Am. J. Physiol. 269, R969–R977.
[5] Zheng, H. et al. (1995) Immunity 3, 9–19.

Other names
Interleukin-1 type 1 receptor, interleukin-1 p80 receptor

Gene symbol
Il1r1

Accession number
MGI: 96545

Area of impact
Immunity and inflammation

General description

IL-1α and IL-1β are cytokines with major roles in the acute phase and inflammatory responses. They are also potent stimulators of bone resorption *in vitro* and *in vivo*, and IL-1 expression is selectively induced in multisystem organ failure during acute pancreatitis. There are two receptors for IL-1 that mediate its actions. IL-1R1 (80 kDa) is expressed on fibroblasts and T cells and binds to IL-1α and IL-1β with equal affinity. IL-1R2 (60 kDa) is expressed on B cells and macrophages and exhibits differing affinities for IL-1α and IL-1β.

KO strain 1 construction[1]

The targeting vector replaced exons 1 and 2 (which encode amino acids 4–146 of the IL-1R1 extracellular domain) with a neomycin-resistance cassette. The deleted sequences are crucial for IL-1 binding. The targeting vector was electroporated into 129Sv-derived ES cells and positive clones were injected into C57BL/6 blastocysts. Chimeric offspring were bred to C57BL/6 mice.

KO strain 2 construction[2]

A replacement type targeting vector containing PGK*neo* and *tk* was designed to delete about 1 kb of DNA including the exon encoding the signal peptide of the IL-1R1 gene (derived from 129Sv). The only potential upstream ATG would be out-of-frame when fused to downstream exons. The targeting vector was electroporated into W9.5 ES cells and positive clones were used to generate chimeric mice that transmitted the mutated allele upon mating to either 129Sv or C57BL/6 females. All mutant mice investigated were of a mixed 129Sv and C57BL/6 background.

Phenotype

IL-1R1$^{-/-}$ mice were viable, overtly normal and fertile. Serum Ig levels and primary and secondary antibody responses were normal. B cells from KO strain 1 IL-1R1$^{-/-}$ mice activated *in vitro* with anti-IgM were able to proliferate in response to IL-4 but not to IL-1. Detectable levels of IL-6 in the serum of these mice were observed after LPS treatment but not after injection of IL-1α. LPS treatment induced normal acute-phase protein mRNA induction. KO strain 1

IL-1R1$^{-/-}$ mice showed normal susceptibility to either a lethal challenge with D-galactosamine plus LPS, or high-dose LPS. KO strain 1 IL-1R1$^{-/-}$ mice on the hybrid background were susceptible to challenge with *Listeria monocytogenes* but once backcrossed to the C57BL/6 background, they were as resistant as control mice.

KO strain 2 IL-1R1$^{-/-}$ mice failed to respond to IL-1 in a variety of assays, including IL-1-induced IL-6 and E-selectin expression. The acute-phase response to turpentine was decreased, similar to IL-1β-deficient mice. KO strain 2 IL-1R1$^{-/-}$ mice were highly susceptible to infection by *L. monocytogenes* and showed decreased delayed-type hypersensitivity.

At 11–12 weeks of age, KO strain 1 IL-1R1$^{-/-}$ mice had a 30% reduction in body weight compared to normal mice. However, calvariae and humeri of mutant mice were normal with respect to cortical thickness, growth plate widths, osteoclast number and surface, and trabecular bone volume[3].

When acute pancreatitis was induced in KO strain 1 IL-1R1$^{-/-}$ mice, the severity of the disease was decreased compared to the wild type. However, increased levels of IL-1 mRNA were detected in affected tissues. Thus, the deletion of IL-1R1 appeared to induce overproduction of IL-1 mRNA in organs known to produce cytokines during pancreatitis[4].

Comments

IL-1R1 is essential for all IL-1-mediated signaling events but not for normal development, homeostasis or bone development. A negative feedback loop may exist between IL-1R1 and IL-1 gene expression. IL-1 signaling may be involved in provoking lethal systemic toxic effects.

Acknowledgements
James Norman
Department of Surgery, University of South Florida, Tampa, FL, USA

References

1 Glaccum, M.B. et al. (1997) J. Immunol. 159, 3364–3376.
2 Labow, M. et al. (1997) J. Immunol. 159, 2452–2461.
3 Socorro, J.V. et al. (1996) J. Bone Mineral Res. 11, 1736–1744.
4 Norman, J.G. et al. (1996) J. Surgical Res. 63, 231–236.
5 Acton, R.D. et al. (1996) Arch. Surg. 131, 1216–1221.

Other names
Interleukin 1 receptor antagonist, IRAP

Gene symbol
Il1rn

Accession number
MGI: 96547

Area of impact
Immunity and inflammation, cytokines

General description

IL-1ra is a variably glycosylated competitive inhibitor of IL-1α and IL-1β. It is the first known example of a specific receptor antagonist of any cytokine or hormone-like molecule. It has no known agonist function, and thus can effectively block the functions of IL-1 when present in high concentrations (in various experimental systems, from 10–1000-fold excess over IL-1). It is expressed by inflammatory cells (primarily macrophages) as part of their response to inflammatory stimuli. There are three forms of IL-1ra, two intracellular and one secreted, derived from the same single copy gene. The intracellular forms are generated by differential splicing of alternative 5′ exons I and II.

KO strain construction

A 9.0 kb *Hin*dIII/*Eco*RI restriction fragment of genomic DNA from murine strain 129 was disrupted by insertion of a neomycin cassette into the third exon. The construct contained the four exons and three introns of the secreted form of IL-1ra, as well as 2.0 kb of upstream and 2.5 kb of downstream untranslated regions. Alternative exons I and II of the intracellular IL-1ra isoforms were not present in the targeting vector. The HSV-*tk* cassette was added at the 3′ end of the construct.

CCE ES cells (from 129Sv mice) were electroporated with the targeting vector. Positive clones were microinjected into C57BL/6J blastocysts. Chimeras were mated with C57BL/6J mice. The mutant strain was backcrossed to C57BL/6J mice for 2–3 generations at the time of publication of the phenotypes.

The null mutation was confirmed by Northern blotting of total lung and liver RNA and by serum ELISA after LPS treatment.

Phenotype

IL-1ra null mutants were born at the expected Mendelian frequency and showed normal development upon gross examination, except for a decrease in body weight of 20–30%. This weight deficit became apparent at 6 weeks of age and persisted into adulthood. IL-1ra$^{-/-}$ mice developed a non-specific illness consisting of further weight loss, decreased activity, apparently labored breathing, and ultimately death in some cases. This illness of unknown etiology was variable in incidence, age of onset, duration and severity.

IL-1ra$^{-/-}$ mice were more susceptible than wild-type mice to the lethal effects of LPS, but were relatively protected from mortality due to infection with *Listeria monocytogenes*. Unexpectedly, since prior evidence had suggested that IL-1 induces its own expression, the expression of IL-1 α and β in the serum was decreased following a systemic dose of LPS. Furthermore, in a separate line of IL-1ra-overexpressing mice, a complementary observation was made: that IL-1ra overexpression led to increased levels of IL-1 in serum.

Comments

IL-1ra appears to have an pivotal role in the regulation of IL-1: inhibition of IL-1 by IL-1ra may protect an organism from a too vigorous a response to infection, but at the risk of impeding the organism's ability to fight infection. In the case of LPS exposure (a model for septic shock), a robust inflammatory cascade leads to hypotension and organ failure. This inflammatory cascade is dependent on the function of IL-1. A relative increase in IL-1 activity in IL-1ra$^{-/-}$ mice leads to an increase in mortality following exposure to LPS. In listeriosis, however, effective elimination of infectious organisms is dependent upon the proper expression of IL-1. Increased IL-1 activity due to elimination of the receptor antagonist in IL-1ra$^{-/-}$ mice increases the host's ability to combat infection.

In addition to its role as an inhibitor of IL-1 activity, IL-1ra also appears to have an unexpected function as a positive regulator of IL-1 expression in serum. Finally, the observation that IL-1ra$^{-/-}$ mice have lower body weights than their wild-type littermates and develop a non-specific illness, suggests a novel role for IL-1ra (and by implication, IL-1) in body mass determination and homeostasis.

Acknowledgements
David Hirsh
Department of Biochemistry and Molecular Biophysics, Columbia University, New York, NY, USA

Reference

1 Hirsch, E. et al. (1996) Proc. Natl Acad. Sci. USA 93, 11008–11013.

Other names
Interleukin 2

Gene symbol
Il2

Accession number
MGI: 96548

Area of impact
Immunity and inflammation, cytokines

General description

IL-2 is a lymphocytotrophic cytokine with a key regulatory role in both the specific immune and inflammatory responses. It is produced by activated T cells and mediates its effects on the growth, differentiation and function of lymphocytes via the multi-subunit IL-2 receptor. *In vitro*, IL-2 promotes T cell proliferation, the differentiation of B cells, and the activation of B cells and NK cells.

KO strain construction

For the targeted disruption of the IL-2 gene, the *neo* cassette of pMCI*neo*polyA was inserted in reverse orientation into *Bgl* III site of the mouse genomic clone (129Ola). This 1.1 kb insertion interrupted the reading frame of the third exon and introduced several stop codons, removing all biological IL-2 activity. The targeting vector was electroporated into E14 ES cells originally derived from 129Ola mice. Positive clones were injected into C57BL/6 blastocysts. Chimeras were mated with C57BL/6 mice and heterozygotes were crossed to obtain IL-$2^{-/-}$ homozygous mutants. The mutation has since been backcrossed into BALB/c, C57BL/6, CH3 and 129Ola strains and is available from Jackson Laboratories.

The null mutation was confirmed by functional assays in which cells from mutant thymus, lymph node and spleen both failed to produce detectable IL-2, and responded poorly to polyclonal T cell activators.

Phenotype

Mutant mice were born at the expected Mendelian frequency and developed normally until 3–4 weeks of age. The ontogeny of the immune system was not affected by the IL-2 deficiency, as seen from the normal development of the lymphoid organs and the presence of all major lymphocyte subsets. In response to viral challenge *in vivo*, young IL-$2^{-/-}$ mice generated cytotoxic T cell responses and helper-dependent and -independent T cell responses which were only partly reduced. NK cell responses were markedly reduced but remained inducible. Although activated B cells, elevated Ig secretion, anti-colon antibodies and aberrant expression of MHC class II molecules were observed, the primary alteration of the immune system was an uncontrolled activation and proliferation of $CD4^+$ T cells which led to autoimmune disease.

Depending on their genetic background, $IL-2^{-/-}$ mice develop either an ulcerative colitis-like inflammatory bowel disease (IBD) or hemolytic anemia. Despite the relatively normal development of the immune system in young mice, 50% of $IL-2^{-/-}$ mice with the mixed 129Ola:C57BL/6 genetic background died by 9 weeks of age from a disease of unknown etiology characterized by splenomegaly, lymphoadenopathy and severe anemia. Surviving mice developed IBD with 100% penetrance which resulted in death by 10–25 weeks. An antigenic stimulation by non-pathogenic intestinal flora was required to trigger IBD, since mice kept under germ-free conditions did not develop the disease. In contrast, $IL-2^{-/-}$ mice of the BALB/c genetic background developed generalized autoimmune disease characterized primarily by severe hemolytic anemia[5]. Uncontrolled activation and proliferation of both T and B cells and inflammatory lesions on several organs were observed. Death occurred within 5 weeks of age. Daily administration of human recombinant IL-2 prevented both the activation of the immune system and the development of autoimmune disease if started before day 10 after birth.

Mice deficient for both IL-2 and IL-4 showed increased proliferation of T cells but all major T cell subsets and B cells were normal[7]. Also, adoptive transfer of lymphocytes from IL-2-treated $IL-2^{-/-}$ animals conferred protection to $IL-2^{-/-}$ mice[8], suggesting that IL-2 induces a post-natal differentiation of regulatory cells necessary for self- and non-self discrimination.

Comments

IL-2 is not critical for normal ontogeny of the immune system. The essential role of IL-2 which is not compensated by other cytokines is a negative regulatory function required for the maintenance of self-tolerance.

Acknowledgements
Ivan Horak
Institute of Molecular Pharmacology, Berlin, Germany

References

1 Schorle, H. et al. (1991) Nature 352, 621–624.
2 Sadlack, B. et al. (1993) Cell 75, 253–261.
3 Kündig, T.M. et al. (1993) Science 262, 1059–1063.
4 Horak, I. et al. (1995) Immunol. Rev. 48, 35–44.
5 Sadlack, B. et al. (1995) Eur. J. Immunol. 25, 3053–3059.
6 Sadlack, B. et al. (1994) Eur. J. Immunol. 24, 281–284.
7 Klebb, G. et al. (1996) Clin. Immunol. Immunopathol. 81, 282–286.

IL-2Rα

Other names
Interleukin 2 receptor α chain, CD25, p55, Tac antigen

Gene symbol
Il2ra

Accession number
MGI: 96549

Area of impact
Immunity and inflammation, cytokine

General description

The IL-2Rα chain complexes with the IL-2Rβ and γc common chains to form the high affinity receptor for IL-2. Signaling through the high-affinity IL-2 receptor promotes cell cycle progression in activated lymphocytes, as well as effector function in T and B cells. IL-2 receptor signals also activate T cell death.

KO strain construction

The neomycin resistance gene was cloned into a genomic portion of the IL-2Rα gene, resulting in a deletion of approximately 5 kb of genomic DNA. The neomycin gene replaced portions of exons 2 and 3, including the region which includes the ligand-binding site of the receptor. The vector was electroporated into the J1 ES cell line (129 strain), and the resulting ES cells were injected into C57BL/6 blastocysts. Chimeric mice were bred with C57BL/6 mice. The deletion of the IL-2Rα gene was confirmed by Southern blotting.

Phenotype

Mutant mice developed normally and were normal in appearance. The mice exhibited phenotypically normal development of normal B and T cells, but they uniformly exhibited a 7- to 10-fold expansion of the peripheral lymphoid compartment by the age of 6 weeks. This expansion was polyclonal, including all subsets of T and B cells. T cells exhibited a mature phenotype, expressing high levels of CD44 and low levels of CD62L. Switched isotypes of immunoglobulin were also elevated, but with normal levels of IgM.

The mice developed an antibody-mediated hemolytic anemia, which was lethal in approximately 25% of the mice between 8 and 20 weeks. Older mice developed inflammatory bowel disease characterized by massive inflammatory cell infiltration of the colon only, with diarrhea and wasting.

The proliferative responses of T cells *in vitro* were diminished, but could be rescued with high doses of IL-2, consistent with the presence of low-affinity IL-2 receptors but a lack of high-affinity IL-2 receptors. T cells exhibited a defect in activation-induced peripheral deletion in response to staphylococcal enterotoxin B.

Comments

IL-2 signaling is important in negative regulation of the peripheral T cell compartment, possibly by controlling activation-induced cell death. The massive polyclonal peripheral lymphoid expansion seen in the IL-2Rα$^{-/-}$ mice may indicate that IL-2Rα has a negative regulatory role in controlling the overall size, as well as the content, of the peripheral lymphoid compartment.

Acknowledgements
Dennis Willerford
Division of Hematology, University of Washington School of Medicine, Seattle, WA, USA

Reference

1 Willerford, D.M. et al. (1995) Immunity 3, 521–530.

Other names
Interleukin 2 receptor β chain, p75–85 subunit of the IL-2 receptor, CD122

Gene symbol
Il2rb

Accession number
MGI: 96550

Area of impact
Immunity and inflammation, cytokine

General description

IL-2Rβ, a component of the IL-2 receptor complex, is indispensable for IL-2-mediated signal transduction in the mouse. IL-2Rβ is also a component of the IL-15 receptor. IL-2Rβ is expressed on T cells, B cells, NK cells, $NK1^+$T cells, dendritic epidermal T cells (DETCs), neutrophils, monocytes, and large granular lymphocyte-like cells in decidua during early pregnancy. Generally, IL-2Rβ is expressed at low density on $CD8^+$ resting T cells, and is upregulated on both $CD4^+$ and $CD8^+$ T cells following T cell activation.

KO strain construction

The pMC1*neo*polyA cassette was inserted into exon 6 of the IL-2Rβ gene. Exon 6 corresponds to a part of the extracellular domain close to the transmembrane region. The total size of the KO construct was approx. 4.1 kb (2.3 kb long arm, 1.1 kb *neo* cassette, 0.7 kb short arm). The targeting vector was introduced into D3 ES cells derived from the 129 strain. Chimeric mice were crossed with C57BL/6 females. The progeny of heterozygous F1 (B6/129) were backcrossed to C576BL/6 5–6 times to generate mice homologous for the disrupted gene. The null mutation was confirmed by PCR and Southern hybridization.

Phenotype

Mutant mice showed normal growth until approx. 3 weeks after birth, after which they were generally smaller in size than normal or heterozygous littermates. Mutant mice had fuzzy hair, poorly developed external genitalia, and slow locomotion. Hemolysis was a primary cause of progressive anemia in young mutant mice, coupled with poor erythropoiesis in older mice. Most mutants died before 3 months of age of severe anemia and generalized autoimmune disease.

No significant defects were observed in the differentiation of either T cells (both TCRαβ and TCRγδ) or B cells. However, T cells of both the $CD4^+$ and $CD8^+$ lineages were spontaneously activated and expressed high levels of CD69 and CD44. Activated T cells appeared before 1 week of age and increased in number with advancing age. These activated T cells produced elevated levels of many cytokines including IL-2, IL-4, IFN-γ and IL-10. B cells were able to differentiate into plasma cells but antibody production was abnormal, resulting in a 10–100-fold elevation of serum IgG1 and IgE. Autoantibodies such as anti-nuclear antibody and anti-DNA antibody were also present. Mutant B cells were unable to mount either T cell-dependent or independent Ig responses. The number of B cells declined with age because of exhaustive plasma cell differentiation and reduced B cell lymphopoiesis. No functional $CD8^+$ response could be generated against lymphocytic choriomeningitis virus (LCMV) infection in mutant mice. Granulocytopoiesis was markedly increased in bone marrow and spleen, depressing the hematopoiesis of other cell lineages.

In contrast to T and B cell differentiation, NK cell differentiation was severely damaged in IL-2R$\beta^{-/-}$ mice. Almost no mature NK cells were present and numbers of $NK1^+$T cells were also significantly decreased. DETCs were totally absent from the skin. Numbers of extrathymically differentiated intestinal intra-epithelial lymphocytes (IELs) with a surface phenotype of $CD8\alpha^+\beta^-$and $TCR\gamma\delta^+$ were significantly reduced in IL-2R$\beta^{-/-}$ mice.

Comments

The B cell pathology and autoimmune disease in IL-2R$\beta^{-/-}$ mice are dependent on the presence of activated $CD4^+$ T cells. IL-2Rβ not only mediates signals for T cell activation but is also required for hematopoietic homeostasis. IL-2Rβ is not required for thymocyte development but is involved in the development of extrathymic lymphocyte subsets.

Acknowledgements
Haruhiko Suzuki
Department of Immunology, Nagoya University School of Medicine, Nagoya, Japan
Tak W. Mak
Ontario Cancer Institute and Amgen Institute, Toronto, ON, Canada

References

[1] Suzuki, H. et al. (1995) Science (Washington DC) 268, 1472–1476.
[2] Suzuki, H. et al. (1997) J. Exp. Med. 185, 499–505.

Other names
Common cytokine receptor γ chain, common γ chain, γc, interleukin 2 receptor γ chain

Gene symbol
Il2rg

Accession number
MGI: 96551

Area of impact
Immunity and inflammation, cytokines

General description

IL-2Rγ(γc) is a shared cytokine receptor chain used by the receptors for IL-2, IL-4, IL-7, IL-9 and IL-15. It demonstrates no ligand-binding activity, but increases the affinity of the receptor complexes, and most importantly, functions in signal transduction through its liaison with the tyrosine kinase Jak3. Located on the murine X chromosome, γc is widely expressed in the hematopoietic system (HSC, lymphocytes, myelocytes, mast cells, etc.) and has been detected in epithelial cell lines and in such diverse tissues such as heart and lung. Absence of γc in humans results in severe combined immunodeficiency (SCIDX1).

KO strain 1 construction[1]

A *loxP*/Cre approach was used to create a defined deletion in the γc locus encompassing exons 2–6, encoding the extracellular and transmembrane domains. A single *loxP* site was inserted into intron 1 and a *loxP*-flanked neomycin-resistance cassette was inserted into intron 6. (This targeting strategy also permitted the derivation of a mouse strain carrying a "floxed" γc locus. These mice are immunologically normal but allow *in vivo* conditional gene targeting approaches to be used for additional analyses.)

The targeting vector was electroporated into E14 ES cells which are of 129Ola origin. Positive clones were injected into CB20 blastocysts to generate chimeras in which the phenotype resulting from the inactivation of the γc gene could be examined directly. The mutation was also germ line transmitted and backcrossed onto a number of inbred backgrounds, including BALB/c, C57BL/6 and C3H, with no change in phenotype. The null mutation was confirmed by flow cytometry.

KO strain 2 construction[2]

The targeting vector was designed to replace part of exon 3 and all of exons 4–8 with the neomycin-resistance cassette, deleting much of the γc extracellular domain as well as the entire transmembrane and cytoplasmic domains. The HSV-*tk* cassette was appended at the 3′ end of the genomic sequences. J1 ES cells were transfected with the targeting vector and positive clones were injected into C57BL/6 blastocysts. Chimeric progeny were bred to C57BL/6 mice to generate heterozygotes. Wild-type males were crossed with heterozygous females to generate γc-deficient males. The null mutation in γc-deficient males was confirmed by Northern and Western blotting of mRNA and protein in spleen and thymus.

KO strain 3 construction[3]

Because previous work *in vitro* by this group established that the intracytoplasmic domain of the γc chain was indispensable for growth signals, a targeting vector was designed that deleted only this region from the γc locus. Mutant mice therefore expressed a γc chain that could still bind ligands but which could not transduce signals. The neomycin-resistance cassette pMC1*neo*polyA replaced the 3′ two-thirds of exon 7, the intron, and the 5′ half of exon 8. The diphtheria toxin A gene was added to the 3′ end of the genomic sequence.

The targeting vector was electroporated into E14-1 or CCE ES cells. Positive clones were injected into C57BL/6 blastocysts. Heterozygotes were intercrossed with C57BL/6 mice to generate mutants homozygous for the truncated γc chain. Expression of only the truncated γc chain in cells of mutant mice was confirmed by RT-PCR on spleen mRNA.

Phenotype of KO strains 1 and 2 (null mutants)

The γc-deficient male mice appeared normal at birth and developed normally. However, the absence of the γc chain, which results in a failure to respond to the key growth and differentiation factors IL-2, IL-4, IL-7, IL-9 and IL-15, perturbed lymphoid development in mutant mice. Mutant mice had severely hypoplastic thymi with basically normal architecture and containing Hassall's corpuscles. Mutant mice showed an absence of NK cells, TCRγδ cells (including γδ i-IELs and DETCs), peripheral lymph nodes and all gut-associated lymphoid tissues, including the Peyer's patches. Monocytes/macrophages and granulocytes were normal or even increased in number. The block in early B cell development was incomplete, since small numbers of mature B cells were found in the spleen. Peritoneal B1 cells were normal. While serum IgM levels were normal, all other Ig isotypes were decreased in mice of greater than 5 weeks of age.

Although the T cell compartment as a whole was decreased, γc-deficient mice showed a surprising degree of TCRαβ cell development. These T cells were primarily $CD4^+$, had an activated phenotype, and provoked a number of "autoimmune" manifestations, including increased extramedullary hematopoiesis, colitis and B cell loss. Although no defects in intrathymic selection mechanisms were detected for $CD8^+$ cells, it was likely that the resultant $CD4^+$ T cells were not self-tolerant and were autoreactive. *In vitro*, there was no activity by mutant cells against the NK-sensitive YAC target cell line, and γc-deficient thymocytes and splenic T cells showed a defective proliferative response to ConA or PMA that was not overcome by the addition of IL-2, IL-7 or IL-4. However, γc-deficient thymocytes were able to respond normally to PMA plus ionomycin, or to anti-CD3 plus anti-CD28; splenocytes showed a similar but more moderate response. Splenocytes were also able to respond to LPS, but not to control levels. When treated with anti-CD3 plus anti-CD28, γc-deficient splenocytes (only) showed drastically decreased production of IFN-γ, and only half-normal stimulation of IL-2 and IL-4. Double mutants in which both c-*kit* and γc were deficient had essentially no thymic development.

Phenotype of KO strain 3 (truncated γc chain mutant)

Mutant mice with the truncated γc chain showed an atrophied thymus and a spleen of doubled size but lacking white pulp. Peyer's patches and peripheral lymph nodes were barely detectable even at 8 weeks after birth. Peripheral blood lymphocyte numbers decreased by 10-fold while peripheral blood and splenic monocytes/macrophages were increased 3-fold. However, thymocyte differentiation was not completely arrested, in contrast to human SCIDX1. Serum levels of IgG and IgG3 were reduced, but serum IgM was increased in 8-week-old mutant mice in spite of a drastic decrease in $CD45R^+$ $sIgM^+$ B cells. Spleen B cell development was blocked after the pro-B stage. Spleen T cells were dramatically reduced in number but showed normal proportions of $CD4^+$ and $CD8^+$ T cells. An apparent increase in the $CD34^+$ c-*kit*$^+$ Sca-1^+ stem cell population was observed. Colony-forming assays showed that 15-fold greater numbers of hematopoietic precursors occurred in mutant spleens. The $TCR\alpha\beta^-/NK1.1^+$ population was not found in either peripheral blood or spleens of mutant mice. An inflammatory bowel disease-like phenotype was noted.

Comments

γc has an important role in peripheral lymphoid maturation and homeostasis. While TCR γδ cells and NK cells appear to be strictly γc-dependent (probably for survival), TCRαβ cells have alternative pathways to "rescue" their development in the absence of γc, possibly involving the c-*kit* receptor. Because γc-deficient mice lack NK cells, this mouse strain should prove instrumental in deciphering the role *in vivo* of NK cells in various immune responses. γc-deficient mice also provide a small animal model for the evaluation of alternative therapies for human γc deficiency (SCIDX1).

The phenotype of the truncated γc mutants was similar to that of the null mutants in many respects, including a complete absence of NK cells. However, increased hematopoietic stem cells, and increased serum IgM were found only in the case of the truncated γc mutant. An exclusive impairment in signal transduction by the truncated IL-2Rγ may account for these differences.

It is interesting to note that γc-deficient mice show some T and B cell development with a total absence of NK cells, while SCID mice show almost no T or B cell development but normal or increased NK cell activity.

Acknowledgements
James DiSanto
INSERM U429 Hôpital Necker-Enfants Malades, Paris, France
Kazuo Sugamura
Tohoku University School of Medicine, Sendai, Japan

References

[1] DiSanto, J.P. et al. (1995) Proc. Natl Acad. Sci. USA 92, 377–381.
[2] Cao, X. et al. (1995) Immunity 2, 223–238.
[3] Ohbo, K. et al. (1996) Blood 87, 956–967.
[4] DiSanto, J.P. et al. (1995) Immunol. Rev. 148, 19–34.
[5] DiSanto, J.P. et al. (1996) J. Exp. Med. 183, 1111–1118.
[6] Sharara, L.I. et al. (1997) Eur. J. Immunol. 27, 990–998.
[7] Rodewald, H.R. et al. (1997) Immunity 6, 265–272.
[8] Sugamura, K. et al. (1996) Annu. Rev. Immunol. 14, 179–205.
[9] Ikebe, M. et al. (1998) Int. J. Exp. Pathol. 78, 133–148.

Other names
Interleukin 4

Gene symbol
Il4

Accession number
MGI: 96556

Area of impact
Immunity and inflammation, cytokines

General description

Interleukins are cytokines secreted by cells of the immune system. IL-4 has been shown to affect the proliferation and differentiation of both B and T cells *in vitro*. In B cells, IL-4 induces the expression of MHC class II and CD23, the low-affinity receptor for IgE. IL-4 also directs isotype switching to IgG1 and IgE in LPS-treated B cells. *In vitro*, IL-4 promotes the growth and differentiation of thymocytes, Th2 helper T cells, mast cells, macrophages and hematopoietic progenitors, and has been shown to enhance CTL responses. IL-4 is produced primarily by mast cells and Th2 cells. Th2 cells, involved in antibody and delayed-type hypersensitivity responses to certain pathogens, secrete a cytokine profile characterized by IL-4, IL-5 and IL-10.

KO strain 1 construction[1]

The targeting vector used to disrupt the mouse IL-4 gene contained a genomic fragment encompassing exons 1 and 2. A neomycin-resistance gene and a translational stop codon were inserted into exon 1. The HSV-*tk* gene was included at the 3′ end of the genomic sequences. The targeting vector was electroporated into E14-1 ES cells and positive clones were injected into C57BL/6 blastocysts. Male chimeras were mated to C57BL/6 females, and heterozygous offspring were intercrossed to obtain IL-4$^{-/-}$ mice which were of (129Ola × C57BL/6)F2 background.

The functional disruption of the IL-4 gene was confirmed by analysis of lymphokine production in supernatants of cultured ConA-stimulated spleen cells derived from animals infected with the nematode *Nippostrongylus brasiliensis*. No IL-4 activity was detected in supernatants of mutant cells, but normal amounts of IFN-γ and IL-2 were detected.

KO strain 2 construction[2]

The targeting vector was designed such that a neomycin-resistance cassette was inserted into exon 3 of the IL-4 gene. The HSV-*tk* cassette was appended at the 3′ end of the genomic sequences. The targeting vector was electroporated into D3 ES cells and chimeras were generated by standard procedures. F1 mice (129Sv × C57BL/6) heterozygous for the mutated allele were intercrossed to obtain homozygous IL-4$^{-/-}$ mice.

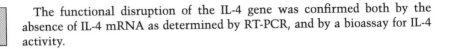

The functional disruption of the IL-4 gene was confirmed both by the absence of IL-4 mRNA as determined by RT-PCR, and by a bioassay for IL-4 activity.

Phenotype

IL-4$^{-/-}$ mice were born at the expected Mendelian frequency and were overtly normal. The development of both B and T cells was normal, but 6-week-old mutant mice had twice the normal number of thymocytes. Serum IgE was undetectable and serum IgG1 was decreased to one-sixth to one-twelfth of the control level, but other isotypes were present at normal concentrations. T helper function was not impaired but IgG1 dominance was absent in a T-dependent immune response. IgE was not detectable in the serum following a nematode infection. Naive IL-4$^{-/-}$CD4^{+} T cells did not produce Th2 cytokines in response to stimulation *in vitro*. Levels of IL-5, IL-9 and IL-10 produced by IL-4$^{-/-}$ CD4^{+} T cells following nematode infection were significantly reduced. IL-5 production was reduced in IL-4$^{-/-}$ mice, corresponding to a 2–3-fold reduction in helminth-induced eosinophilia. Delayed-type hypersensitivity (DTH) reactions to lymphocytic choriomeningitis virus (LCMV) injection were not impaired.

Double IL-4$^{-/-}$IL-2$^{-/-}$ mutants

In mice doubly mutated to lack both IL-2 and IL-4[3,4], all major T and B cell subsets were normal. The proliferation of T cells was unexpectedly increased in IL-4$^{-/-}$IL-2$^{-/-}$ mice. Primary CTL responses against LCMV and vaccinia virus were reduced but still protective. In virus-specific restimulation experiments, IL-4$^{-/-}$ IL-2$^{-/-}$ spleen cells did not generate CTL responses against virus-infected target cells; responses could be restored by the addition of IL-2. T-dependent and T-independent B cell responses were essentially normal.

Comments

IL-4 is required for the production of Th2-derived cytokines and its absence impairs immune responses dependent on these cytokines. IL-4 plays an important role in isotype switching and is essential for the induction of IgE responses *in vivo*. IL-4 is responsible for IgG1 dominance in T-dependent immune responses but not IgG1 production. IL-4 is also required for the production of IL-5 by CD4^{+} T cells.

Neither IL-4 nor IL-2 is essential for the development of the immune system. Mice lacking both IL-4 and IL-2 show a surprising degree of immunocompetence, emphasizing the intricacy and redundancy of the cytokine network.

References

1 Kuhn, T. et al. (1991) Science 254, 707–710.
2 Kopf, M. et al. (1993) Nature 362, 245–248.
3 Sadlack, B. et al. (1994) Eur. J. Immunol. 24, 281–284.
4 Bachmann, M. et al. (1995) J. Virol. 69, 4842–4846.

Other names
Interleukin 5, eosinophil differentiation factor (EDF), T cell replacement factor, B cell growth factor (BCGFII)

Gene symbol
Il5

Accession number
MGI: 96557

Area of impact
Immunity and inflammation, cytokines

General description

IL-5 is produced mainly by Th2 cells and was originally recognized by its activity as a B cell growth factor, an IgA-enhancing factor, and as a differentiation factor for the eosinophil lineage. The involvement of eosinophils in allergic diseases such as asthma and in parasitic infections has created considerable interest in defining the biological role of IL-5, because it appears to most specifically regulate the maturation and differentiation of the eosinophil lineage.

KO strain construction

The *neo* gene was inserted into exon 3 of a 3.7 kb fragment of the IL-5 gene, into a codon for one of the cysteine residues required for IL-5 activity. The targeting vector was electroporated into BL-III ES cells (C57BL/6-derived), and the chimeric mice were bred with C57BL/6 females.

Phenotype

The IL-5-deficient mice were healthy and fertile, with no gross abnormalities. In contrast to previous studies, no obligatory role for IL-5 was demonstrated in the regulation of conventional B cells, in normal T cell-dependent antibody responses, or in cytotoxic T cell development. However, IL-$5^{-/-}$ mice were impaired in the generation of $CD5^+$ B (B-1) cells, and did not develop blood and overt tissue eosinophilia after nematode infection or during inflammatory responses. In addition, using an artificially induced mouse asthma model, it was demonstrated that the eosinophilia, airway hyperreactivity, and lung damage seen in this disease model were abolished by the lack of IL-5 in IL-$5^{-/-}$ mice.

Comments

The results demonstrate that IL-5 plays an important role in the development of $CD5^+$ B-1 cells, as well as in eosinophilia. However, IL-5 is not obligatory for the development of normal T cell-dependent antibody responses or cytotoxic T cell responses. IL-5 also plays a crucial role in the development of lung damage in a mouse model for asthma.

Acknowledgements
Klaus Matthaei
Division of Biochemistry and Molecular Biology, Australian National University, Canberra, Australia
Manfred Kopf
Basel Institute for Immunology, Basel, Switzerland

References

[1] Kopf, M. et al. (1996) Immunity 4, 15–24.
[2] Foster, P.S. et al. (1996) J. Exp. Med. 183, 195–201.
[3] Mould, A.W. et al. (1997) J. Clin. Invest. 99, 1064–1071.
[4] Simeonovic, C.J. et al. (1997) J. Immunol. 158, 2490–2499.
[5] Takamoto, M. et al. (1997) Immunology 90, 511–517.
[6] Matthaei, K.I. et al. (1997) Mem. Inst. Oswaldo Cruz 92, 63–68.

Other names
Interleukin 6

Gene symbol
Il6

Accession number
MGI: 96559

Area of impact
Immunity and inflammation, cytokines

General description

IL-6 is a pleiotropic cytokine that plays key roles in the inflammatory response, hematopoiesis, and the immune response. IL-6 induces acute-phase response genes in the liver and the terminal differentiation of B cells, and stimulates the differentiation of cytotoxic T cells and the proliferation of hematopoietic progenitors in bone marrow. It is also thought to mediate the effects of estrogens on bone. IL-6 is not detectable in the blood or tissues of healthy animals, but intense and rapid production of IL-6 is induced by both inflammatory and pathological stimuli. IL-6 is implicated in several autoimmune diseases, multiple myeloma and osteoporosis. It binds to the signal-transducing receptor gp130, a receptor shared by several other cytokines which appear to have some redundancy of function *in vitro*.

KO strain 1 construction[1]

The targeting vector was designed such that a 2.1 kb fragment of the IL-6 gene (derived from BALB/c) containing the proximal promoter region and exons 1–3 was replaced with a neomycin-resistance cassette. The deletion eliminated the amino terminal half of the protein, essential for biological activity. An HSV-*tk* expression cassette was included 5′ of the genomic sequences. The targeting vector was electroporated into CCE ES cells (derived from strain 129Sv/Ev) and positive clones were injected into C57BL/6 mice. Male chimeras were mated to MF1 females and female heterozygous offspring were bred with 129Sv/Ev mice. Heterozygous offspring of this breeding were intercrossed to obtain homozygous IL-6$^{-/-}$ mutant mice. Mutants have since been backcrossed to the C57BL/6 strain, DBA1J (5 generations) and BALB/cAn (14 generations).

The null mutation was confirmed by Northern blotting of splenic RNA from LPS-treated mice; ELISA determination of serum IL-6; and a bioassay for IL-6 activity. No IL-6 transcripts or functional IL-6 protein were observed.

KO strain 2 construction[2]

The targeting vector was designed to disrupt exon 2 (the first coding exon) of the IL-6 gene by the insertion of a neomycin-resistance cassette. The long and short arms of homology were 1.4 kb and 8.5 kb, respectively. An HSV-*tk* expression cassette was appended 3′ of the genomic sequences. The targeting vector was introduced into D3 ES cells. F1 mice (C57BL/6 × 129Sv) heterozygous for the mutated allele were intercrossed to obtain IL-6$^{-/-}$ mutants. The

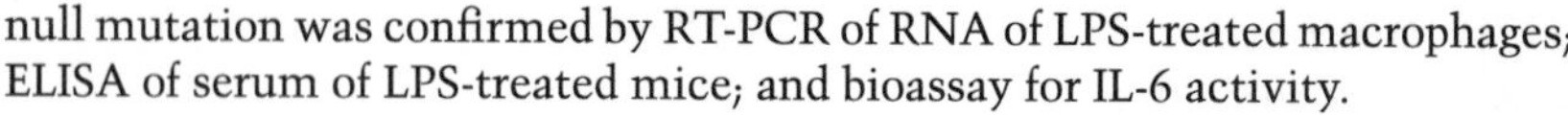

null mutation was confirmed by RT-PCR of RNA of LPS-treated macrophages; ELISA of serum of LPS-treated mice; and bioassay for IL-6 activity.

Phenotype

Mutant mice were born at the expected frequency and were of normal size and development. Both males and females were fertile. However, mutant mice had distinct phenotypes in several compartments. IL-$6^{-/-}$ female mice had higher bone turnover than IL-$6^{+/+}$ littermates, regardless of estrogen status. While no significant changes to the amount of trabecular bone were observed, cortical bone volume was markedly decreased compared to wild type. IL-$6^{-/-}$ female mice were resistant to bone loss caused by estogen depletion because the animals failed to increase the number of bone marrow osteoclast precursors (GM-CFU)[1].

Numbers of thymocytes and peripheral T cells were reduced by 20–40% but expression of surface markers was normal. Bone marrow and splenic B cells were also normal. While the generation and activity of CTLs against lymphocytic choriomeningitis virus (LCMV) was normal, mutant mice were unable to control infection by either vaccinia virus or *Listeria monocytogenes*, and the T cell-dependent antibody response against vesicular stomatitis virus (VSV) was impaired. Mucosal IgA responses were defective[2].

Mutant mice were unable to mount a normal inflammatory response to turpentine injection. The induction of acute-phase proteins was dramatically decreased, body weight was maintained, and only mild anorexia and hypoglycemia were noted. In contrast, systemic inflammation in response to LPS injection occurred at essentially the same intensity as in wild-type mice[2,3]. Under these conditions, TNFα production in IL-$6^{-/-}$ mice was increased 3-fold over normal, suggesting that TNFα might be compensating for the absence of IL-6. Corticosterone induction was normal in IL-$6^{-/-}$ mice[3]. IL-$6^{-/-}$ mice were also highly susceptible to infection with the yeast *Candida albicans*, when either a virulent strain or an attenuated live vaccine was used.

Comments

IL-6 is not required for normal embryogenesis (despite its reported expression at the 8-cell stage) or normal lymphoid development. The regulation of local turnover of bone depends on IL-6. IL-6 mediates the generation of bone loss associated with estrogen deficiency. It is also a critical mediator of localized inflammation, but not systemic inflammation. IL-6 is required for optimal defense against certain infectious organisms.

Acknowledgements
Valeria Poli
Instituto di Richerche di Biologia Molecolare P. Angeletti, Rome, Italy

References

[1] Poli, V. et al. (1994) EMBO J. 13, 1189–1196.
[2] Kopf, M. et al. (1994) Nature 368, 339–342.
[3] Fattori, E. et al. (1994) J. Exp. Med. 180, 1243–1250.

Other names
Interleukin 7

Gene symbol
Il7

Accession number
MGI: 96561

Area of impact
Immunity and inflammation, cytokines

General description

IL-7 is a cytokine produced by cells of the immune system that preferentially affects immature lymphocytes. It is a powerful stimulus for the proliferation of early T and B lymphocyte populations *in vitro*, often without a requirement for costimulatory agents. Both early B cells and thymocytes have been shown to be dependent on IL-7 for survival in cell culture. IL-7 binds to a receptor that contains the IL-7α chain and the IL-2Rγ (common γ) chain, shared with the receptors for IL-2, IL-4, IL-9 and IL-15.

KO strain construction

The targeting vector was designed to replace exon 4 and 300 bp of the flanking introns of the IL-7 gene (derived from 129Sv) with a neomycin-resistance cassette. This deletion removed two of the four α-helical bundles of the cytokine protein. The 5′ and 3′ homologous regions were both 2.3 kb in length. The HSV-*tk* gene was appended to the 3′ end of the genomic sequences. The targeting vector was electroporated into E14.1 ES cells and positive clones were injected into C57BL/6 blastocysts. Male chimeras were mated to C57BL/6 females and heterozygous progeny were intercrossed to obtain IL-7$^{-/-}$ mutant mice.

RT-PCR analysis of thymic stromal RNA showed that no wild-type IL-7 mRNA was present in mutant mice, indicating that the disruption resulted in mRNA instability.

Phenotype

Mutant mice were born at the expected frequency, were apparently healthy at birth, developed normally and were fertile. The thymus and spleen were reduced in size and weight but were of essentially normal gross architecture. Lymph nodes and Peyer's patches were absent. Mutant animals showed a significant decrease in white cell count that was due to a severe reduction in lymphocytes. Other blood parameters and cell populations were normal. Thymus and spleen of IL-7$^{-/-}$ mice were decreased in cellularity to 5% and 15% of wild type, respectively, and numbers of mature splenic B and T cells were dramatically reduced. No increase in peripheral lymphocytes was observed in IL-7$^{-/-}$ mice of up to 4 months of age. While normal numbers of cells were present in mutant bone marrow, almost no B220$^+$ IgM$^+$ cells were present. Further investigation showed that there was a significant block in the transition from the pro-B to the pre-B cell stage (fraction D). The small remaining populations of thymocytes present in the mutant thymus and splenic T cells in the spleen showed normal ratios of T cell subpopulations, indicating that IL-7 is important for proliferation and expansion of T cells but not their differentiation. Mitogenic responses of residual thymocytes and splenic T cells were normal. Fetal $\gamma\delta$ T cell maturation was impaired in IL-7$^{-/-}$ mice, resulting in a dramatic reduction in adult thymic $\gamma\delta$ T cells and splenic and intestinal intra-epithelial $\gamma\delta$ T cells. In contrast, NK cell maturation was only mildly affected. The expression of c-*kit* (CD117) on IL-7$^{-/-}$ CD4^{low} and pro-T cells was also decreased.

Comments

Normal B and T cell development are dependent on IL-7, particularly with respect to lymphoid lineage expansion. IL-7 function cannot be replaced by other cytokines. Differentiation of fraction B/C B cells to fraction D is absolutely dependent on IL-7. IL-7 is critical for thymocyte expansion but not for T cell differentiation. IL-7 may influence the expression of other cytokine receptors involved in early hematopoiesis.

IL-7-deficient mice are unusual in that they show severe perturbations of the lymphoid system that are due to the absence of a single cytokine. Interestingly, IL-7$^{-/-}$ mice showed a later block in B cell development than did IL-7R$\alpha^{-/-}$ mice, possibly due to the use by other cytokines of the IL-7Rα chain. The phenotype of IL-2R$\gamma^{-/-}$ mice, similar in many aspects to that of IL-7$^{-/-}$ mice, is primarily due to the absence of IL-7 signaling, even though the IL-2Rγ receptor chain is shared by four other cytokines.

References

1 von Freeden-Jeffry, U. et al. (1995) J. Exp. Med. 181, 1519–1526.
2 Moore, T.A. et al. (1996) J. Immunol. 157, 2366–2373.

Other names
Interleukin 7 receptor α subunit, IL-7R, high-affinity IL-7R

Gene symbol
Il7r

Accession number
MGI: 96562

Area of impact
Immunity and inflammation, cytokines

General description

IL-7 was initially identified as a soluble factor produced by bone marrow stromal cells that supported the short-term proliferation of B cell progenitors. IL-7 is mitogenic for double negative thymocytes, comitogenic for mature T cells, induces CTL and LAK generation, and induces tumoricidal activity and pro-inflammatory cytokine synthesis by monocytes. IL-7 exerts its effects by binding to the high-affinity IL-7 receptor, which consists of a unique α chain and the common γ chain shared by the IL-2, IL-4, IL-9 and IL-15 receptors.

KO strain 1 construction[1]

The targeting vector was of the replacement type. An MC1*neo* cassette was inserted into exon 3 of the IL-7Rα gene at approximately amino acid 90 of the 180 amino acid extracellular domain; the mutant allele therefore encoded less that half of the IL-7R extracellular domain. The thymidine kinase cassette MC1*tk* was inserted into the 3′ end of the vector. The IL-7R mutation was originally generated in 129-derived AB1 ES cells. Positive clones were injected into C57BL/6 blastocysts and chimeras were bred to 129J females. The mutation has since been moved onto C57BL/6 to B6N5 by successive backcrossing.

RT-PCR analysis confirmed the lack of intact IL-7R mRNA, although the expected IL-7R/*neo* hybrid transcript was present.

KO strain 2 construction[2]

A disruption vector with a PGK*neo* cassette and the MC1*tk* cassette was used to replace the exon 2 of the IL-7Rα gene. The targeting vector was electroporated into E14-1 ES cells and positive clones were injected into C57BL/6N blastocysts. Chimeras were bred to C57BL/6J mice and onto B6N2. The null mutation was confirmed by RT-PCR analysis of mutant spleen mRNA and flow cytometry of bone marrow and spleen cells.

Phenotype

IL-7R$^{-/-}$ mice were born at the expected Mendelian frequency, bred normally, had normal lifespans and did not display any overt external phenotypic anomalies. IL-7R$^{-/-}$ mice had reduced numbers of mature peripheral B cells and αβ T cells, and lacked γδ T cells. Levels of NK and myeloid lineage cells were unaffected by the mutation. The stages of B and T lymphopoiesis that were affected in IL-7R$^{-/-}$ mice were defined by flow cytometric analyses of bone marrow and thymus, respectively. Late pro-B cells and their descendants were dramatically under-represented. Thymocyte cellularity in IL-7R$^{-/-}$ mice ranged from10^4 to 10^7. Under specific pathogen free (SPF) conditions, ~70% of IL-7R$^{-/-}$ mice had thymic cellularities of less than 10^6. The basis for this variation was not defined and was observed on both 129 and C57BL/6 genetic backgrounds. CD4$^-$CD8$^-$ thymocytes expressing CD25 were dramatically underrepresented in IL-7R$^{-/-}$ mice. γδ T cell development in fetal and adult thymi was completely blocked. Peripheral T cells in IL-7R$^{-/-}$ mice were also reduced in numbers, were dysfunctional, and failed to respond appropriately to a number of activation regimens. While T cell-independent B cell responses appeared to be normal, T cell-dependent B cell responses were defective, and the ability to reject allogeneic tumors was impaired.

Comments

Signaling through the IL-7R is indispensable for thymic and extrathymic γδT cell development, and for functional T cell maturation. IL-7Rα may be involved in signaling for the expansion and/or survival of the earliest T cell precursors in the thymus without being essential for the induction of αβTCR rearrangement.

The phenotypic similarities between IL-7R$^{-/-}$ mice and IL-2Rγ-deficient humans and mice indicate that the T cell developmental defect due to IL-2Rγ deficiency can in large part be explained by an inability to respond to IL-7. These comparisons also reveal that in the mouse, the IL-7R may be an obligate receptor for an additional cytokine that does not utilize the IL-2Rγ chain, and that the cytokine and cytokine receptor requirements for B lymphopoiesis differ in humans and in mice.

Acknowledgements
Jacques Peschon
Immunex Corporation, Seattle, WA, USA

References

[1] Peschon, J.J. et al. (1994) J. Exp. Med. 180, 1955–1960.
[2] Maki, K. et al. (1996) Proc. Natl Acad. Sci. 93, 7172–7177.
[3] He, Y. and Malek, T.R. (1996) J. Exp. Med. 184, 289–293.
[4] Maraskovsky, E. et al. (1996) J. Immunol. 157, 5315–5323.

Other names
Interleukin 8 receptor

Gene symbol
Cmkar2

Accession number
MGI: 105303

Area of impact
Immunity and inflammation, cytokines

General description

IL-8 is an important chemoattractant and activator of neutrophils that is required for the transvenule migration of these cells at sites of inflammation. Unlike humans, which have two high-affinity IL-8 receptors, the murine genome contains a single IL-8R gene encoded by a single exon. IL-8R plays a role not only in the normal inflammatory response by activating neutrophils to migrate to the site of acute injury, but also in the differentiation or expansion of myeloid and lymphocytic cells. The possible involvement of IL-8R in the regulation of hematopoiesis and myelopoiesis was an unexpected result of experiments in which targeted gene deletion was used to further study IL-8R gene function.

KO strain construction

The single exon containing the open reading frame of IL-8R was replaced with a neomycin-resistance gene under control of the PGK promoter. The targeting vector was electroporated into D3 ES cells. Positive clones were injected into C57BL/6 blastocysts to generate chimeras. Chimeras were mated with C57BL/6J females and heterozygotes were intercrossed to obtain IL-8R$^{-/-}$ animals. Mice were backcrossed onto BALB/c and C57BL/6 genetic backgrounds.

The IL-8R gene was completely eliminated after homologous recombination, as confirmed by Southern blot using the complete exon as a probe.

Phenotype

IL-8R$^{-/-}$ mice were initially indistinguishable from the wild type, but over time, the mutant mice failed to gain normal weight and their fur appeared ruffled. Spleens ranged from 2 to 4 times normal size and cervical lymph nodes were enlarged 3- to 10-fold. Most other lymph nodes were enlarged to varying degrees, with the exception of the inguinal and popliteal nodes, which were grossly normal. Histopathology showed splenomegaly resulting from expansion of the splenic white pulp due to proliferation of myeloid elements and megakaryocytes. Longitudinal sections of the femur and tibia of IL-8R$^{-/-}$ mice revealed grossly white marrow instead of the normal red. Histologically, there was a great increase in bone marrow cellularity which was composed of the normal myeloid maturation series. In addition, IL-8R$^{-/-}$ bones were more fragile. In lymph nodes, the medullary cords were expanded by abundant foci of myelopoiesis, Russell bodies, and plasma cells, and they compressed the adjacent medullary sinuses. The nearly 10-fold increase in B cells was confirmed by flow cytometric analysis. The number of circulating neutrophils was increased 12-fold. However, the acute migration ability of the neutrophils was compromised and these cells could not respond to an inflammatory challenge of thioglycollate injection. *In vitro* migration and intracellular calcium flux response to receptor ligands IL-8, MIP-2, and KC were significantly negatively affected.

Mice reared in conventional animal breeding facilities that are specific pathogen free (SPF) are still subject to bacterial and fungal exposure. When the IL-8R KO mice were rederived in a germ-free (GF) setting so that the mice were negative for cultured flora and showed a completely naive immune system, the neutrophilia and previously described phenotypes were ameliorated. Upon reintroduction into an SPF breeding facility, the neutrophilia, enlarged spleen and enlarged lymph nodes returned. In addition, homozygous breeding of IL-8R$^{-/-}$ mice in an SPF facility ranged from extremely difficult to impossible, but completely normal breeding occurred under GF conditions.

Comments

IL-8R is critical for the inflammatory and migratory response of neutrophils. IL-8R may be involved in negative regulation of neutrophil production. The apparent involvement of IL-8R in the regulation of myelopoiesis was unexpected.

Acknowledgements
Grace Cacalano
Genentech, S. San Francisco, CA, USA

References

1 Cacalano, G. et al. (1994) Science 265, 682–684.
2 Lee, J. et al. (1995) J. Immunol. 155, 2158–2164.
3 Moore, M.W. et al. (1995) Science 269, 1591.
4 Broxmeyer, H. et al. (1996) J. Exp. Med. 184, 1825–1832.

Other names
IL-11Ra

Gene symbol
Il11ra1

Accession number
MGI: 107426

Area of impact
Development and hematopoiesis

General description

The interleukin 11 (IL-11) receptor α chain is a member of the hematopoietin family of receptors. It complexes with a β chain, gp130, to form a functional high-affinity receptor for IL-11. Two genes encode IL-11 receptor α chains in the mouse: *Il11Ra1* and *Il11Ra2*. *Il11Ra1* is widely expressed whereas the expression of *Il11Ra2* is restricted. *Il11Ra2* is absent in some mouse strains and is not found in humans.

KO strain construction

The targeting vector was designed such that PGK*neo*polyA replaced parts of exons 8 and 13 and all of exons 9–12 of the *Il11Ra1* gene. This resulted in the deletion of part of the hematopoietin subdomain, including the WS × WS region, the transmembrane region and the membrane-proximal portion of the cytoplasmic tail. PGK*tk* was included 5′ of the genomic sequences. The targeting vector was electroporated into W9.5 ES cells (129Sv). Positive clones were used to generate chimeric mice which were mated with C57BL/6 females. Heterozygous offspring were intercrossed to obtain IL-11Ra1$^{-/-}$ mice. The mutation has also been derived on the 129Sv background.

The null mutation was confirmed by expression analysis and by the absence of response of IL-11Ra1$^{-/-}$ bone marrow cells in hematopoietic colony assays.

Phenotype

Mice with a null mutation of the *Il11Ra1* gene appeared phenotypically normal. Examination of peripheral blood, bone marrow and spleen, CFU-S assays and clonal cultures of hematopoietic progenitor cells did not identify any perturbation of hematopoiesis in these mice. The hematopoietic recovery after cytoablation and hemolysis was also normal.

Unexpectedly, homozygous mutant females were infertile due to altered decidualization – the process of uterine stromal cell differentiation into specialized decidual cells in response to implantation. In IL-11Ra1$^{-/-}$ uteri, the anti-mesometrial decidual response to the implanting blastocyst was reduced and mesometrial decidua failed to develop, allowing invasion by increased numbers of trophoblastic giant cells. This resulted in the death of the embryo at 8 days pc. Oil-induced deciduomata formation was also abnormal in the IL-11Ra1$^{-/-}$ uteri and was not rescued by administration of progesterone. *In situ* hybridization studies showed IL-11 expression in the primary decidual zone adjacent to the implanting blastocyst and IL-11Ra1 expression in the decidual cells. IL-11 expression in the pregnant uterus followed the time course of the decidualization response.

Comments

IL-11 receptor signaling is not required for adult hematopoiesis. The ablation of IL-11 action by disruption of the IL-11 receptor α chain alters the maternal decidual transformation in response to the implanting blastocyst, resulting in female infertility.

Acknowledgements
Lorraine Robb
The Walter and Eliza Hall Institute of Medical Research, The Royal Melbourne Hospital, Victoria, Australia

References

[1] Nandurkar, H.H. et al. (1997) Blood 90, 2148–2159.
[2] Robb, L. et al. (1998) Nature Med. 4, 303–308.

Other names
Interleukin 12α, p35

Gene symbol
Il12a

Accession number
MGI: 96539

Area of impact
Immunity and inflammation, cytokines

General description

IL-12 is a heterodimeric cytokine composed of a 35 kDa and a 40 kDa subunit (p35 and p40). IL-12 affects the Th1/Th2 balance by inducing Th1 differentiation, as well as the expression of IFNγ. IL-12α (p35) is constitutively expressed in immune cells but is not secreted in the absence of IL-12β (p40). IL-12β can be expressed as a homodimer, which can function as an antagonist of intact, heterodimeric IL-12. In immune cells, the expression of IL-12β enables formation and secretion of bioactive IL-12 p75 heterodimers.

KO strain construction

Upon homologous recombination with the replacement targeting vector, exons 1 and 2 of the IL-12α gene were replaced with the *neo* cassette. Correctly targeted ES cell clones from the W9.5 ES cell line derived from 129Sv mice were injected into host C57BL/6J blastocysts which were transferred into (C57BL/6J × CBA/J) F1 females. Chimeric males were mated with 129Sv females and the progeny were intercrossed. Homozygous deletion of the IL-12α gene was confirmed by Southern blotting.

Phenotype

IL-12α homozygous mutant mice developed normally and were fully fertile, with no gross abnormalities. The mice were unable to restrict the progression of *Leishmania major* infection and had lost the ability to mount a *Leishmania*-specific delayed-type hypersensitivity response. IL-4 levels were elevated in mutant lymph node $CD4^+$ cells after infection. When lymph node cells were stimulated *in vitro* with *Leishmania* antigen, IFNγ expression was much lower in mutant $CD4^+$ T cells than in either 129Sv or BALB/c mice, while IL-4 production was elevated in mutant 129Sv mice relative to wild-type mice.

Comments

These results are consistent with a central role for IL-12 as a Th1 cytokine, controlling the Th1/Th2 balance during the course of an *in vivo* infection. In addition, the influence of IL-12 on Th1 responses during *L. major* infection appears to be crucial for the ability of resistant mice to control the infection.

IL-12-deficient mice also represent a useful tool for studying the *in vivo* role of Th1 vs. Th2 responses to different types of infection.

Acknowledgements
Gottfried Alber
University of Leipzig, Leipzig, Germany

Reference

1 Mattner, F. et al. (1996) Eur. J. Immunol. 26, 1553–1559.

Other names
Interleukin 12β, p40

Gene symbol
Il12b

Accession number
MGI: 96540

Area of impact
Immunity and inflammation, cytokines

General description

IL-12 is a heterodimeric cytokine composed of a 35 kDa and a 40 kDa subunit (p35 and p40). IL-12 affects the Th1/Th2 balance by inducing Th1 differentiation, as well as the expression of IFNγ. It is expressed by antigen-presenting cells such as macrophages, monocytes and dendritic cells. IL-12α (p35) is constitutively expressed in immune cells, but is not secreted in the absence of IL-12β (p40). IL-12β is expressed upon induction, and can be expressed as a homodimer (p80) which can function as an antagonist of intact, heterodimeric IL-12. In immune cells, the expression of IL-12β enables formation and secretion of bioactive IL-12 p75 heterodimers.

KO strain 1 construction

Upon homologous recombination with the replacement targeting vector a portion of exon 3 of the IL-12β gene was replaced with a *neo* cassette. Correctly targeted ES cell clones from the W9.5 ES cell line derived from 129Sv mice were injected into host C57BL/6J blastocysts which were transferred into C57BL/6J females. Chimeric males were mated with 129Sv females and the progeny were intercrossed. Deletion of the IL-12β gene was confirmed by Southern blotting and by an IL-12 bioassay.

Phenotype

IL-12β homozygous mutant mice developed normally and were fully fertile, with no gross abnormalities. The mice had an impaired ability to produce IFN-γ in response to LPS stimulation *in vivo,* and IL-4 production was enhanced. Secretion of IL-2 and IL-10 in response to antigen stimulation was normal, as were allogeneic CTL responses. The mice were unable to restrict the progression of *Leishmania major* infection and had lost the ability to mount a *Leishmania*-specific delayed-type hypersensitivity (DTH) response. IL-4 levels were elevated in mutant lymph node $CD4^+$ cells after infection. When lymph node cells were stimulated *in vitro* with *Leishmania* antigen, IFN-γ expression was much lower in mutant $CD4^+$ T cells than in either 129Sv or BALB/c mice, while IL-4 production was elevated in mutant 129Sv mice relative to wild-type mice.

Comments

These results indicate that IL-12 plays an essential role in regulating IFN-γ production and in facilitating normal DTH responses. However, IL-2 secretion and CTL generation were not compromised by the absence of IL-12. The influence of IL-12 on initiating a Th1 response can be crucial to the successful defense against some types of infections.

Acknowledgements
Gottfried Alber
University of Leipzig, Leipzig, Germany

Reference

[1] Magram J. et al. (1996) Immunity 4, 471–481.

α-Inhibin

Gene symbol
Inha

Accession number
MGI: 96569

Area of impact
Reproduction, hormone

General description

Activins and inhibins are members of the TGFβ growth factor family. There are two inhibin heterodimers, inhibin A (α:βA) and inhibin B (α:βB). These forms have an inhibin-specific α subunit but share β subunits with the activin β:β dimers. There are three activins: activin A (βA:βA), activin B (βB:βB) and activin AB (βA:βB). Activins are activated by follicle-stimulating hormone (FSH) and have functions that are antagonistic to inhibins.

Inhibins are synthesized and secreted in the placenta during embryogenesis, and in the gonads (Sertoli cells in males and the granulosa cells in females), pituitary gland, adrenal gland, spleen, and nervous system in adults. In males, inhibin abrogates spermatogonia DNA synthesis and spermatogonia numbers. Inhibin and luteinizing hormone (LH) induce steroidogenesis in Leydig cells. In females, inhibin can increase numbers of ovarian follicles, inhibit oocyte meiosis, and induce steroidogenesis in thecal cells. Inhibin secretion is inhibited by pituitary FSH. Inhibins regulate synthesis of human chorionic gonadotrophin (hCG) in the placenta.

KO strain construction

Exons 1 and 2 were replaced by a PGK*hprt* cassette. For negative selection, two MC1*tk* expression cassettes were placed at the 5′ and 3′ ends of the construct. Hprt-mutant AB2.1 ES cells were targeted and injected into C57BL/6 mice.

Phenotype

Male and female α-inhibin$^{-/-}$ mice developed normally. However, as early as 4 weeks of age these mice developed multifocal or intratubular uni- or bilateral ovarian and testicular stromal tumors, implying that α-inhibin is a negative regulator of gonadal stromal cell proliferation. Serum levels of FSH were increased in α-inhibin$^{-/-}$ male and female mice.

Reference

[1] Matzuk, M.M. et al. (1992) Nature 360, 313–319.

Other names
Inducible nitric oxide synthase, nitric oxide synthase 2 (NOS2)

Gene symbol
Nos2

Accession number
MGI: 97361

Area of impact
Immunity and inflammation

General description
iNOS belongs to a tripartite multigene family of flavoenzymes (NOSs; EC 1.14.13.39) catalyzing the 5-electron oxidation of L-arginine to L-citrulline + NO. The iNOS gene is induced and expressed to high levels in a wide variety of cells and tissues upon exposure to inflammatory stimuli, including cytokines such as IFN-γ, IL-1β, TNFα and LPS. This induction results in a high output over prolonged periods of nitric oxide, a radical-generating gas. Nitric oxide produced by iNOS has been shown to be beneficial through its anti-tumor and anti-microbial activities. Once activated, it is capable of restricting the growth of taxonomically unrelated intracellular pathogens and extracellular tumors. Studies in animal models of inflammatory bowel disease, cerebral ischemia, arthritis, diabetes and immune complex glomerulonephritis, as well as in experimental allergic encephalomyelitis, have suggested a role for NO in the pathogenesis of inflammation and pain.

KO strain 1 construction[1]
The gene replacement vector (pINOS-RV1) was constructed to delete the proximal 585 bases of the iNOS promoter (previously shown to be required for mouse iNOS expression) as well as exons 1–4. The vector thus consisted of a 1.034 kb 5′ short arm, a 6.7 kb 3′ long arm, and an antisense PGK*neo*^{r} cassette insertion. The targeting vector was electroporated into AB2.1 ES cells (129SvEv) and iNOS$^{-/-}$ mutants were derived on 129SvEv, C57BL/6, B10.RIII, PL/J, NOD, MRL-*lpr/lpr*, and DBA1/J backgrounds. The null mutation was confirmed by Northern analysis of RNA from mice injected with LPS. iNOS$^{-/-}$ macrophages lacked expression of iNOS mRNA (as determined by Northern blot), iNOS protein (as determined by Western blot) and iNOS enzymatic activity (as determined by assay for NO generation) when treated with LPS ± IFN-γ, TNF or IFN-α/β. The expression of other NOS genes was not affected.

KO strain 2 construction[2]
The targeting construct (pSPKO-NOS) consisted of an *Sph*I/*Eco*RI 9 kb iNOS genomic fragment (originally derived from strain 129) in which the *Apa*I/*Kpn*I calmodulin-binding domain (exons 11–13) was replaced by the neomycin-

resistance gene. The HSV-*tk* gene was also included to allow positive–negative selection. The targeting vector was electroporated into the 129-derived ES cell line E14TG2a. Positive clones were injected into C57BL/6J (B6) blastocysts. Chimeric males were mated with B6 females and the B6/129 F1 heterozygotes were crossed to generate homozygous iNOS$^{-/-}$ mutants. This iNOS knockout has been backcrossed onto C57BL/6J for ten generations. The null mutation was confirmed by Northern analysis of RNA from peritoneal macrophages treated with LPS/IFN-γ, by immunoblot, and by enzymatic assay for NOS activity.

KO strain 3 construction[3]

The targeting construct, based on a 129Sv genomic clone, was of the replacement type. The expected deletion of exons 1–5 did not occur during production of the vector, but the rearranged mutant allele that was obtained did give rise to iNOS$^{-/-}$ mice. The mutant allele was electroporated into the CGR8 ES cell line. Chimeras were mated with outbred MF1 mice and homozygous mutants were generated by interbreeding. Loss of the wild-type allele was confirmed by Southern blotting and by Northern analysis of RNA from LPS/IFN-γ-induced peritoneal macrophages. Although a transcript of abnormal size was present in mutant mice, no iNOS protein could be detected by Western blot, and nitrite production was at background levels.

Phenotype

Mutant mice were developmentally normal, had normal lifespans and were reproductively fertile. No gross or histologic changes in tissues of mutant mice were observed at either 8 weeks or 14 months of age. Peritoneal leukocyte migration, macrophage oxidative burst and MHC class II expression were normal, as were splenic and thymic Ig^{+} B cell, CD4^{+}CD8^{-} and CD4^{-}CD8^{+} T cell populations.

To test the *in vivo* contribution of iNOS in mouse models of inflammatory diseases, iNOS$^{-/-}$ mice were bred to mouse backgrounds shown to be susceptible to induced and spontaneous inflammatory diseases. The role of iNOS was examined in murine models of collagen-induced arthritis (CIA), experimental autoimmune encephalomyelitis (EAE), type I diabetes in the NOD genetic background, and inflammatory bowel disease. Disease susceptibility and progression were unaffected in CIA (in both the B10.RIII and DBA1 backgrounds), in diabetes (NOD background), in inflammatory bowel disease, and in joint disease and nephritis in the MRL-lpr/lpr background. However, disease susceptibility and progression were exacerbated in a model of EAE. Mutant mice also showed a reduced non-specific inflammatory response to carrageenin, while leukocyte rolling and adhesion was elevated when given LPS.

Only when infectiously or tumorigenically challenged did iNOS$^{-/-}$ mice exhibit defects in cell-mediated immunity compared to wild-type littermates. iNOS$^{-/-}$ mice failed to effectively eliminate infections by gram-positive bacilli (*Listeria monocytogenes*), mycobacteria, ectromelia, or neurovirulent herpes simplex 1 virus *in vivo*, or restrict the growth of L1210 tumor cells *in vitro*. The immunologic lesions were related to the absence of iNOS, since anti-listerial T cell memory, tuberculostatic IFN-γ and TNFα release, and virus-specific CTL and NK cell activities were intact. Mutant mice were highly susceptible to *L. major* infection and developed significantly stronger Th1-type immune responses compared to wild-type or heterozygous mice. iNOS$^{-/-}$ mice were not protected from multi-organ failure, inflammatory cytokine release (IL-1α, IL-1β, TNFα and IL-6), or mortality in two models of *Escherichia coli* LPS-induced sepsis, despite an attenuated drop in mean arterial blood pressure.

Comments

The loss of iNOS increased the susceptibility of mutant mice to listeria, leishmania, tuberculosis, and other infections. However, these studies of iNOS$^{-/-}$ mice may prompt the re-evaluation of the role of iNOS in inflammatory diseases. In addition, these results show that while iNOS may predominantly govern LPS-induced hypotension, iNOS-independent routes to endotoxic death exist. There is also evidence that these mice will have altered responses to organ rejection and ischemia-reperfusion injury. iNOS$^{-/-}$ mice will be important in the study of inflammatory disorders, and of cancer growth and metastasis.

Acknowledgements
John Mudgett
Merck Research Laboratories, Rahway, NJ, USA
John MacMicking
Laboratory of Immunology, Howard Hughes Medical Institute, The Rockefeller University, New York, NY, USA
Victor Laubach
University of Virginia, Department of Surgery, Charlottesville, VA, USA

References

1. MacMicking, J.D. et al. (1995) Cell 81, 641–650.
2. Laubach, V.E. et al. (1995) Proc. Natl Acad. Sci. USA 92, 10688–10692.
3. Wei, X. et al. (1995) Nature 375, 408–411.
4. MacMicking, J. et al. (1997) Annu. Rev. Immunol. 15, 323–350.
5. MacMicking, J. et al. (1997) Proc.Natl Acad. Sci. USA 94, 5243–5248.
6. Karupiah, G. et al. (1993) Science 261, 1445–1448.
7. Hickey, M.J. et al. (1997) FASEB J. 11, 955–964.
8. Gilkenson, G.S. et al. (1997) J. Exp. Med. 186, 365–373.
9. Fenyk-Melody, J. et al. (1998) J. Immunol. 160, 2940–2946.
10. Diefenbach, A. et al. (1998) Immunity 8, 77–87.

Integrin α1

Gene symbol
Itga1

Accession number
MGI: 96599

Area of impact
Development

General description

Integrins are cell surface receptors for extracellular matrix molecules. They exist as heterodimers of an α and a β chain. Integrin α1 is a receptor for laminin and collagen which is expressed widely and dynamically in embryogenesis, and has been implicated in various developmental processes including establishment of the placenta and formation of the central and peripheral nervous systems. In the adult it is the sole collagen receptor in smooth muscle and liver and is thought to be important for stability of these tissues.

KO strain construction

A PGK*neo* cassette was inserted into a unique *Cla*I site in an exon near the N-terminus. The targeting vector was introduced into J1 129 ES cells, which were injected into BALB/c blastocysts. The mutant strain was maintained on a 129 × BALB/c hybrid background.

Phenotype

Mice homozygous for this mutation are viable and fertile and have no overt phenotype, demonstrating that the molecule is not required for development. Embryonic fibroblasts derived from mutant animals are unable to spread on or migrate into substrate of collagen IV and are deficient in spreading on and migrating into laminin. Further *in vitro* analysis of cell spreading and migration suggests that α1β1 is not required for binding to collagen 1, and implicates a third receptor, possibly integrin α3β1, in collagen 1 binding.

Acknowledgements
Humphrey Gardner
Scripps Research Institute, La Jolla, CA, USA

Reference

[1] Gardner, H. et al. (1996) Dev. Biol. 175, 301–313.

Integrin α3

Gene symbol
Itga3

Accession number
MGI: 96602

Area of impact
Development

General description

Integrins are cell surface receptors for extracellular matrix molecules. They exist as heterodimers of an α and a β chain. Integrin α3β1 is a receptor for laminin 5 and also binds to a variety of other extracellular matrix proteins.

KO strain construction

A PGK*neo* cassette was inserted in a *Bcl* site within an exon near the 5′ end of the gene. The construct was transfected into J1 ES cells (129Sv) and resulting targeted clones were injected into recipient C57BL/6 blastocysts. The mutant phenotype was analyzed on a 129 × B6 hybrid background.

Phenotype

Mice homozygous for the mutation show perinatal lethality. Histological abnormalities in kidney and lung development and small skin blisters are observed. The basement membranes of the glomerulus and the epidermal–dermal junction are disorganized, indicative of extracellular matrix disruption.

Acknowledgements
Jordan Kreidberg
Children's Hospital and Harvard Medical School, Boston, MA, USA

References

[1] Kreidberg, J.A. et al. (1996) Development 122, 3537–3547.
[2] DiPersio, C.M. et al. (1997) J. Cell Biol. 137, 729–742.

Integrin $\alpha4$

Gene symbol
Itga4

Accession number
MGI: 96603

Area of impact
Development

General description

Integrins are cell surface receptors for extracellular matrix molecules. They exist as heterodimers of an α and a β chain. Integrin $\alpha4$ complexes with two different integrin β subunits. $\alpha4\beta1$ and $\alpha4\beta7$ are receptors for both an alternatively spliced form of fibronectin and for the Ig superfamily molecule VCAM-1. $\alpha4\beta7$ also binds MAdCAM-1.

KO strain construction

A PGK*neo* cassette was inserted in place of most of the first exon, including the ATG and signal sequence. The construct was transfected into D3 ES cells (129Sv) and targeted clones were injected into C57BL/6 recipient blastocysts. Chimeras were bred to C57BL/6, and the mutant phenotype examined on a 129 $\times$ B6 background.

Phenotype

Homozygotes show embryonic lethality. Approximately half the embryos fail to form a placenta because of failure of fusion between the allantois (VCAM-1^+) and the chorion ($\alpha4^+$). The rest die by cardiac hemorrhage resulting from detachment and rupture of the epicardium ($\alpha4^+$) and the coronary vessels which lie below it. Chimeric mice made from $\alpha4$-null ES cells allow dissection of the role of $\alpha4$ integrins in development of lymphocytes and leukocytes[2] and of muscle. Integrins $\alpha4$ are required for post-natal lymphopoiesis in the bone marrow but are not required for fetal development of T cells or for monocyte or NK cell development. Integrins $\alpha4$ are necessary for lymphocyte homing to Peyer's patches but not to many other lymphoid organs. They appear to be dispensible for myogenesis or muscle development.

Acknowledgements
Richard Hynes
Massachusetts Institute of Technology, Cambridge, MA, USA

References

1 Yang, J.T. et al. (1995) Development 121, 549–560.
2 Arroyo, A.G. et al. (1996) Cell 85, 997–1008.
3 Yang, J.T. et al. (1996) J. Cell Biol. 135, 829–835.

Integrin α5

Other names
Fibronectin receptor α

Gene symbol
Itga5

Accession number
MGI: 96604

Area of impact
Development

General description

Integrins are cell surface receptors for extracellular matrix molecules. They exist as heterodimers of an α and β chain. Integrin α5β1 is a major receptor for fibronectin and mediates fibronectin matrix assembly and fibronectin-dependent adhesion, spreading, migration and intracellular signaling.

KO strain construction

A PGK*neo* cassette was inserted in place of most of the first exon, including the ATG and signal sequence. The construct was transfected into D3 ES cells (129Sv) and targeted clones were injected into C57BL/6 recipient blastocysts. Chimeras were bred to C57BL/6. The mutant phenotype was analyzed on a 129 × B6 hybrid background.

Phenotype

Homozygous embryos die around days 10–11 of gestation with vascular defects, both extra-embryonic and embryonic. The anterior part of the embryo develops relatively normally (up to the forelimb level) but the posterior develops very poorly, producing axial deformities and shortening. Surprisingly, the embryos do assemble fibronectin-rich matrices as do cells cultured from null embryos. Integrin α5-null cells also perform many fibronectin-dependent functions (adhesion, spreading, cytoskeletal assembly, focal contact formation, migration) normally.

Comments

The phenotype is milder than that of fibronectin-null embryos. Together with the fibronectin-responsiveness of α5-null cells, this shows that other fibronectin receptors must be active during development. Further analysis implicates αvβ1 integrin.

Acknowledgements
Richard Hynes
Massachusetts Institute of Technology, Cambridge, MA, USA

References
[1] Yang, J.T. et al. (1993) Development 119, 1093–1105.
[2] Yang, J.T. et al. (1996) Mol. Biol. Cell 7, 1737–1748.

Integrin α6

Other names
ITGAG in human

Gene symbol
Itga6

Accession number
MGI: 96605

Area of impact
Development

General description

Integrins are cell surface receptors for extracellular matrix molecules. They exist as heterodimers of an α and a β chain. Integrin α6 can associate with either β1 chain or β4 chain. Integrins α6β1 and α6β4 are receptors for laminins. α6β1 is widely expressed in the embryonic epithelia. α6β4 is associated with hemidesmosomes of stratified epithelia, Schwann cells and other sites.

KO strain construction

The targeting vector results in the deletion of 7 kb comprising three exons in the 5′ region, and replacement of the deleted region by a PGK*neo* cassette. The targeting vector was introduced into 129Sv ES cells, which were injected into C57BL/6 blastocysts. The mutant phenotype was analyzed on a 129 × B6 background.

Phenotype

Homozygous embryos die neonatally. There is detachment of the epidermis and other epithelia (oral cavity, esophagus) and absence of hemidesmosomes in basal keratinocytes. Other phenotypes may exist and are under investigation.

Comments

The skin phenotype is similar to that of the integrin β4 knockout. The phenotype also resembles the human condition junctional epidermolysis bullosa.

Acknowledgements
Elisabeth Georges-Labouesse, IGBMC, Illkirch, France

References

[1] Georges-Labouesse, E.N. et al. (1996) Nature Genet. 13, 370–373.
[2] Van der Neut, R. et al. (1996) Nature Genet. 13, 367–369.
[3] Dowling, J. et al. (1996) J. Cell Biol. 134, 559–572.
[4] Borradori, L. et al. (1996) Curr. Opin. Cell Biol. 8, 647–656.
[5] Fassler, R. et al. (1996) Curr. Opin. Cell Biol. 8, 641–646.

Integrin α9

Gene symbol
Itga9

Accession number
MGI: 104756

Area of impact
Development

General description

Integrins are cell surface receptors for extracellular matrix molecules. They exist as heterodimers of an α and a β chain. The α9 subunit is part of a single integrin heterodimer, α9/β1, that is widely expressed in epithelia and smooth muscle. This integrin recognizes tenascin and an N-terminal fragment of osteopontin as ligands, but its function is largely unknown.

KO strain construction

A replacement construct was used to insert a neomycin-resistance cassette into a deleted region of a critical exon encoding the ligand-binding region. 129 ES cells (RF 8 from Robert Farese) were targeted. Chimeras were backcrossed to either 129 or C57BL/6.

Phenotype

Mice homozygous for the null mutation were born at the expected frequency, but died within the first 7–10 days of life from respiratory failure due to bilateral chylothorax.

Comments

The observations suggest that α9/β1 plays a critical role in the development of the thoracic duct.

Acknowledgements
Dean Sheppard
University of California San Francisco, San Francisco, CA, USA

References

1 The integrin $\alpha 9^{-/-}$ mouse is unpublished
2 Palmer, E.L. et al. (1993) J. Cell Biol. 123, 1289–1297.
3 Yokosaki, Y. et al. (1994) J. Biol. Chem. 269, 26691–26696.
4 Smith, L.L. et al. (1996) J. Biol. Chem. 271, 28485–28491.

Integrin αV

Other names
CD51

Gene symbol
Itgav

Accession number
MGI: 96608

Area of impact
Development

General description

Integrins are cell surface receptors for extracellular matrix molecules. They exist as heterodimers of an α and a β chain. Integrin αV combines with several different β subunits to form a family of αV integrins. These integrins all recognize the sequence RGD in a variety of extracellular matrix proteins, including vitronectin, fibronectin, thrombospondin and fibrinogen.

KO strain construction

A PGK*neo* cassette was inserted in place of exon 1. The construct was transfected into D3 ES cells (129Sv) and targeted clones were injected into C57BL/6 recipient blastocysts. Chimeras were bred to C57BL/6.

Phenotype

Homozygous embryos show a variable phenotype. Eighty per cent of embryos die between E10.5 and 12.5, while 20% pass through this phase and develop to term. All of these show cerebral hemorrhage and underdevelopment of the brain and die soon after birth. The results suggest that some aspects of vasculogenesis and angiogenesis are independent of αV integrins.

Acknowledgements
Richard Hynes
Massachusetts Institute of Technology, Cambridge, MA, USA

Reference

[1] Bader, B. et al. (unpublished results).

Integrin β1

Other names
Fibronectin receptor β

Gene symbol
Itgb1

Accession number
MGI: 96610

Area of impact
Development

General description

Integrins are cell surface receptors for extracellular matrix molecules. They exist as heterodimers of an α and a β chain. The β1 subunit dimerizes with a number of different α subunit to form one major family of integrin receptors, which are widely expressed in different tissues and show different ligand specificities.

KO strain 1 construction[1]

The targeting vector inserts a *neo* cassette into the first coding exon to disrupt translation. The vector was introduced into JM-1 129SvJ ES cells, which were injected into C57BL/6 blastocysts. The mutant phenotype was analyzed on a 129 × outbred Swiss Webster Black background.

KO strain 2 construction[2]

Two mutations were made. The first inserted a promoterless β*geo* gene in-frame to the start codon of β1 integrin. The second contained only a promoterless neo^r gene at the same site. Both vectors were introduced into D3 or R1 129 ES cells, which were injected into C57BL/6 blastocysts. The phenotype was examined on either a mixed 129 × B6 or inbred 129 background.

Phenotype

Homozygous mutant embryos undergo implantation in the uterus but fail to develop normally beyond E5.5. Blastocyst outgrowths from mutant embryos show normal trophoblast development but poor growth and differentiation of the inner cell mass. These data suggest that β1 integrins are required for normal morphogenesis of the inner cell mass. In chimeric mice, β1 integrin$^{-/-}$ cells were capable of contributing to all tissues except liver and spleen. However, very large contributions of mutant cells led to abnormalities of development.

References

1 Stephens, L.E. et al. (1995) Genes Dev. 9, 1883–1895.
2 Fassler, R. and Meyer, M. (1995) Genes Dev. 9, 1896–1908.

Integrin β3

Other names
Beta 3 integrin, β3

Gene symbol
Itgb3

Accession number
MGI: 96612

Area of impact
Hematopoiesis

General description

Integrin β3 combines with αv to form the widely expressed integrin αvβ3, believed to be important in angiogenesis and in osteoclast function. β3 also combines with αIIb to form the platelet-specific integrin αIIbβ3, which is a receptor for fibrinogen, fibronectin, von Willebrand factor and thrombospondin. αIIbβ3 plays a key role in hemostasis.

KO strain construction

A PGK*neo* cassette was inserted in place of the first two exons of the β3 gene. Fragments of 2.0 kb and 3.5 kb of genomic DNA were included 5′ and 3′ of the selection cassette, and PGK*tk* cassettes were added at both ends of the construct. The construct was transfected into D3 ES cells (129Sv) and targeted clones were injected into C57BL/6 recipient blastocysts. Chimeras were bred to C57BL/6.

Phenotype

The β3-null mice were viable and outwardly normal. Further details of their phenotype are under investigation.

Comments

These mice will be a model for the human bleeding disease Glanzmann's thrombasthenia, which arises from mutations in the human β3 gene.

Acknowledgements
Richard Hynes
Center for Cancer Research, Massachusetts Institute of Technology, Cambridge, MA, USA

Reference
[1] The integrin $\beta3^{-/-}$ mouse is not yet published.

Integrin β6

Gene symbol
Itgb6

Accession number
MGI: 96615

Area of impact
Immunity and inflammation

General description

The β6 subunit is a component of a single integrin heterodimer (αvβ6) that is restricted in its distribution to epithelial cells, particularly during development. It is also expressed in inflamed and/or injured epithelial organs, and in epithelial tumors. This integrin protein is a receptor for fibronectin and tenascin. Heterologous expression augments epithelial cell proliferation in polarized cultures.

KO strain construction

The targeting construct contained a neomycin-resistance cassette inserted so as to replace a portion of an exon in the region of the protein known to bind ligands. The genomic sequences were flanked by a thymidine kinase cassette for negative selection. RF8 ES cells (129 strain) were electroporated. The ES cells were injected into C57BL/6 blastocysts, and the resulting chimeras were backcrossed to 129 and C57BL/6 mice. Successful knockout of the endogenous gene was confirmed by Southern blot and PCR analyses.

Phenotype

Mice homozygous for the null mutation developed and reproduced normally. However, they developed inflammatory baldness in areas of low-grade trauma, accompanied by infiltration of macrophages into the dermis of the affected areas. In addition, the mice spontaneously developed lymphocytic inflammation around the conducting airways of the lungs, as well as activation of alveolar macrophages. This effect principally occurred in mice maintained in unventilated cages, suggesting a role for environmentally induced injury in this phenotype. The knockout mice also manifested persistent airway hyperresponsiveness to the bronchoconstrictor acetylcholine, but were protected against bleomycin-induced pulmonary fibrosis.

Comments

The phenotype suggests that signals from epithelial cells can initiate airway inflammation and hyperresponsiveness (key features of human asthma) and can contribute to the development of lung fibrosis. Furthermore, these pro-inflammatory responses of epithelial cells appear to be downregulated by signals initiated through the epithelial-restricted integrin αvβ6.

Acknowledgements
Dean Sheppard
University of California, San Francisco, CA, USA

References

[1] Huang, X.-Z. et al. (1996) J. Cell Biol. 133, 921–928.
[2] Agrez, M. et al. (1994) J. Cell Biol. 127, 547–556.
[3] Weinacker, A. et al. (1994) J. Biol. Chem. 269, 6940–6948.
[4] Breuss, J.M. et al. (1995) J. Cell Sci. 108, 2241–2251.

Other names
Type 1 inositol 1,4,5-trisphosphate receptor 1, P400 protein

Gene symbol
Itpr1

Accession number
MGI: 96623

Area of impact
Neurology

General description

Numerous hormones and transmitters trigger hydrolysis of phosphatidylinositol bisphosphates through a G protein-coupled receptor, leading to the production of diacylglycerol and inositol 1,4,5-trisphosphates (IP3). IP3 is a key intracellular second messenger. After binding of IP3 to its receptor (IP_3R), calcium is released into the cytoplasm from its internal store in the ER. The efflux of calcium triggers various cellular functions. IP_3R exists on ER, and serves as a calcium-releasing channel. At present, IP_3R can be classified into at least three groups: type 1 (IP3R1), type 2 (IP3R2) and type 3 (IP3R3). IP3R1 is highly expressed in cerebellar Purkinje cells and hippocampus.

KO strain construction

A neomycin-resistance gene was inserted just after the ATG translation initiation site of the IP3R gene, so that no functional protein was expected from the mutated allele. The targeting vector consisted of a 7.6 kb 5′-homologous region, a 1.8 kb neomycin-resistance gene, a 1.2 kb 3′ homologous region and the diphtheria toxin A (DT-A) gene for negative selection of ES cells. The strain of ES cells used for establishing IP3R1 gene-disrupted recombinants was J1 (129Sv). The recombinant was injected into C57BL/6-derived blastocytes to produce germ line chimeras and the F1 generation. Thus, the mutant strain had a genetic background of 129Sv and C57BL/6.

Phenotype

Most IP3R1-deficient mice generated by gene targeting died *in utero*. Animals which were born had severe ataxia and tonic–clonic seizures, and died by the weaning period. An electroencephalogram showed that they suffered from epilepsy, indicating that IP3R1 is essential for proper brain function. However, observation by light microscope of hematotoxylin-eosin staining of the brain and peripheral tissues of IP3R1-deficient mice showed no abnormalities, and the unique electrophysiological properties of the cerebellar Purkinje cells of IP3R1-deficient mice were not severely impaired.

Comments

IP3R1 is essential for proper brain function.

Acknowledgements
Eiichiro Nagata and Katsuhiko Mikoshiba
Department of Molecular Neurobiology, Institute of Medical Science, University of Tokyo, Tokyo, Japan

Reference

1 Matsumoto, M. et al. (1996) Nature 379, 168–171.

Other names
Insulin receptor

Gene symbol
Insr

Accession number
MGI: 96575

Area of impact
Hormone

General description

The metabolic effect of insulin is viewed as a set of branching pathways, with some actions serving to regulate energy metabolism and others to regulate cellular growth and development. The insulin receptor belongs to the receptor tyrosine kinase family. The *Insr* gene encodes a polypeptide precursor that generates the signal peptide, and the α and β subunits of the receptor following proteolytic cleavage. Insulin binding to the extracellular domain (α subunit) activates the intracellular kinase domain (β subunit). The activated kinase is responsible for mediating insulin action by activating intracellular substrates. In humans, complete lack of insulin receptors due to mutations of the insulin receptor gene results in severe growth retardation and insulin-resistant diabetes. In mice, targeted inactivation of insulin receptor substrate 1, an important substrate of the insulin receptor kinase, leads to inhibition of growth and mild resistance to the metabolic actions of insulin.

KO strain 1 construction[1]

A *neo*-resistance replacement targeting vector from isogenic DNA was used to disrupt the coding sequence of the *Insr* gene at amino acid 306 in exon 4. J1 ES cells were targeted and injected into C57BL/6 blastocysts. The resulting allele gave rise to mRNA that was probably unstable, since mRNA was not detectable in several tissues and developmental stages in knockout animals. The mutation is currently on a mixed 129 × C57BL/6J strain.

KO strain 2 construction[2]

Exon 2 was replaced by a *neo* cassette. The HSV-*tk* gene was added at the 3′ end. Targeted ES cells were injected into C57BL/6 blastocysts.

Phenotype

Unlike human patients lacking insulin receptors, mice homozygous for a null allele of the insulin receptor gene were born at term with normal intrauterine growth and development. Within hours of birth, however, homozygous null mice developed severe hyperglycemia and hyperketonemia, and died as the result of diabetic ketoacidosis in 48–72 hours. These data are consistent with a model in which the insulin receptor functions primarily to mediate the metabolic actions of insulin. Mice heterozygous for the null allele were hyperinsulinemic, suggesting that they might be insulin resistant. On a mixed 129 × C57BL background, 10% of the heterozygous mice developed diabetes.

Comments

This model is important for the evaluation of insulin action *vis-à-vis* the action of other growth factors of the insulin family, such as IGF-I and IGF-II, as well as the role of signaling molecules such as IRS-1 in mediating the metabolic actions of insulin. Intercrosses of mice heterozygous for an IR-null allele with mice heterozygous for an IRS-1-null allele develop insulin-resistant diabetes with high frequency (~40%), which largely exceeds the frequency of diabetes in the IR heterozygous animals (~10%) and in the IRS-1 heterozygous animals (0%)[3]. The double heterozygous insulin receptor/IRS-1 mouse provides a model for epistatic genetic interaction underlying the pathogenesis of common non-Mendelian disorders such as type II diabetes.

Acknowledgements
Domenico Accili
National Institute of Health, Bethesda, MD, USA

References

1 Accili, D. et al. (1996). Nature Genet. 12, 106–109.
2 Joshi, R.L. et al. (1996) EMBO J. 15, 1542–1547.
3 Bruening, J.C. et al. (1997) Cell 88, 561–572.
4 Di Cola, G. et al. (1997) J. Clin. Invest. 99, 2538–2544.

Other names
Interferon regulatory factor 1, IFN-stimulated gene factor 2 (ISGF-2)

Gene symbol
Irf1

Accession number
MGI: 96590

Area of impact
Immunity and inflammation, transcription factors

General description

IRF-1, a member of the IRF transcription factor family, binds to the interferon (IFN)-stimulated regulatory elements (ISREs) found in the promoters of many IFN-induced genes. cDNA expression studies have suggested that IRF-1 is a transcriptional activator for IFNs and IFN-inducible genes. The IRF-1 gene itself can be induced by IFNs, suggesting that IRF-1 is directly involved in IFN-mediated responses. IFN-γ and LPS stimulation of macrophages results in the induction of both inducible nitric oxide synthase (iNOS) and IRF-1, due to the presence of an ISRE in the iNOS promoter. IRF-1 also shows anti-oncogene activity, counteracting the oncogenic influence of IRF-2 in NIH 3T3 cells. The human IRF-1 gene, which maps to 5q31.1, is deleted, inactivated, or both at one or both alleles in leukemia or pre-leukemic myelodysplastic syndrome.

KO strain 1 construction[1]

The targeting construct (cloned into PUC18) contained 4.9 kb of homologous IRF-1 genomic BALB/c DNA, and a deletion of 1.2 kb into which pMC1*neo*-polyA gene was introduced in the same orientation as IRF-1. This design deleted amino acids 63–223 of the IRF-1 protein, including an essential part of the DNA-binding domain. The targeting construct was electroporated into D3 ES cells (129Sv). Positive clones were injected into C57BL/6J blastocysts. Chimeras were crossed with (C57BL/6J × DBA/2)F1 or C57BL/6J mice. Analysis of RNA from splenocytes induced with IFN-β showed a shorter IRF-1-related transcript in mutant mice. Gel-shift assays showed that no functional IRF-1 was present in mutant cells, and no truncated IRF-1 protein was observed.

KO strain 2 construction[2]

The targeting vector was designed such that 308 bp of mouse IRF-1 coding sequence were replaced by the PGK*neo*/UMS (transcription termination sequence) cassette. The targeting construct was electroporated into GS1 ES cells (129Sv) and positive clones were injected into C57BL/6 blastocysts. Chimeras were mated with C57BL/6 females and heterozygotes were crossed to generate homozygous IRF-1$^{-/-}$ mutants. Northern blot analysis revealed an aberrant mRNA in mutant mice which lacked sequences corresponding to exons 4–6. Reverse band shift assays showed that the DNA-binding site had been eliminated from the truncated protein found in mutant mice.

Phenotype

IRF-1$^{-/-}$ mice were born at the expected Mendelian frequency, showed no abnormalities in size, behavior or fertility, and had normal lifespans. No gross anomalies of internal organs were observed. *In vivo*, no differences were observed in the inducibility of type I IFN genes by either Newcastle disease virus (NDV) or polyI:polyC as determined by anti-viral activity, and by steady-state levels of IFN mRNA in various organs[2]. However, IRF-1-deficient mice exhibited a profound reduction of TCR$\alpha\beta^{+}$CD4^{-}CD8^{+} T cells, with a thymocyte developmental defect[1].

In vitro, type I IFN could not be induced in IRF-1$^{-/-}$ fibroblasts by (polyI:-polyC) but was induced to normal levels by NDV infection, indicating that there are at least two distinct pathways of type I IFN induction, one of which is IRF-1 dependent[1]. IRF-1 deficiency impaired the induction by IFN-γ of the anti-viral state against encephalomyocarditis virus, but not against vesicular stomatitis virus[3]. Nitric oxide was not produced by IRF-1$^{-/-}$ peritoneal macrophages after IFN or IFN-γ plus LPS induction[4].

IFNs and double-stranded RNA are known to augment NK cell activity. In the absence of IRF-1, NK cell development was affected in a gene-dosage-dependent manner. A reduced number of NK cells was present in IRF-1$^{-/-}$ mice, but these cells were functionally inactive, exhibiting impaired cytotoxicity and IFN-γ production in response to IL-12[5,6].

IRF-1 is induced not only by IFNs but also by T cell mitogens. Normally, peripheral T cells stimulated with mitogens undergo apoptosis when exposed to DNA-damaging agents. However, IRF-1$^{-/-}$ peripheral T cells were resistant to cell death induced by DNA-damaging agents[7]. Interestingly, the cell death IL-1β-converting enzyme (ICE) protease was not induced in IRF-1$^{-/-}$ peripheral T cells[7]. Oncogene-induced transformation has been documented in IRF-1$^{-/-}$ embryonic fibroblasts. These cells were also resistant to γ-irradiation and chemotherapeutic agents, and deficient in DNA damage-induced cell cycle arrest[8]. Furthermore cooperation of IRF-1 and p53 in the induction of p21 (WAF1, CIP1) was observed.

Finally, IRF-1$^{-/-}$ mice showed decreased incidence and severity of antigen-induced autoimmune disease compared to heterozygous mice[9]. IFN-γ induction by cognate antigen, but not by anti-CD3 cross-linking, was significantly reduced in mutant mice.

Comments

IRF-1 is critical for T cell development but is dispensable *in vivo* for normal inducibility of type I IFN genes and the establishment of the anti-viral state against some viruses. IRF-1 is an essential transcriptional regulator of iNOS in murine macrophages.

Recently, it has been revealed that the IRF family consists of at least seven members. Since the IRF members studied so far bind to similar motifs, redundancies and interactions between IRF-1 and other family members are likely to exist, and to depend on the genes involved and/or the physiological settings.

Acknowledgements
Toshifumi Matsuyama
Department of Oncology, Nagasaki University, School of Medicine, Nagasaki City, Japan
Tak W. Mak
Ontario Cancer Institute and Amgen Institute, Toronto, ON, Canada

References

1. Matsuyama, T. et al. (1993) Cell 75, 83–97.
2. Reis, L.F.L. et al. (1994) EMBO J. 13, 4798–4806.
3. Kimura, T. et al. (1994) Science 264, 1921–1924.
4. Kamijyo, R. et al. (1994) Science 263, 1612–1615.
5. Taki, S. et al. (1997) Immunity 6, 673–679.
6. Duncan, G.S. et al. (1996) J. Exp. Med. 184, 2043–2048.
7. Tamura, T. et al. (1995) Nature 376, 596–599.
8. Tanaka, N. et al. (1996) Nature 382, 816–818.
9. Tada, Y. et al. (1997) J. Exp. Med. 185, 231–238.

Other names
Interferon regulatory factor 2, IFN-stimulated gene factor 1 (ISGF-1)

Gene symbol
Irf2

Accession number
MGI: 96591

Area of impact
Immunity and inflammation, transcription factors

General description

IRF-2, a member of the IRF transcription factor family, binds to DNA elements termed IRF-E (IRF-element)/ ISRE (IFN-stimulated response element) found in the promoters of type I IFN and many IFN-inducible genes. IRF-2 functions as an antagonistic repressor for the transcription activator IRF-1. Overexpression of IRF-2 results in the transformation of NIH 3T3 cells, which can be reverted by IRF-1 overexpression. IRF-2 also positively regulates additional target genes such as the histone H4 gene.

KO strain construction

The targeting construct contained 4 kb of the endogenous IRF-2 gene derived from a library of DBA/2 mouse DNA. Exon 3, encoding a part of the DNA-binding domain, was replaced with the neomycin-resistance cassette. The construct was designed such that even if the mutant allele were to be aberrantly spliced, a frameshift occurring after amino acid 30 would produce a nonsense protein.

The targeting vector was electroporated into D3 ES cells derived from 129Sv mice. Positive clones were injected into C57BL/6J blastocysts. Heterozygotes were obtained by breeding chimeric mice with (C57BL/6J × DBA/2) F1 or C57BL/6J mice. The null mutation was confirmed by gel-shift assay of proteins in extracts of cells treated with IFN-β, or infected with Newcastle disease virus (NDV).

Phenotype

IRF-2$^{-/-}$ and IRF-2$^{+/-}$ mice showed no apparent abnormalities in either reproductive ability or behavior. However, IRF-2$^{-/-}$ mice showed increased physical vulnerability; a substantial proportion of mutant mice died within 8 months after birth, and females often died after giving birth. Old mice showed erosions and ulcers around the neck, back and abdomen, which, in severe cases, resulted in patchy scars and post-inflammatory hair loss. Induction of type I IFN genes by double-stranded RNA was normal, whereas that by NDV infection was slightly upregulated in embryonic fibroblasts (EFs) from IRF-2$^{-/-}$ mice. Induction of several IFN-inducible genes (including the 2′-5′ oligoadenylate synthetase gene, the double-stranded RNA-activated kinase gene, and the inducible NO synthase gene) by type I and type II IFNs was not affected in IRF-$2^{-/-}$ EFs. Anti-viral states against EMCV (encephalomyocarditis virus) induced by type I and type II IFN treatment were established normally in IRF-$2^{-/-}$ EFs *in vitro*.

IRF-$2^{-/-}$ mice mounted normal CTL responses against LCMV (lymphocytic choriomeningitis virus) and normal humoral responses against VSV (vesicular stomatitis virus) *in vivo*. In spite of normal CTL activity, mutant mice succumbed to LCMV infection. IRF-$2^{-/-}$ mice also exhibited weak bone marrow suppression of hematopoiesis and B lymphopoiesis. Unlike IRF-$1^{-/-}$ EFs, IRF-$2^{-/-}$ EFs were not transformed by the introduction of a single oncogene.

Comments

The IFN system is only weakly affected by IRF-2 gene disruption, suggesting a redundancy among related factors such as ICSBP. For unknown reasons, IRF-2 appears to be essential for recovery from viral infection. The mechanism by which IRF-2 affects hematopoiesis remains to be elucidated.

Acknowledgements
Tohru Kimura
Tokyo Medical and Dental University, Toyko, Japan
Tak W. Mak
Ontario Cancer Institute and Amgen Institute, Toronto, ON, Canada

References

1 Matsuyama, T. et al. (1993) Cell 75, 83–97.
2 Kamijo, R. et al. (1994) Science 263, 1612–1615.
3 Tanaka, N. et al. (1994) Cell 77, 829–839.
4 Kimura, T. et al. (1994) Science 264, 1921–1924.

Other names
Interferon regulatory factor 4, lymphocyte-specific interferon regulatory factor (LSIRF), PIP, ICSAT, NF-EM5

Gene symbol
Irf4

Accession number
MGI: 1096873

General description

IRF-4 is a member of the IRF family of transcription factors which are defined by a characteristic DNA-binding domain and the ability to bind to the ISRE (interferon-stimulated response element). IRF-4 is expressed in B cells and activated T lymphocytes, and is induced by the crosslinking of the TCR or BCR. The function of IRF-4 is not entirely clear: in B cells, it binds an element within the enhancers of the Ig light chains. Other binding sites in B and T cells remain to be determined.

KO strain construction

The targeting vector was designed such that exons 2 and 3 of the IRF-4 gene (derived from 129J) were replaced by a neomycin resistance gene. The short arm of the construct contained intron 1, while the long arm contained exons 4, 5 and 6. The targeting vector was introduced into E14 ES cells (129Ola) and positive clones were injected into CD1 blastocysts. Chimeric offspring were mated with C57BL/6 mice. The mutation has been backcrossed into both C57BL/6 (F5) and BALB/c (F5).

Phenotype

Mice deficient in IRF-4 showed normal distribution of B and T lymphocytes at 4–5 weeks of age, but developed progressive generalized lymphadenopathy. IRF-4$^{-/-}$ mice exhibited a 500-fold reduction in serum immunoglobulin concentrations and did not mount detectable antibody responses. T lymphocyte function was also impaired *in vivo* since these mice could not generate cytotoxic or anti-tumor responses.

Comments

IRF-4 is essential for the function and homeostasis of both mature B and mature T lymphocytes.

Acknowledgements
Hans-Willi Mittrücker
Max-Planck Institute for the Biology of Infection, Berlin, Germany
Tak W. Mak
Ontario Cancer Institute and Amgen Institute, Toronto, ON, Canada

Reference

1 Mittrücker, H.-W. et al. (1997) Science 275, 540–543.

Other names
Insulin receptor substrate 1, pp185

Gene symbol
Irs1

Accession number
MGI: 99454

Area of impact
Hormone, signal transduction

General description

The 185 kDa protein IRS-1 is the principal substrate for insulin and insulin-like growth factor I (IGF-I) receptors. IRS-1 is tyrosine-phosphorylated in physiological target tissues of insulin such as liver, muscle and adipocytes, and serves as an intracellular substrate for the insulin receptor or IGF-I receptor. In addition, tyrosine phosphorylation of IRS-1 has been reported in response to several interleukins, interferons, angiotensin II and growth hormone. Tyrosine phosphorylation of IRS-1 provides binding sites for several distinct Src homology 2 (SH2) proteins which mediate multiple signaling pathways.

KO strain 1 construction[1]

Base pairs −77 to +179 (including the start codon) were replaced by a neomycin-resistance cassette. Targeted D3 ES cells were injected into C57BL/6 blastocysts.

KO strain 2 construction[2]

A neomycin-resistance gene with an MC1 promoter and without a polyA$^+$ addition signal was inserted into the *Eco*RI site located about 2.6 kb 3′ to the ATG translation initiation codon. A diphtheria toxin A fragment gene with an MC1 promoter was ligated to the 3′ terminus across the vector backbone for negative selection. TT2 ES cells derived from a B6 × CBA hybrid were targeted. Positive clones were injected into eight-cell embryos from ICR mice and transferred into pseudopregnant ICR females to generate offspring. Male chimeras were bred to C57BL/6J females.

Phenotype

The growth of the homozygous IRS-1 knockout mice was retarded. At 3, 8 and 15 weeks of age, they weighed about 30% less than their normal littermates. Serum IGF-I, IGF-II and growth hormone levels were normal in the mutant mice. The growth retardation of the embryos was apparent at E15.5 and the weight of homozygous IRS-1 knockout embryos was about 80% of that of normal and heterozygous embryos at E18.5. The growth retardation of the IRS-$1^{-/-}$ mice was milder than that seen in insulin receptor knockout mice, in leprechaunism (the human insulin receptor-deficient state), and in IGF-I receptor knockout mice. Adult IRS-$1^{-/-}$ mice were fertile and had normal litter sizes. These observations suggest the existence of both IRS-1-dependent and IRS-1-independent pathways for signal transduction through insulin and IGF-I receptor tyrosine kinases.

The drop in blood glucose levels after the injection of human insulin was significantly smaller in homozygous mutant mice than in heterozygous mutant or wild-type mice. Similar results were observed with IGF-I or IGF-II injection, indicating that IRS-$1^{-/-}$ mice are resistant to IGF-I and IGF-II in addition to insulin. The stimulation of glucose transport activity of isolated adipocytes with insulin and of soleus muscle with insulin and IGF-I was significantly impaired. Although there were no significant differences in blood glucose among the three genotypes, serum insulin levels before and after glucose load were significantly higher in the homozygous mutant mice. Thus, IRS-1 knockout mice showed mild to moderate post-receptor insulin resistance in the muscle, yet normal glucose tolerance by compensatory hyperinsulinemia from pancreatic β cells.

Histologically, mutant mice exhibited β cell hyperplasia accompanied by partial differentiation of non-endocrine tissues into insulin-positive cells. Insulin secretory profile and glucose sensitivity to insulin secretion were indistinguishable between wild-type and IRS-$1^{-/-}$ mice when the same numbers of islets were compared. Thus, moderate insulin resistance appeared to induce hyperplasia and neogenesis of pancreatic β cells rather than β cell hyperfunction, leading to a moderate increase in insulin levels *in vivo.*

Comments

The IRS-1 mutation has led to the identification of IRS-2. IRS-2 was phosphorylated and associated with phosphatidylinositol-3-OH kinase after insulin/IGF-I activation. IRS-1 and IRS-2 appear to mediate overlapping but distinct signals. Mice heterozygous for IRS-1 and insulin receptor (IR) null mutations develop diabetes at 4–6 months of age. IRS-$1^{-/-}$IR$^{-/-}$ double knockout mice die of uncontrolled diabetes within 72 hours after birth[6].

Acknowledgements
Takashi Kadowaki
Third Department of Internal Medicine, Faculty of Medicine, University of Tokyo, Tokyo, Japan

References

1 Araki, E. et al. (1994) Nature 372, 186–190.
2 Tamemoto, H. et al. (1994) Nature 372, 182–186.
3 Tobe, K. et al. (1995) J. Biol. Chem. 270, 5698–5701.
4 Yamauchi, T. et al. (1996) Mol. Cell. Biol. 16, 3074–3084.
5 Terauchi, Y. et al. (1997) J. Clin. Invest. 99, 861–866.
6 Brüning, J.E. et al. (1997) Cell 88, 561–572.

ISL-1

Other names
Insulin-related protein, LIM homeobox transcription factor, islet-1

Gene symbol
Isl1

Accession number
MGI: 101791

Area of impact
Transcription factors

General description

ISL-1 is a LIM homeobox transcription factor. It is expressed during embryonic development in several neuronal and non-neuronal cell types, including motor neurons, retinal ganglion cells, sensory neuron ganglia, sympathetic ganglia, basal forebrain neurons, pancreatic islet cells and splanchnic mesoderm.

KO strain construction

A 7 kb *Xba*I fragment of genomic DNA was cloned from a 129Sv library. The second LIM-encoding exon was replaced with a *neo* cassette for targeting in ES cells (129Sv). Chimeras were crossed to 129 and BL6.

Phenotype

Homozygous ISL-1 knockout mice died at E9.5. Motor neuron and pancreatic islet cells failed to differentiate. At E9.5, apoptotic cells were detected in tissues where ISL-1 is normally expressed.

Acknowledgements
Samuel L. Pfaff
Salk Institute, La Jolla, CA, USA

References

1 Pfaff, S. et al. (1996) Cell 84, 309–320.
2 Ahlgren, U. et al. (1997) Nature 385, 257–260.
3 Karlsson, O. et al. (1990) Nature 344, 879–882.
4 Thor, S. et al. (1991) Neuron 7, 881–889.
5 Ericson, J. et al. (1992) Science 156, 1555–1560.

Other names
Intestinal trefoil factor

Gene symbol
Tff3

Accession number
MGI: 104638

Area of impact
Metabolism, immunity and inflammation

General description

ITF is a member of the trefoil family of proteins. It is expressed abundantly in the small intestine and colon, where it is secreted to the mucosal surface by goblet cells. ITF is also expressed at sites of injury, inflammation and dysplasia, including ulcers, inflammatory bowel diseases and adenomatous polyps.

KO strain construction

The entire second exon of the *Tff3* gene, which encodes most of the protein, was replaced by a neomycin-resistance gene cassette in J1 ES cells. HSV-*tk* was placed at the 3′ end. Targeted ES cells were injected into C57BL/6 blastocysts.

Phenotype

Homozygous mutant mice lacking ITF grew normally but demonstrated increased proliferation of cells in the colonic crypts and abnormal migration of cells to the apical surface of the lumen. The mice also showed increased susceptibility to colonic injury. Mutant mice succumbed due to extensive mucosal ulceration following oral administration of dextran sodium sulfate, an agent which causes mild epithelial injury in wild-type mice. ITF-deficient mice lacked histologic evidence of epithelial regeneration following injury. Rectal instillation of ITF was able to protect against acetic acid injury in these mutant mice by promoting epithelial regeneration.

Comments

These data show that ITF has a protective role in the maintenance and repair of the intestinal mucosal barrier. Because these proteins are resistant to high concentrations of acid and proteolytic enzymes, they represent potentially novel orally administered therapy for various forms of colonic injury, including inflammatory bowel diseases.

Acknowledgements
Hiroshi Mashimo
Massachusetts General Hospital, Boston, MA, USA

Reference

[1] Mashimo, H. et al. (1996) Science 274, 262–265.

Other names
Janus kinase 3

Gene symbol
Jak3

Accession number
MGI: 99928

Area of impact
Immunity and inflammation, signal transduction

General description

Jak3 is a member of the Janus family of non-receptor tyrosine kinases. These genes are characterized by distinct Jak homology regions which contain two unique tandemly arranged domains: the kinase, and the kinase-like, domains. Jak3 is expressed predominantly in hematopoietic and lymphoid tissues (but also in lung, kidney, and thymic and glomerular epithelial cells). Jak3 is constitutively associated with the common gamma chain (γc), which is part of the receptors for IL-2, IL-4, IL-7, IL-9 and IL-15. *In vitro*, Jak3 is activated by transphosphorylation following ligand binding and signaling through these receptors. Jak3 phosphorylates signal transducers and activators of transcription (STATs), and other signal transduction proteins.

KO strain 1 construction[1]

A neomycin-resistance gene under the control of the mouse PGK promoter replaced a 0.6 kb genomic fragment of the *Jak3* gene. The region replaced encompasses part of exon 16 and all of exon 17, and encodes the conserved protein kinase subdomains I–IV. The targeting vector was introduced into J1 ES cells (derived from the 129 strain) and positive clones were injected into C57BL/6 blastocysts; these KO mice are therefore of mixed background. The null mutation was confirmed by protein immunoblot of total lysates from mutant spleen, thymus and bone marrow. Neither intact Jak3 protein, nor a protein fragment containing the N-terminal portion of Jak3, could be detected.

KO strain 2 construction[2]

The construct consisted of a targeting vector based on pBluescript II which introduced the hygromycin-resistance gene into the 5′ region of the *Jak3* gene just downstream of the ATG, disrupting the first coding exon. The HSV-*tk* cassette was inserted into the 3′ end of the vector. 129-derived E14 ES cells were electroporated with the construct. Positive clones were injected into C57BL/6 blastocysts. The chimeric mice were bred to C57BL/6 mice to generate heterozygotes which were separately bred to obtain homozygous mutants. No Jak3 protein was detectable in extracts of mutant bone marrow, spleen or thymus.

KO strain 3 construction[3]

The targeting vector based on pBluescript was constructed by replacing a 7.8 kb genomic region of the *Jak3* gene, containing the Jak homology regions, the kinase-like domain, and the 5′ end of the kinase domain, with the neomycin-resistance cassette. The HSV-*tk* cassette was located at the 3′ end of the genomic sequences. The construct was injected into C57BL/6 blastocysts. Chimeras were bred to C57BL/6 mice, and heterozygous progeny were intercrossed to obtain homozygous mutants. The null mutation was confirmed by RT-PCR analysis of total RNA of mutant bone marrow cells.

Phenotype

Jak3$^{-/-}$ mice were born at the expected Mendelian ratio and had normal appearance, weight and fertility. However, mutant mice showed severe defects in the development and function of the immune system. Although myeloid progenitors were normal, B cell development was blocked at the pre/pro-B cell stage, and very few mature B220$^+$IgM$^+$ B cells were observed in the periphery. Defects in T cell development were manifested by the presence of a very small thymus (1–10% of normal number of cells) with a relatively normal distribution of thymic subsets (as judged by CD3, CD4, CD8, CD25, CD44, Thy-1, and c-*kit* markers). Peripheral T cells were present at normal or even elevated numbers, but the CD4:CD8 ratio was significantly increased. Mutant T cells appeared to be phenotypically activated as assessed by the expression of surface markers, including CD44, CD62L and CD69. Activation apparently occurred in the periphery, after thymic export, and was dependent on signaling through the TCR.

Measurements of functional responses *in vitro*, including proliferation and IL-2 secretion, were limited by the reduced survival of Jak3$^{-/-}$ T cells in culture. In short term cultures (4–5 hours), Jak3$^{-/-}$ T cells produced some IL-2, IL-10 and IFNγ in response to PMA plus ionomycin treatment, but not in response to TCR plus coreceptor crosslinking. Peripheral T cells were able to proliferate *in vivo*, and this expansion was dependent on antigenic signals through the TCR. Lymph nodes, except for the mesenteric lymph nodes, were absent or nearly undetectable. Jak3$^{-/-}$ mice lacked $\gamma\delta^+$ T cells and NK cells, but had monocytes and dendritic cells.

These Jak3 KO mice were were severely immunocompromised, and even under virus-antibody-free housing conditions, sporadically developed autoimmune diseases such as inflammatory bowel disease.

Additional features of KO strain 2 mice

Jak3$^{-/-}$ mice of KO strain 2 showed a similar phenotype to KO strain 1 mice except that 50–200-fold reductions in the numbers of both thymocytes and peripheral T cells were observed. Comparable deficiencies were present in the B cell population. In addition, mesenteric and gut-associated lymphoid tissues were much smaller than those of the wild type. Mutant thymocytes failed to respond to ConA or PMA in combination with anti-CD3, or ConA plus IL-2. Bone marrow cells did not respond to IL-7.

Additional features of KO strain 3 mice

In addition to phenotypic features described for KO strains 1 and 2, analysis of Jak3$^{-/-}$ mice of KO strain 3 showed reductions in spleen size and splenocyte number. No formation of cytokeratin networks and no immunoreactivity for Hassall's corpuscles were found in mutant thymic medulla. Thymic epithelial cells expressed normal levels of MHC class II. T cells appeared in mutant spleens only with aging; no NK cells were found. Lin^{-}c-*kit*$^{+}$ Sca-1^{+} stem cells were present in normal numbers in mutant bone marrow and spleen, and formed normal numbers of colonies when assayed *in vitro*, but Lin^{-}c-*kit*$^{+}$ thymocyte precursors were absent. No dendritic epidermal T cells (DETCs) or intestinal intra-epithelial lymphocytes (i-IEL cells) were detected, and CD8$\alpha^{+}\beta^{-}$ and CD8$\alpha^{+}\beta^{+}$ cells were dramatically reduced. IL-7R^{+} cells could not develop in Jak3$^{-/-}$ mice. Thymocyte progenitors seeded into the thymus of Jak3$^{-/-}$ mice were able to develop normally.

Comments

The phenotype of Jak3$^{-/-}$ mice is similar in many respects to that of mice lacking the γc chain, IL-7 or the IL-7Rα chain (except for aspects of thymic and splenic architecture). Jak3 appears to have a crucial and non-redundant function for the delivery of developmental signals in lymphoid cells, particularly for IL-2 receptor signaling in primary T cells, and for the progression of B cell development in the bone marrow. Jak3 is essential for the development of both NK cells, and intra- and extrathymic $\gamma\delta^{+}$ i-IEl, but is not required (or is functionally redundant) for the development of myeloid lineages.

Acknowledgements
Leslie Berg
Department of Molecular and Cellular Biology, Harvard University, Cambridge, MA, USA
James Ihle
Howard Hughes Medical Institute, Department of Biochemistry, St. Jude's Children's Research Hospital, Memphis, TN, USA

References

1 Thomis, D.C. et al. (1995) Science 270, 794–797.
2 Nosaka, T. et al. (1995) Science 270, 800–802.
3 Park, S.Y. et al. (1995) Immunity 3, 771–782.
4 Eynon, E.E. et al. (1996) J. Interferon Cytokine Res. 16 , 677–684.
5 Thomis, D.C. and Berg, L.J. (1997) J. Exp. Med. 185, 197–206.

Keratin 14

Gene symbol
Krt1-14

Accession number
MGI: 96688

Area of impact
Epithelial integrity

General description

Keratins are divided into two classes, I and II, and usually exist as specific pairs *in vivo*. Keratin 14 is a class I keratin and is found associated with keratin 5 in mitotically active keratinocytes of all stratified squamous epithelia. It has been suggested that the function of the keratin filaments in the epidermis is to impart mechanical integrity to the cells, without which the cells become fragile and prone to rupturing.

KO strain construction

The targeting vector was constructed as follows: the 5′ arm was the 3.4 kb *Eco*R1/*Nco*I fragment ending at the initiation ATG; the 3′ arm was the 3.4 kb *Eco*R1 fragment beginning in intron I resulting in a deletion of 1.2 kb of the keratin 14 gene from the initiation ATG. Inserted between the arms was a human vimentin cDNA with human growth hormone sequences as a 3′ UTR and the neomycin-resistance gene driven by the PGK1 promoter. The HSV-*tk* gene driven by the PGK1 promoter was downstream of the 5′ arm. The targeting vector was introduced into R1 ES cells, derived from strain 129Sv. C57BL/6 blastocysts were used to produce chimeric animals. The transgenic human vimentin was never expressed, showing that the resultant animals were indeed keratin 14 knockouts.

Phenotype

Homozygotes were born normally and appeared normal until 2 days after birth, when they became frail and began to show signs of gross blistering over their body surfaces, especially the paws, legs and ears. Light microscopy revealed blister formation in the basal layer of the mutant epidermis, with the suprabasal layers remaining intact. Ultrastructural analysis of skin showed a paucity of keratin filament bundles in the basal layer. There was a residual minor network of keratin filaments attached to hemidesmosomes and desmosomes in mutant cells. These filaments were sparse and wispy rather than in bundles. The suprabasal layers were indistinguishable from normal skin.

Protein analysis of 2-day-old mutant animals showed no expression of keratin 14 in the skin but the presence of keratin 5 at a high level. However, no other keratin proteins were upregulated to compensate for the loss of keratin 14. Immunohistochemical analysis revealed the presence of keratin 15 in the basal cells of both normal and mutant skin. Immunoelectron microscopy also revealed that keratin 15 was associated with keratin 5 in the residual filaments of the mutant basal cells. It was also found to be associated with keratin 5 in normal cells. Protein analysis showed that the levels of keratin 15 remained the same in normal and mutant tissues. Other stratified squamous epithelia in 2-day-old mutant animals were examined but only the cornea showed major and consistent signs of basal cell cytolysis. Protein analysis revealed that keratin 15 was expressed in other stratified tissues at varying levels. In tissues where there was no apparent cytolysis, there tended to be an increased level of keratin 15 relative to keratin 14, compared to other tissues.

The level of keratin 15 expression varied with the maturation of the animals. Two mutant animals survived past 2 days of age (one to 2 months, the other to 4 months). The esophagus of the 2-day-old mutant mouse had very high levels of keratin 15 relative to keratin 14 and showed no cytolysis. However, in a 2-month-old mutant animal, the level of keratin 15 was comparable to that of keratin 14. The esophagus was found to be fragile and cytolysis was observed.

Comments

These results confirm that the function of the epidermal keratin filaments is to impart mechanical integrity to cells. Keratin 15 was found to be basal specific and interacted with keratin 5. The expression levels of keratin 15 varied between different stratified squamous epithelial tissues and with neonatal development. The formation of a wide variety of suprabasal keratin networks was not dependent on the presence of keratin 14 in the basal layer.

Acknowledgements
Catriona Lloyd
Department of Anatomy and Human Biology, University of Western Australia, Perth, Australia

Reference

[1] Lloyd, C. et al. (1995) J. Cell Biol. 129, 1329–1344.

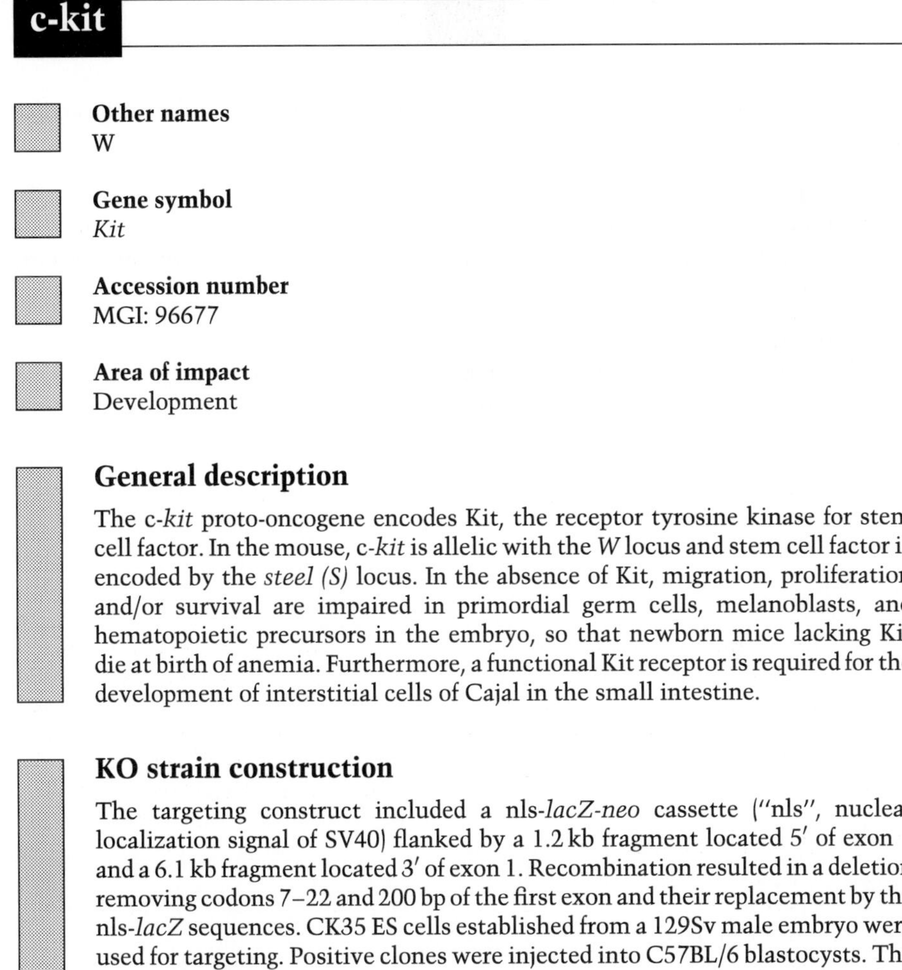

Other names
W

Gene symbol
Kit

Accession number
MGI: 96677

Area of impact
Development

General description

The c-*kit* proto-oncogene encodes Kit, the receptor tyrosine kinase for stem cell factor. In the mouse, c-*kit* is allelic with the *W* locus and stem cell factor is encoded by the *steel (S)* locus. In the absence of Kit, migration, proliferation and/or survival are impaired in primordial germ cells, melanoblasts, and hematopoietic precursors in the embryo, so that newborn mice lacking Kit die at birth of anemia. Furthermore, a functional Kit receptor is required for the development of interstitial cells of Cajal in the small intestine.

KO strain construction

The targeting construct included a nls-*lacZ-neo* cassette ("nls", nuclear localization signal of SV40) flanked by a 1.2 kb fragment located 5′ of exon 1 and a 6.1 kb fragment located 3′ of exon 1. Recombination resulted in a deletion removing codons 7–22 and 200 bp of the first exon and their replacement by the nls-*lacZ* sequences. CK35 ES cells established from a 129Sv male embryo were used for targeting. Positive clones were injected into C57BL/6 blastocysts. The mutation, called *WlacZ*, was maintained in the 129Sv genetic background.

Phenotype

Normal expression of the c-*kit* gene in *WlacZ*/+ heterozygous embryos was reflected by their *lacZ* expression. By comparing the patterns of *lacZ*-expressing cells between *WlacZ*/+ and *WlacZ*/*WlacZ* embryos, it was possible to detect where and when melanoblasts, primordial germ cells and hematopoietic progenitors failed to survive in the absence of Kit. Interstitial cells of Cajal (ICC) were found to be identical in *WlacZ*/+ and *WlacZ*/*WlacZ* embryos, suggesting that ICC do not depend on Kit expression during embryogenesis. However, Kit has been found to be required for the pacemaker activity of the intestine in the adult mouse[2].

Comments

The function of the c-*kit* gene is required only for post-natal development of the ICC. Unexpected sites of c-*kit* expression were uncovered in normal embryos, including in endothelial, epithelial and endocrine cells. However, none of these cells were dependent on Kit expression for their migration, proliferation and/or survival during embryogenesis.

Acknowledgements
Jean-Jacques Panthier
URA INRA de Genetique Moleculaire, Ecole Nationale Veterinaire d'Alfort, Maisons-Alfort, France

References

[1] Bernex, F. et al. (1996) Development 122, 3023–3033.
[2] Huizinga, J.D. et al. (1995) Nature 373, 347–349.

Krox-20

Other names
Early growth response 2, zinc-finger transcription factor, Zfp-6

Gene symbol
Egr2

Accession number
MGI: 95296

Area of impact
Neurology

General description

Krox-20 is a general zinc-finger transcription factor. It is expressed in the hindbrain rhombomeres 3 and 5. From E10.5 to E14, expression of Krox-20 is confined primarily to the motor and sensory roots of spinal and cranial nerves. From E15.5 to the adult, expression extends to Schwann cells along the entire length of peripheral nerves (no expression is observed within the cranial or dorsal root ganglia or within the sympathetic system). Krox-20 has been shown to directly regulate the *Hoxb2* gene.

KO strain construction

A *neo*^{r} /*lacZ* fusion was inserted into the coding exon of Krox-20. The vector was electroporated into D3 ES cells (129) and clones showing homologous recombination were injected into C57BL/6 blastocysts. Chimeric offspring were crossed with C57BL/6 × DBA/2 F1 mice.

Phenotype

The skeleton of Krox-20$^{-/-}$ mice was much more transparent to X-rays, indicating a severe calcium deficiency. Long bones of Krox-20$^{-/-}$ mice were thinner and shorter than those of their control littermates and very poorly stained by alizarin; skull bones were much less affected by the mutation. Staining with anti-osteocalcin, an early marker of bone differentiation, revealed a very strong reduction in the number of osteocalcin-positive cells in the endosteal part of mutant bones. These mutants exhibited defects in PNS myelination, with Schwann cells initiating only 1–2 wraps of myelin around individual axons. Krox-20$^{-/-}$ mice expressed early myelin markers such as MAG and beta S-100, but did not express late myelin markers such as MBP and P0. Krox-20 may therefore play a role in the progression of myelination within the PNS. Mutant mice exhibited signs of trembling at 10–15 days post-natal.

Comments

Krox-20 controls myelination and plays a role in the development of the hindbrain.

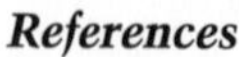

References

[1] Schneider-Maunoury, S. et al. (1993) Cell 75, 1199–1214.
[2] Topilko, P. et al. (1994) Nature 371, 796–799.
[3] Nonchev, S. et al. (1996) Development 122, 543–554.

Other names
Ku86

Gene symbol
Xrcc5

Accession number
MGI: 104517

Area of impact
Immunity and inflammation

General description

DNA-dependent protein kinase (DNA-PK) is a mammalian serine/threonine kinase implicated in DNA replication, transcription, the repair of double-stranded DNA breaks, and V(D)J recombination. DNA-PK is composed of a catalytic subunit called DNA-PK_{cs}, and a DNA-binding subunit called Ku. Ku is a heterodimer consisting of a 70 kDa subunit and an 80 (or 86) kDa subunit. The binding of Ku to altered DNA structures (such as nicks, hairpins and double-stranded breaks) activates the kinase activity of DNA-PK.

KO strain 1 construction[1]

A replacement-type vector containing the neomycin-resistance cassette was constructed using a 1.2 kb genomic fragment (derived from 129SvJ) 5′ of the translation start site, and a 9 kb fragment extending from intron 2 to 6 of the Ku80 gene. The result was a 3.4 kb deletion of the Ku80 locus, including 100 bp of the promoter and exons 1 and 2. The targeting vector was electroporated into R1 ES cells and positive clones were injected into C57BL/6 blastocysts. Chimeric mice were used to generate Ku80$^{-/-}$ mutants. Ku80 was undetectable in mutant ES cells and fibroblasts; Ku70 was present at greatly reduced levels.

KO strain 2 construction[2]

The targeting vector was designed to delete a 6 kb region which included two exons (nucleotides 701–964; amino acids 229–313). This fragment (derived from strain 129) was replaced with the PGK*puro*bpA cassette, resulting in a frameshift after this point. The targeting vector was electroporated into ES cells and positive clones were injected into blastocysts to generate chimeras. Chimeras were bred to C57BL/6 females to produce Ku80 (Ku86)$^{-/-}$ mutant mice.

Phenotype

Ku80$^{-/-}$ mice were viable and fertile but were 40–60% the size of littermate controls. Ku80$^{-/-}$ embryonic fibroblasts exhibited premature senescence with an early loss of proliferating cells and an extended doubling time. Cell cycle checkpoints were intact and prevented cells with damaged DNA from entering the cell cycle. T and B cell development was arrested at early progenitor stages. There was a profound deficiency in V(D)J recombination in cells of $Ku80^{-/-}$ mice, characterized by impaired formation of both signal and coding joints. This stands in contrast to the defect in *scid* mice, in which coding joints fail to form but signal joints are normal. The accumulation of both blunt full-length signal ends and hairpin coding ends was observed in $Ku80^{-/-}$ cells.

Comments

Ku80 is required for the processing of V(D)J recombination intermediates, apparently after the cleavage step. Ku80 does not appear to be essential for the protection of V(D)J recombination intermediates. The reduced size of the $Ku80^{-/-}$ mice suggests that there may be an important link between Ku80 and growth control.

Acknowledgements
André Nussenzweig
Memorial Sloan-Kettering Cancer Center, New York, NY, USA

References

[1] Nussenzweig, A. et al. (1996) Nature 382, 551–555.
[2] Zhu, C. et al. (1996) Cell 86, 379–389.

Kv3.1

Other names
Voltage-gated potassium channel, NGK2/Kv4

Gene symbol
Kcnc1

Accession number
MGI: 96667

Area of impact
Neurology

General description

Kv3.1 is a high-threshold, voltage-gated potassium channel with fast activation and deactivation kinetics. It is expressed in cerebellar granule cells, the thalamic reticular nucleus, a subset of cells in cerebral cortex and hippocampus, and several brainstem nuclei involved in auditory signal processing. It appears that, with the exception of cerebellar granule cells, Kv3.1 channels are expressed in parvalbumin-containing, fast-spiking GABAergic neurons. Outside the CNS, Kv3.1 is expressed in skeletal muscle and in T lymphocytes.

KO strain construction

A replacement vector was constructed with 6 kb homology of isogenic 129SvEMS genomic DNA including the second coding exon. PGK*neo* was inserted in the S2–S3 linker between the *Eco*RI and the *Msc*I sites (corresponding to codons 273 and 285), replacing 35 bp of genomic DNA. The mouse strain 129SvEMS was used for the initial tests. The disrupted Kv3.1 locus is currently being backcrossed to C57BL/6 (N7 generation as of January 1998).

Phenotype

Homozygous Kv3.1$^{-/-}$ mice were viable and fertile but had significantly reduced body weights compared to their wild-type and Kv3.1$^{+/-}$ littermates. Wild-type, heterozygous and homozygous Kv3.1 channel-deficient mice exhibited similar spontaneous locomotor and exploratory activity. In a test for coordinated motor skill, however, Kv3.1$^{-/-}$ mice performed significantly worse than their heterozygous or wild-type littermates. Both fast and slow skeletal muscles of Kv3.1$^{-/-}$ mice were slower to reach peak force and to relax after contraction, consequently leading to tetanic responses at lower stimulation frequencies. Both muscle types generated significantly smaller contractile forces during a single twitch and during tetanic conditions. The behavioral motor skill deficit of Kv3.1$^{-/-}$ mice was probably related to the slowing of muscle contraction and relaxation. Although Kv3.1$^{-/-}$ mice displayed a normal auditory frequency range, they showed significant differences in their acoustic startle responses. Contrary to expectation, Kv3.1$^{-/-}$ mice did not have increased spontaneous seizure activity.

It has been shown that the manifestation of a particular mutation may depend on the genetic background in which it is expressed. Initially, a series of behavioral tests were performed with $Kv3.1^{-/-}$ mice generated by intercrosses of (129SvEMS × C57BL/6)F1 hybrids. During these initial studies, it was suspected that the behaviour of the Kv3.1-deficient mice might be affected by the genetic background. Although the results were qualitatively similar to the ones obtained subsequently with Kv3.1 mutants on pure 129SvEMS background, larger variations were observed among trials and among different animals of the same Kv3.1 genotype on mixed background. In contrast to mice of mixed background, both wild-type $Kv3.1^{+/+}$ and $Kv3.1^{-/-}$ 129SvEMS mice improved their performances only marginally in two different assays for learning and memory.

The reticular nucleus of the thalamus expresses high levels of Kv3.1 and participates in synchronized thalamocortical oscillations characteristic of slow-wave sleep and generalized absence seizures[2]. The very high rates of spike discharge of reticular neurons during rhythmic spike-bursts are thought to synchronize slow oscillations in thalamocortical relay cells and reinforce slow-wave activity in the cortical EEG. GABAergic reticular neurons deficient in Kv3.1 may not achieve high discharge rates during spike-bursts, resulting in a failure to effectively synchronize slow oscillations. The cortical EEGs of freely moving wild-type and $Kv3.1^{-/-}$ mice were recorded over a period of 24 hours. No differences were found between groups in the time spent in waking, slow-wave sleep or REM sleep. Period amplitude analysis was used to quantify the somatomotor cortical EEG. Significant differences were observed in the amplitude and incidence of waves in the delta band (0.8–4.1 Hz) through all 6 hour blocks of the light and dark cycle. The delta amplitude was nearly 40% lower and incidence 20% lower in the $Kv3.1^{-/-}$ mice[3].

Comments

These findings strongly support the hypothesis that absence of Kv3.1 K^+ channels decreases neuronal synchronization and slow-wave activity in the EEG. In as much as neuronal synchronization and slow-wave activity are instrumental to the function of slow-wave sleep, further study of Kv3.1-deficient mice may help to elucidate the biological significance of this state of arousal. In addition, the $Kv3.1^{-/-}$ mice may be a useful model to study the regulation of thalamocortical synchronization. Because Kv3.1 channels are overexpressed in double-negative T lymphocytes in mice with autoimmune disease, the Kv3.1-deficient mouse may also be a useful model to study pathogenesis of autoimmunity.

Acknowledgements
Rolf Joho
Department of Cell Biology and Neuroscience, The University of Texas Southwestern Medical Center, Dallas, TX, USA

References

[1] Ho, C.S. et al. (1997) Proc. Natl Acad. Sci. USA 94, 533–538.
[2] Steriade, M. et al. (1993) Science 262, 679–685.
[3] Joho, R. et al. (1997) Eur. J. Physiol. 434, R90–91.
[4] Drewe, J.A. et al. (1992) J. Neurosci. 12, 538–548.
[5] Chandy, K.G. et al. (1993) Semin. Neurosci. 5, 125–134.
[6] Grissmer, S. et al. (1994) Mol. Pharmacol. 45, 1227–1234.
[5] Weiser, M. et al. (1995) J. Neurosci. 15, 4298–4314.

Kvbeta1.1

Other names
Potassium voltage-gated channel, shaker-related subfamily, beta member 1 voltage-gated potassium channel, mKv(beta)1

Gene symbol
Kcnab1

Accession number
MGI: 109155

Area of impact
Neurology

General description

Kvbeta1.1 is derived from alternative splicing of the Kvbeta1 gene (the alternative isoforms differ in their N-terminal sequences). It is a β subunit of voltage-gated potassium channels which associates with Kv1 α subunits. Kvbeta1.1 confers A-type inactivation on non-inactivating delayed rectifier-type α subunits. It appears to be expressed exclusively in brain where it is preferentially expressed in the hippocampal CA1 area and in striatum.

KO strain construction

The Kvbeta1.1 targeting construct contained a PGK*neo*bpolyA insertion in the first coding exon of Kvbeta1.1, which is specific for the Kvbeta1.1 isoform (insertion after the 24th codon in the same transcriptional orientation as the Kvbeta1 gene). The insertion of the *neo* gene resulted in a 43 bp deletion in the exon. The construct consisted of 11 kb of homologous sequences. R1 ES cells (129) were used for the Kvbeta1.1 targeting. The genetic background in which the mutants were studied was either F2(129Sv × C57BL/6) or four crosses into C57BL/6 (approx. 93.75% C57BL/6 and the rest, 129 background from the R1 ES cells).

Phenotype

The loss of Kvbeta1.1 did not lead to abnormal expression of other voltage-gated potassium channel subunits. At the light microscope level, there were no morphological abnormalities in the mutant brain. Furthermore, the mutants did not show any obvious aberrant behavior. In hippocampal CA1 pyramidal neurons, the loss of Kvbeta1.1 led to a reduction of the A-type potassium current as well as to a reduction in the ratio between A-type potassium current and non-inactivating potassium current. Therefore, distinct A-type potassium channels seemed to be transformed into delayed rectifier channels in the Kvbeta1.1-deficient neurons. As a consequence of this change, frequency-dependent spike broadening was reduced in the mutant hippocampal CA1 neurons. Furthermore, the calcium-dependent slow afterhyperpolarization (sAHP) was reduced in the mutants, probably because there was less calcium influx during a spike train in these mice. No abnormalities in synaptic plasticity (paired-pulse facilitation, long-term potentiation, depotentiation) were found in the mutant hippocampal CA1 area.

The Kvbeta1.1-deficient mice were normal in contextual conditioning. They were impaired in reversal learning in the Morris water maze, but not in the initial spatial learning (under distributed training conditions). Furthermore, the mutants were impaired in the social transmission of food preference task 21 hours after the interaction with demonstrator mice.

Comments

These results demonstrate that A-type potassium channels control spike dynamics by influencing spike duration during a spike train, thereby regulating calcium influx during the train and affecting the calcium dependent sAHP. Furthermore, these results suggest that the observed alterations in spike dynamics lead to behavioral impairments.

Acknowledgements
Karl P. Giese
Cold Spring Harbor Laboratory, Cold Spring Harbor, NY, USA

References
[1] The Kvbeta1.1$^{-/-}$ mouse is not yet published.
[2] Rettig, J. et al. (1994). Nature 369, 289–294.
[3] Heinemann, S.H. et al. (1996) J. Physiol. (Lond.) 493, 625–633.
[4] Rhodes, K.J. et al. (1996) J. Neurosci. 16, 4846–4860.

Other names
Leukocyte-type 12/15-lipoxygenase, Alox12l

Gene symbol
Alox12l

Accession number
MGI: 87997

General description

L-12LO catalyzes the first step in the biosynthesis of 12-hydroperoxy-eicosatetraenoic acid (12-HPETE) and small amounts of 15-HPETE from arachidonic acid substrate. In the mouse, this enzyme is found in highest abundance in peritoneal macrophages. The functions of this pathway are poorly understood.

KO strain construction

Two tandem copies of a neomycin-resistance cassette were inserted into the *Stu*I site of exon 3 of the L-12LO gene derived from strain 129Sv. An HSV-*tk* cassette was included 5′ of the genomic sequences. No deletions were made but any aberrant transcript, if made, would be shifted out-of-frame and a premature stop codon introduced 109 bp downstream. It was expected that this construct would result in a null allele since the non-heme iron atom ligands essential for activity are located in exons 3′ of this point. The targeting vector was introduced into D3H ES cells and positive clones were injected into C57BL/6 blastocysts. Male chimeras were mated with C57BL/6 females. All mice studied were of mixed B6/129Sv background. No evidence of L-12LO mRNA was detected in peritoneal macrophages of mutant mice.

Phenotype

L-12LO$^{-/-}$ mice appeared overtly normal and were fertile. There were no gross anomalies or histological defects in either external or internal organs. Macrophages appeared morphologically normal but had altered arachidonic acid metabolism in that 5-HETE became a major metabolite in mutant cells. The mechanism behind this increase remains to be clarified. Responses of mutant mice in well-characterized inflammatory models were normal.

Comments

The L-12LO pathway is not important for many functions usually carried out by macrophages. It may play a role in the protection against, or enhancement of, atherosclerosis.

Acknowledgements
Colin Funk
University of Pennsylvania, Philadelphia, PA, USA

Reference
1 Sun, D.-X. and Funk, C.D. (1996) J. Biol. Chem. 271, 24055–24062.

Other names
α-Lactalbumin

Gene symbol
Lalba

Accession number
MGI: 96742

Area of impact
Metabolism

General description

α-lac is a major mammalian milk protein, expressed only in lactating mammary gland epithelial cells and at the end of gestation. It interacts with UDP galactosyltransferase to induce lactose synthesis.

KO strain 1 construction[1]

A 2.7 kb fragment comprising the entire structural gene was replaced by PGK-*hprt*. PGK*tk* was used for negative selection. The 5′ homology was 4.3 kb in length and the 3′ homology was 0.85 kb. The construct was electroporated into HM-1 ES cells. The background strain used was 129Ola × BALB/c.

KO strain 2 construction[2]

The genomic *Lalba* gene was cloned from a BALB/c mouse. Exons 1 and 2 of the transcription unit and 0.56 kb 5′ flanking sequence were replaced with an HSV-*neo* cassette. An HSV-*tk* cassette was added at the 5′ end for additional selection. The construct was introduced into D3 ES cells. Recipient blastocysts were from C57BL/6 mice.

Phenotype

α-lac$^{-/-}$ mice had no detectable lactose in their milk. The milk was highly viscous (>10-fold increase) and had high concentrations of protein and fat. High milk viscosity disabled normal milk ejection and pups were unable to remove milk from the lactating mammary gland. The volumes of mammary epithelial cell cytoplasm, secretory vesicles and the Golgi apparatus were reduced. Female mutant mice could not sustain pups.

Comments

α-lac has a key role in lactogenesis such that milk composition and lactation can be dramatically altered by changes in α-lac expression. The KO mice may offer new insights into manipulation of milk composition.

Acknowledgements
Jean-Luc Vilotte
Laboratoire de Génétique Biochimique et de Cytogénétique, Jouey-en-Josas, France
Alex Kind
PPh Therapeutics, Edinburgh, Scotland

References

[1] Stacey, A. et al. (1994) Mol. Cell. Biol. 14, 1009–1016.
[2] Stinnakre et al. (1994) Proc. Natl Acad. Sci. USA 91, 6544–6548
[3] Stacey, A. et al. (1995) Proc. Natl Acad. Sci. USA 92, 2835–2839.
[4] Soulier, S. et al. (1997) J. Dairy Res. 64, 145–148.

Gene symbol
Lag3

Accession number
MGI: 106588

Area of impact
Immunity and inflammation

General description

LAG-3 is a transmembrane protein of the Ig superfamily, sharing several common features with the CD4 protein. It is expressed on activated T and NK cells.

KO strain construction

A targeting vector was constructed which replaced exons 1 to 3 of the *Lag3* gene with the neomycin-resistance gene. D3 ES cells (129Sv strain) were electroporated with the vector. Mice used for analysis were of a (129Sv × C57BL/6)F2 background.

Phenotype

Development and functions of T cells were entirely normal in the mutant mice. However, killing of certain tumor targets by NK cells was inhibited or even abolished, while lysis of cells displaying major histocompatibility complex (MHC) class I disparities remained intact.

Comments

LAG-3 appears to be a receptor or coreceptor that defines different modes of natural killing. Killing of targets due to absence or disparity of MHC molecules seems to occur independently of LAG-3, but killing of tumor targets is at least partially dependent upon the influence of LAG-3. These two forms of recognition do not seem to be mutually exclusive. The role of LAG-3 on activated T cells, if any, is as yet unknown.

Acknowledgements
Toru Miyazaki
Basel Insitute for Immunology, Basel, Switzerland

References
[1] Miyazaki, T. et al. (1996) Science 272, 405–408.
[2] Miyazaki, T. et al. (1996) Int. Immunol. 8, 725–729.
[3] Triebel, T. et al. (1990) J. Exp. Med. 171, 1393.

Other names
Lambda5, BL-CAM

Gene symbol
Lgl-5

Area of impact
Immunity and inflammation, B cell

General description

The λ5 gene is a homolog of the Ig Jλ-Cλ genes, and is expressed specifically in pro- and pre-B cells along with the V_{preB} gene, a homolog of the Ig Vλ gene. λ5 and V_{preB} proteins, which together are called the "surrogate light chain", bind to Igμ heavy chains to form an IgM-like complex (now called the pre-B cell receptor) on the surface of pre-B cells. The function of the λ5 protein and the pre-B cell receptor was unknown until its mutation in the knockout mouse was analyzed.

KO strain construction

The neomycin-resistance gene was inserted into exon 1 of the λ5 gene at a *Stu*I site. A promoter and transcription initiation site 5′ of the *neo* gene insertion site were deleted by the removal of a *Bam*HI/*Bgl*II fragment. D3 ES cells (129Sv strain) were transfected by electroporation, and homologous recombinants were selected by G418 selection, followed by PCR screening. Recombinant ES cell clones were injected into C57BL/6 blastocysts, and chimeric mice were bred to C57BL/6 backgrounds. The original strain of the homozygous mutant mice was of mixed background (129Sv × C57BL/6) but mice have since been backcrossed to either C57BL/6 or 129 strains and are available from the Jackson Laboratory. Pre-B cell lines were generated from the homozygous KO mice, and the presence of the *neo* insertion was confirmed by Northern blot analysis. In addition, the lack of λ5 protein was confirmed by immunoprecipitation.

Phenotype

$\lambda5^{-/-}$ mice showed a severe impairment in B cell development, with most developing B cells arrested at the large pre-B cell stage. This selective lack of expansion of large pre-B cells in the mutant bone marrow was presumably caused by the absence of a signal through the pre-B cell receptor, resulting in very few small pre-B cells and B cells being found in the bone marrow. Very few B cells were found in the periphery of young mice, although the lack of B cells grew more leaky as the mice aged. By 4 months of age, mutant mice showed one-third the number of splenic B cells found in normal mice. Numbers of peritoneal B cells became equal to those found in normal mice by 5 weeks of age. T cell-independent immune responses of the mutant mice were equivalent to those of wild-type mice, but T cell-dependent responses were 5–10 times lower than those of wild-type mice. Allelic exclusion of the heavy chain gene locus was abrogated in the pre-B cell stage, as revealed by single-cell DNA analysis, but was found to be maintained in the peripheral B cells when examined at the protein level. In addition, the λ5 gene was shown to be involved in reading frame II suppression during D_HJ_H gene rearrangement in pro-B cells, possibly by pairing with the truncated Dμ protein.

Comments

The phenotype of the $\lambda 5^{-/-}$ mouse is consistent with the proposed role of the λ5 and $V_{pre\text{-}B}$ surrogate light chain in pairing with the newly rearranged μ heavy chain in pre-B cells. The μ heavy chain/surrogate light chain molecule is thought to signal the pre-B cell to continue through the subsequent stages of B cell development, including light chain rearrangement, as well as being important in the proliferation and survival of pre-B cells. In addition, λ5 appears to be involved in heavy chain allelic exclusion and heavy chain reading frame usage. The $\lambda 5^{-/-}$ mice show a slow accumulation of mature B cells in the periphery, indicating that this mutation is leaky.

Acknowledgements
Daisuke Kitamura
Research Institute for Biological Sciences, Science University of Tokyo, Noda City, Chiba, Japan

References

[1] Kitamura, D. et al. (1992) Cell 69, 823–831.
[2] Ehlich, A. et al. (1993) Cell 72, 695–704.
[3] Loeffert, D. et al. (1994) Immunol. Rev. 37, 135–153.
[4] Karasuyama, H. et al. (1994) Cell 77, 133–143.
[5] Loeffert, D. et al. (1996) Immunity 4, 133–144.

Other names
Lecithin : cholesterol acyltransferase

Gene symbol
Lcat

Accession number
MGI: 96755

Area of impact
Metabolism

General description

LCAT is a 63 kDa glycoprotein enzyme that catalyzes intravascular esterification of cholesterol and regulates HDL metabolism. The *Lcat* gene is expressed in the liver, brain and testes and most of the body's LCAT protein is produced and secreted by the liver into the circulation. The majority is associated with the circulating high-density lipoprotein but a small proportion is associated with low-density lipoprotein.

KO strain 1 construction[1]

The backbone vector was pPN2T containing a neomycin-resistance cassette and two tandemly arranged *tk* gene cassettes. The 1.5 kb short arm and 7.5 kb long arm were inserted to flank the *neo* cassette. Recombination with the targeting vector disrupted the first coding exon and part of 5′ upstream sequence of the *Lcat* gene. The targeting vector was introduced into ES cells and positive clones were injected into blastocysts (strain 129). Chimeric mice were bred with DBA × C57BL/6 F1 hybrids.

KO strain 2 construction[2]

Exons 2–5 of the murine *Lcat* gene (derived from 129Sv) were replaced by a neomycin-resistance cassette. HSV-*tk* was placed at the 5′ end of the construct. RW-4 ES cells derived from 129SvJ mice were targeted and injected into C57BL/6J blastocysts.

Phenotype

Mutant LCAT$^{-/-}$ mice were healthy and fertile. The adrenal glands of male LCAT$^{-/-}$ mice were severely depleted of neutral lipid stores. Quantitation of tissue lipids in the adrenal glands showed a 61% reduction in total tissue cholesterol, an 82% reduction in cholesteryl ester (CE) content, and a 30% reduction in unesterified cholesterol (UC) content. This phenotype was associated with a 2-fold upregulation of adrenal scavenger receptor class B type I (SR-BI) expression. Red blood cells in LCAT$^{-/-}$ mice had an elevated surface-to-volume ratio, consistent with the mild anemia observed in these mice. KO strain 1 LCAT$^{-/-}$ mice displayed a 70% and 90% reduction in plasma total and

HDL-cholesterol, respectively. Plasma CE was reduced to 10% of control whereas the level of plasma UC was essentially unchanged. There was a relative enrichment of pre-beta migrating high-density lipoprotein (HDL) in the mutant mice. By negative-staining electron microscopy, HDL particles from wild-type mice appeared to be homogeneous spherical particles. In contrast, HDL particles from KO strain 1 LCAT$^{-/-}$ mice were heterogeneous in morphology but also showed the distinctive discoidal particles which form the classical rouleaux stuctures. No corneal opacity or renal insufficiency was observed in 4-month-old KO strain 2 LCAT$^{-/-}$ mice.

Comments

The LCAT-deficient mouse created by gene targeting reproduces the human LCAT-deficient HDL metabolic abnormalities with high fidelity; however, KO strain 2 LCAT$^{-/-}$ mice did not develop the corneal opacity or renal insufficiency usually associated with human LCAT deficiency. The reduced cholesterol content of the HDL may affect the flux of cholesterol to the adrenals via a selective uptake pathway.

Acknowledgements
Edward M. Rubin
Human Genome Center, University of California, Berkeley, CA, USA

References

[1] Ng, D.S. et al. (1997) J. Biol. Chem. 272, 15777–15781.
[2] Sakai, N. et al. (1997) J. Biol. Chem. 272, 7506–7510.

Other names
Lymphocyte protein tyrosine kinase

Gene symbol
Lck

Accession number
MGI: 96756

Area of impact
Oncogenes, immunity

General description

Lck is a member of the Src-family of tyrosine protein kinases. It is expressed predominantly by thymocytes and peripheral T cells but also by NK cells, B cell lines, and in the brain. Lck is a major signal transduction molecule of T cells. The N-terminal domain of $p56^{lck}$ coprecipitates with the intracytoplasmic domain of CD4 and CD8α. Lck is overexpressed in a murine thymic lymphoma (LSTRA) and in cases of human T cell lymphoblastic leukemia.

KO strain construction

A replacement type vector was used containing 2.3 kb of the genomic murine *Lck* gene from a mouse BALB/c genomic library. A neomycin-resistance gene cassette was inserted in exon 12. D3 (129) ES cells were selected and injected into C57BL/6 blastocysts. Chimeras were backcrossed with C57BL/6 mice.

Phenotype

Thymic atrophy was related to a dramatic reduction in double positive αβ-thymocytes. Almost no mature single positive thymocytes were seen. Increased TCRβ chain surface expression was found among double negative and double positive thymocytes. An increase in the proportion of the $CD44^{-}CD25^{+}$ double negative subset was observed. There was almost complete allelic exclusion of the TCRβ locus. There was almost no phosphorylation of TCRζ chain and ZAP-70. Only 10% of double positive cells were generated in $RAG2^{-/-} \times Lck^{-/-}$ mice after anti-CD3 epsilon treatment. There was an unbiased TCR Vβ and Vα repertoire. There was clonal deletion of superantigen-reactive CD4 T cells but no clonal deletion of superantigen-reactive CD8 T cells. Immaturity of thymic cortical stromal cells was seen.

There were very few peripheral T cells (5–10% lymph node population) and these had decreased surface expression of CD4 and CD8. There were no significant anti-viral effector functions of peripheral T cells, although the proliferative response *in vitro* to CD3 and TCRβ crosslinking was normal. There was no rejection of allogeneic grafts and syngeneic tumors, but NK function was normal. There was a defect in thymic maturation of γδ T cells, but normal development of intra-epithelial intestinal TCR γδ cells. γδ T cells did not protect against *Listeria monocytogenes* infection.

Acknowledgements
Thierry Molina
Department of Pathology, Hôtel Dieu de Paris, Paris, France
Tak W. Mak
Ontario Cancer Institute and Amgen Institute, Toronto, ON, Canada

References

[1] Molina, T.J. et al. (1992) Nature 357, 161–164.
[2] Penninger, J. et al. (1993) Science 260, 358–361.
[3] Molina, T.J. et al. (1993) J. Immunol.151, 699–706.
[4] Wallace, V. et al. (1995) Eur. J. Immunol. 25, 1312–1318.
[5] Wen, T. et al. (1995) Eur. J. Immunol. 25, 3155–3159.

Other names
Low-density lipoprotein receptor

Gene symbol
Ldlr

Accession number
MGI: 96765

Area of impact
Metabolism

General description

The low-density lipoprotein (LDL) receptor (LDLR) regulates plasma cholesterol levels through the clearance of cholesterol-rich intermediate density lipoprotein particles (IDL). Triglycerides are removed from VLDLs via lipoprotein lipase, thus generating IDLs. IDLs are then rapidly cleared in the liver via high-affinity binding of apolipoprotein E (apoE) to the LDLR. Some IDL particles are converted into LDLs which contain apoB100. Since apoB100 has low affinity for the LDLR, LDL particles circulate for a prolonged time in the plasma. Moreover, the LDLR can mediate the uptake of cholesterol-rich remnants of intestinal chylomicrons. Genetic defects in the LDLR lead to familial hypercholesterinemias in humans, rabbits (Watanabe rabbits), and rhesus monkeys. Patients with familial hypercholesterinemia have high levels of IDL and LDL and develop atherosclerosis at an early age.

KO strain construction

A neomycin-resistance cassette was inserted into exon 4. Two copies of the HSV-*tk* gene were placed in tandem at the 3′ end. AB1 ES cells were targeted and mutated ES cells were injected into C57BL/6 blastocysts.

Phenotype

LDLR$^{-/-}$ mice were viable and fertile. Total plasma cholesterol levels (2-fold) and the levels of IDL and LDL (7–9-fold) were increased. HDL and plasma triglyceride levels remained unaffected. The clearance of VLDL and LDL, but not of HDL, was impaired in LDLR-null mice. Dietary cholesterol challenge led to a large increase in IDL and LDL cholesterol levels. Restoration of LDLR expression using an adenovirus encoding the human LDLR restored VLDL clearance.

Comments

The LDLR is important for the clearance of VLDL, IDL, and LDL.

Reference

[1] Ishibashi, S. et al. (1993) J. Clin. Invest. 92, 883–893.

LEF-1

Other names
Lymphoid enhancer factor 1

Gene symbol
Lef1

Accession number
MGI: 96770

Area of impact
Development

General description

LEF-1 is a DNA-binding protein of the high mobility group (HMG) protein family, members of which act as DNA-binding proteins and regulate transcription in concert with other transcription factors. LEF-1 was first identified in lymphoid lineages, but is expressed more widely during embryogenesis, including in the neural crest, branchial arches, limb buds, thymus, lung, kidney, tooth genes, brain, etc.

KO strain construction

The PGK*neo* gene was inserted into the second exon of the HMG domain, which is essential for LEF-1 function. The targeting vector was introduced into D3 129Sv ES cells, which were injected into C57BL/6 blastocysts. The mutant phenotype was examined on a C57BL/6 × 129 hybrid background.

Phenotype

Homozygous mutant mice had reduced viability, with none surviving to weaning. The most obvious phenotype was lack of body hair and vibrissae. Hair follicles were present but in small numbers and poorly developed. Tooth development begins but arrests at the bud stage in E13 embryos. Mammary gland formation also arrests at an early stage. The mesencephalic nucleus of the trigeminal nerve is also missing.

Comments

LEF-1 is essential for normal development in a subset of organs in which it is expressed, all of which depend on inductive interactions. LEF-1 has been shown to be an important component of the *Wnt* signaling pathway, interacting with β-catenin to activate gene transcription.

Reference

[1] vanGenderen, C. et al. (1994) Genes Dev. 8, 2691–2703.

Other names
Leukemia inhibitory factor, CDF, DRF, HSF III, CNDF, DIA, MLPLI, OAF, D11Mit106, D11Mitl6

Gene symbol
Lif

Accession number
MGI: 96787

Area of impact
Immunity and inflammation, neurology

General description

LIF is a pleiotropic cytokine of the neuropoietic family (IL-6, IL-11, OSM, CNTF, CT-1) that can regulate the proliferation and differentiation of many cell types in culture. LIF is known to inhibit the differentiation of ES cells and to promote the survival and maturation of hematopoietic progenitors. During pregnancy, it is expressed in a burst in the maternal endometrial glands, just preceding implantation of the embryo. The murine LIF gene is localized on the short arm of chromosome 11 and consists of three exons and two introns. Alternative splicing of exon 1 leads to the expression of two different transcripts involved in the synthesis of both matrix-associated and diffusible precursor forms of LIF. However, post-translational regulation of these two protein precursors results in a single unique LIF protein sequence.

KO strain 1 construction[1]

Homologous recombination of the *Lif* gene was performed using 5′ (3853 bp) and 3′ (1422 bp) *Lif* gene fragments. Between these fragments, a 3301 bp DNA fragment including all the known introns and exons (and part of the 3′ untranslated region) of the *Lif* gene was deleted and replaced by the *lacZ* gene and a *neo* cassette. Germ line transmission was obtained by mating chimeric males (generated by injection of recombined 129Sv ES clones into C57BL/6 blastocysts) with C57BL/6 × DBA/2 females.

KO strain 2 construction[2]

The targeting vector was constructed by inserting a PGK*neo* cassette into exon 3 of the *Lif* gene. The resulting truncated LIF protein was missing the C-terminal 81 amino acids, including the last nine that are crucial for LIF biological activity. An HSV-*tk* gene was located about 5 kb downstream from the *Lif* sequences. The vector was electroporated into 129Sv-derived ES cells and positive clones were injected into C57BL/6J blastocysts. Chimeras were bred to C57BL/6J mice. The null mutation was confirmed by Northern blotting and by the failure of ES cells to grow on a feeder layer of LIF$^{-/-}$ primary embryonic fibroblasts.

Phenotype

LIF deficiency had pleiotropic effects on several different biological systems *in vivo*. After birth, $LIF^{-/-}$ mice exhibited no overt anatomical abnormalities but were 30% smaller than wild-type animals. Thymus and spleen were normal in size. Male $LIF^{-/-}$ mice were fertile but $LIF^{-/-}$ females were sterile due to their failure to develop sites for blastocyst implantation on the uterus surface. In the spleens of $LIF^{-/-}$ animals, pluripotent stem cells (CFU-S), committed erythroid (BFU-E), and granulocyte-macrophage progenitors were dramatically decreased by more than 70%. In mutant bone marrow, the CFU-S number was significantly decreased by 59%, while the other progenitor cells were less affected. In contrast to the decrease in the number of stem cells, the circulating mature red and white blood cell and blood platelet levels were normal in $LIF^{-/-}$ mice. Thymic T cell response to Con A or allogenic stimulation was dramatically reduced, but peripheral T cell reponses were normal in adult (6–8 weeks old) $LIF^{-/-}$ mice.

No severe neuronal deficiencies or adrenergic to cholinergic switching were observed in LIF-deficient mice[3]. Neuropeptide (VIP, neurokinin-A, galanin) induction was suppressed in LIF-deficient sympathetic neurons of superior cervical ganglia cultured as explants or axotomized *in situ*. Further analysis of the nervous system showed that LIF was required for appropriate peripheral neuron response to injury, and for the initial inflammatory response. The brains of $LIF^{-/-}$ mice appeared quite normal, but there were striking, sexually dimorphic changes in antibody staining for glial fibrillary acidic protein (GFAP). A double $LIF^{-/-}CNTF^{-/-}$ (ciliary neurotropic factor) knockout mouse displayed accelerated motor neuron death with aging or injury[4].

Almost all aspects (reproduction, hematopoiesis, T cell response, neuropeptide induction) of the LIF-deficient phenotype were partially or entirely restored by the addition of exogenous LIF extracellularly. These data suggest that, in each case, the LIF null mutation primarily affected the LIF-producing microenvironment rather than the LIF-target cell. Heterozygous ($LIF^{+/-}$) mice were affected to a lesser extent, indicating a dosage effect of the null mutation and LIF activity *in vivo*.

Comments

Uterine expression of LIF in mice is essential for implantation. In addition, a major function for LIF in the adult nervous system is the regulation of the neural and inflammatory response to injury. LIF appears to be required for the survival of the normal pool of stem cells, but not for their terminal differentiation. The analysis of gp130 (a common signal transducer for the LIF family of cytokines) KO embryos[5] confirmed the crucial role of LIF family cytokines in hematopoiesis and thymic T cell physiology *in vivo*.

Analysis of the double KO $LIF^{-/-}CNTF^{-/-}$ mice demonstrated crucial roles for LIF and CNTF in trophic support of motorneurons *in vivo*, and that this trophic support is multifactorial. Accordingly, the inherited inactivation of the *Cntf* gene which occurs in a high proportion of the Japanese population may predispose individuals to degenerative disorders of human motor neurons.

Acknowledgements

Jean-Louis Escary
Lipid Research Laboratory, VA Wadsworth Medical Center, Los Angeles, CA, USA
Paul Patterson
California Institute of Technology, Pasadena, CA, USA

References

[1] Escary, J.L. et al. (1993) Nature 363, 361–364.
[2] Stewart, C.L. et al. (1992) Nature 359, 76–79.
[3] Rao, M. S. et al. (1993) Neuron 11, 1175–1185.
[4] Sendtner, M. et al. (1996) Curr. Biol. 6, 686–694.
[5] Yoshida, K. et al. (1996) Proc. Natl Acad. Sci. USA 93, 407–411.

Other names
Leukemia inhibitory factor receptor

Gene symbol
Lifr

Accession number
MGI: 96788

Area of impact
Neurology, hematopoiesis

General description

The leukemia inhibitory factor (LIF) is a neurotropic cytokine that supports the survival of motor neurons *in vitro*. LIF binds to the LIF receptor, which associates in a trimolecular complex with gp130 to initiate signal transduction via tyrosine kinases. LIF plays a role in implantation of the embryo and the development of hematopoietic progenitors.

KO strain construction

Seven exons (20 kb) of the *Lifr* genomic sequence were replaced by *lacZ/neo*^{r} containing an internal ribosomal entry site (IRE Sβ-geo^{pA}). This vector was electroporated into CGR8 ES cells. Homologous recombinants were injected into C57BL/6 blastocyts. Heterozygous chimeras were backcrossed to the parental strain and then intercrossed for F2 and F3 generations.

Phenotype

Homozygous animals exhibited a 20% reduction in body mass compared to heterozygous littermates. These animals lacked vigour, were incapable of righting, and exhibited perinatal lethality (all animals died within 24 hours of birth). Expression of the inserted *lacZ* cassette was apparent at E9.5. At E14.5, expression was observed in the subependymal zone, glial limitans, and in neurons of the brainstem motor nuclei, including the hypoglossal and facial nuclei, and also in the nucleus ambiguus. In the spinal cord, staining was most intense in the sensory neurons of the dorsal root ganglion; however, it was also detectable in most spinal neurons. Some staining was also present in Schwann cells of the PNS.

Analysis of motor neuron number at the time of birth revealed that motoneurons in LIF-R$^{-/-}$ animals were reduced by > 35% in the facial nucleus, and by >40% in lumbar spinal motor neurons (L1–L6). Neurons of the nucleus ambiguus were also reduced by more than 50%. Motor neurons within the nucleus ambiguus innervate the esophagus, pharynx and larynx, coordinate suckling and swallowing, and generate respiratory rhythm. It is therefore possible that the loss of these neurons contributed to the early mortality observed in LIF-R$^{-/-}$ mice.

Comments

Animals which contain a null mutation of the *Lifr* gene have also been produced at Immunex (Seattle). Mice which are homozygous for this mutation show an increased resorption rate, defects at the placental interface (a site of LIF-R expression) and abnormal hematopoiesis. Those which are born die at 1 day post-natal. These animals show decreased mineralization of the bone matrix, and a 6-fold increase in the number of osteoclasts.

Reference
[1] Li, M. et al. (1995) Nature 378, 724–727.

Lim1

Gene symbol
Lhx1

Accession number
MGI: 99783

Area of impact
Development

General description

Lim1 is encoded by a homeobox gene which also contains cysteine-rich LIM domains, first recognized in the *lim-11* gene in *Caenorhabditis elegans*, the *Isl-1* gene in vertebrates and the *mec-3* gene in *C. elegans*. The mouse gene is expressed in early mesoderm wings and the pre-chordal mesoderm. Later in development it is expressed in the developing kidney and parts of the CNS.

KO strain construction

The entire Lim1 coding region was replaced with a PGK*neo* cassette. The targeting occurred in 129 AB1 cells, which were injected into C57BL/6 blastocysts. The phenotype was analyzed on a 129 × B6 mixed background.

Phenotype

Lim1 homozygous mutant embryos lack all head structures anterior to the otic vesicle from E9.5 on. The rest of the body axis appeared normal. Most embryos were resorbed by E10.5, but occasionally homozygous fetuses were stillborn at term. These embryos showed absence of head but normal body structures. Internally, kidneys and gonads were missing. Mutant embryos were recognizable as early as E7.5 by a constriction between the embryonic and extra-embryonic regions and delayed formation of the node. Occasional partial secondary axes were also observed.

Comments

It is proposed that Lim1 is an essential component of head organizer activity.

Reference
[1] Shawlot, W. et al. (1995) Nature 374, 424–430.

5-Lipoxygenase

Other names
Alox5, 5LO, 5LX

Gene symbol
Alox5

Accession number
MGI: 87999

Area of impact
Immunity and inflammation

General description

The 5-lipoxygenase (5LO) gene encodes an enzyme required for the biosynthesis of leukotrienes, which are potent mediators of inflammation and hypersensitivity reactions. The 5LO enzyme catalyzes both the oxygenation of arachidonic acid (AA) to generate 5-hydroperoxyeicosatetraenoic acid (5-HPETE), and the subsequent conversion of 5-HPETE into leukotriene A_4 (LTA_4). LTA_4 can then be metabolized into either LTB_4 or the peptidyl-leukotrienes (LTC_4, LTD_4, and LTE_4). Expression of 5LO is restricted to a few cell types, primarily those of myeloid lineage.

KO strain 1 construction[1]

The targeting vector was prepared from genomic clones containing the murine *Alox5* gene and was identical to the corresponding region in the mouse genome, except that a 1.3 kb fragment extending from the *Aat*II site at the 3′ end of intron 9 to the *Sma*I site in exon 10 was replaced by a neomycin-resistance gene. The targeting vector was electroporated into the ES cell line E14TG2A. Chimeras were originally bred to B6D2F1/J hybrid mice and the inbred 129SvEv mouse strain. The 5LO mutation has also been backcrossed into the C57BL/6, DBA/1LacJ, and MRL-*lpr*/*lpr* inbred mouse strains. The null mutation was confirmed by RT-PCR of total cellular RNA from calcium ionophore-stimulated peritoneal macrophages, and the absence of leukotriene synthesis by these cells was verified by enzyme immunoassay.

KO strain 2 construction[2]

The targeting vector was a replacement construct in which the neomycin-resistance cassette was inserted into the *Sca*I site of exon 6 of the *Alox5* gene derived from a mouse 129Sv genomic library. The HSV-*tk* cassette was located at the 5′ end of the construct. No deletions were made but any aberrant transcript synthesized would be shifted out-of-frame and would contain a premature stop codon. The targeting vector was electroporated into D3H ES cells and positive clones were injected into C57BL/6J blastocysts. Chimeric males were mated with C57BL/6J females to generate heterozygotes which were crossed to obtain $5LO^{-/-}$ mutants. The absence of 5LO mRNA, protein and leukotriene synthesis in macrophages and bone marrow-derived mast cells was confirmed by RT-PCR, Western blot, and HPLC chromatography of products formed by peritoneal macrophages challenged with inducers.

Phenotype

$5LO^{-/-}$ mice were born at the expected Mendelian frequency and were normal in growth and size, but were unable to synthesize detectable levels of leukotrienes. When these overtly normal mice were challenged with various inflammatory stimuli, a distinct phenotype was revealed. 5LO-deficient mice were resistant to mortality induced by treatment with platelet-activating factor (PAF), suggesting that leukotrienes are important mediators of the physiological changes characteristic of PAF-induced systemic anaphylaxis. Acute inflammatory responses, including zymosan A-stimulated peritonitis; edema and cellular infiltration induced by topical applications of AA to mouse ear tissue; neutrophil influx in an immune complex-induced peritonitis model; eosinophil infiltration and airway hyperresponsiveness in a mouse model of allergic airway inflammation, were all reduced. However, examination of delayed-type hypersensitivity reactions, IgE-mediated passive anaphylaxis, and endotoxin-induced shock showed no difference between wild-type and $5LO^{-/-}$ mice.

Alterations in the expression and activity of other inflammatory mediators in $5LO^{-/-}$ mice were observed. Calcium ionophore-stimulated peritoneal macrophages derived from $5LO^{-/-}$ mice released increased amounts of prostaglandin E_2 (PGE_2) and thromboxane B_2 (TXB_2) relative to stimulated cells isolated from wild-type mice. Although examination of the $5LO^{-/-}$ mice clearly demonstrated the importance of leukotrienes in eliciting a response to topical AA in mouse ear tissue, an easily measurable response remained in these animals. This residual inflammatory response in the $5LO^{-/-}$ mice, but not in wild-type animals, could be eliminated by cyclo-oxygenase inhibitors. These data suggest that inflammatory responses are modulated by AA metabolites through a variety of interconnected mechanisms. Furthermore, studies of inflammatory responses in several mouse strains showed that the relative effect of 5LO-deficiency was dependent upon genetic background, suggesting that many gene products may influence the role of leukotrienes in inflammatory processes.

Comments

Studies using $5LO^{-/-}$ mice have shown that 5LO is involved in defined inflammatory states and the relative importance of leukotrienes in inflammation and immune responses varies depending upon the stimulus and type of response elicited. However, these studies also indicate a role in these responses for other inflammatory mediators, particularly the products of the cyclooxygenase pathway of AA metabolism. The role of prostanoids in particular aspects of acute inflammatory processes may be enhanced when the 5LO pathway is blocked. Examination of mouse strains deficient in additional inflammatory pathways, in conjunction with the 5LO-deficient animals, should allow further dissection of complex immune processes *in vivo*.

Acknowledgements
Beverly H. Koller
University of North Carolina, Chapel Hill, NC, USA
Colin Funk
University of Pennsylvania, Philadelphia, PA, USA

References

1 Goulet, J.L. et al. (1994) Proc. Natl Acad. Sci. USA 91, 12852–12856.
2 Chen, X.-S. et al. (1994) Nature 372, 179–182.
3 Funk, C.D. et al. (1995) Adv. Prostaglandin, Thromboxane, Leukotriene Res. 23, 145–150.
4 Bozza, P.T. et al. (1996) J. Exp. Med. 183, 1515–1525.
5 Voelkel, N.F. et al. (1996) J. Clin. Invest. 97, 2491–2498.
6 Byrum, R.S. et al. (1997) J. Exp. Med. 185, 1065–1075.
7 Irvin, C.G. et al. (1997) Am. J. Physiol. 272, L1053–L1058.

Other names
Rbtn2

Gene symbol
Lmo2

Accession number
MGI: 102811

Area of impact
Hematopoiesis, oncogenesis

General description

The *RBTN/TTG* family of oncogenes is associated with chromosomal translocations resulting in T cell acute lymphoblastic leukemia (T-ALL) in humans. The *RBTN* genes encode proteins containing the cysteine-rich LIM domains. The physiologic function of the LIM domain is unknown but it appears to bind specifically to zinc. *RBTN2* is located in humans near recurring translocations involving the T cell receptor, and has been shown in transgenic mouse lines to act as an oncogenic protein. LMO2 (Rbtn2) is expressed in the mouse from E14 in a variety of locations, particularly in fetal liver, but also spleen and brain (but not thymus).

KO strain construction

The replacement targeting vector was designed such that a neomycin-resistance cassette was inserted in the 5′ end of exon 2 (derived from strain 129), disrupting the coding sequence of the first LIM domain in the LMO2 protein. Termination codons in all three reading frames were created. An HSV-*tk* cassette was placed at the 5′ end of the vector. The targeting vector was electroporated into E14 ES cells and positive clones were injected into C57BL/6 blastocysts. Chimeras were crossed to MF1 or 129Sv mice to generate heterozygotes on outbred or inbred genetic backgrounds, respectively. Heterozygotes were intercrossed to produce LMO2$^{-/-}$ mice.

Phenotype

Outbred heterozygotes did not display any abnormal phenotype. Hearts of LMO2$^{-/-}$ embryos (on either the outbred or inbred background) ceased to beat at E10.5, demonstrating that the LMO2-null mutation is embryonic lethal. Studies of LMO2 expression in wild-type tissues showed that it is a nuclear protein expressed *in vivo* in cells of the erythroid lineage. Null mutation of *Lmo2* resulted in a failure of yolk sac erythropoiesis, with no blood island development and no circulating nucleated red blood cells in either the yolk sac or the embryo. Cardiac development was not affected. Erythroid differentiation *in vitro* of LMO2$^{-/-}$ yolk sac tissue was specifically blocked, as was the erythroid differentiation of doubly mutated LMO2$^{-/-}$ ES cells. The differentiation of non-erythroid hematopoietic cells was not affected.

Comments

The LIM protein LMO2 is required for erythroid differentiation in the mouse. Strikingly similar effects on erythroid differentiation have been observed in ES cells lacking Tal1/SCL, suggesting that these proteins may act at the same critical stage of erythroid development.

Acknowledgements
Terence Rabbitts
MRC Laboratory of Molecular Medicine, Cambridge, UK

Reference

[1] Warren, A. et al. (1994) Cell 78, 45–57.

Other names
Low molecular weight protein 2, large multifunctional protease 2

Gene symbol
Lmp2

Accession number
MGI: 96797

Area of impact
Immunity and inflammation, antigen processing

General description

LMP-2 is a subunit of the proteasome, a large proteolytic complex involved in the degradation of cytosolic proteins. Proteasomes are important for the generation of MHC class I-binding peptides. LMP-2 is thought to be a catalytic subunit of the proteasome whose expression is induced by IFN-γ. LMP-2, along with LMP-7, is thought to affect the cleavage specificity of the assembled proteasome complex.

KO strain construction

The targeting vector was designed to delete an 800 bp fragment from the *Lmp2* gene, encompassing a portion of exon 2 and intron 2. The vector contained 2.1 kb and 3.4 kb of homology to the *Lmp2* gene at the 5′ and 3′ ends, respectively, and the gene was disrupted by insertion of the neomycin-resistance gene. The vector was electroporated into the E14 ES cell line (strain 129Sv), and chimeric mice were bred with C57BL/6 mice. Disruption of the *Lmp2* gene was confirmed by Southern blotting, and the absence of LMP2 protein was confirmed by 2D-PAGE analysis of immunoprecipitated proteasomes.

Phenotype

No developmental or other non-immunogical defects were observed in these mice. Proteasomes purified from spleen and liver of mutant mice exhibited altered peptidase activities. Specifically, cleavage of hydrophobic and basic fluorogenic peptide substrates was 2–3 times lower compared with wild-type proteasomes, whereas cleavage of an acidic peptide substrate was 2 times higher. Antigen-presenting cells from mutant mice showed reduced capacity (3–4-fold) to stimulate a T cell hybridoma specific for H-2D^b plus a nucleoprotein epitope of an influenza A virus. Mutant mice generated 5- to 6-fold fewer influenza nucleoprotein-specific cytotoxic T lymphocyte precursors, whereas the cytotoxic T cell reponse to the nucleoprotein for Sendai virus was normal. Mutant mice also had reduced levels (60–70% of wild type) of $CD8^+$ T lymphocytes, although the expression of MHC class I molecules was normal.

Comments

These results establish that LMP-2 can influence antigen processing but is not absolutely required for this process. LMP-2 appears to positively influence proteasome cleavage of acidic substrates, but negatively influence cleavage of basic and hydrophobic substrates. This mouse is useful for studies on MHC class I-restricted antigen processing.

Acknowledgements
Luc Van Kaer
Howard Hughes Medical Institute, Department of Microbiology & Immunology, Vanderbilt University School of Medicine, Nashville, TN, USA

Reference

[1] Van Kaer, L. et al. (1994) Immunity 1, 533–541.

Other names
Low molecular weight protein 7, large multifunctional protease 7

Gene symbol
Lmp7

Accession number
MGI: 96798

Area of impact
Immunity and inflammation, antigen processing

General description

Endogenous proteins are intracellularly degraded in the proteasomes, resulting in peptides that associate with newly synthesized MHC class I molecules in the endoplasmic reticulum. Proteasomes are proteolytic complexes of 700 kDa which occur abundantly in the cytoplasm and nucleus of all cell types examined. Each proteasome is composed of 13–15 distinct subunits, including LMP-7 and LMP-2. Like the gene for LMP-2, the LMP-7 gene is located in the MHC class II region and is closely linked to the TAP-1 and TAP-2 genes involved in peptide transport.

KO strain construction

The targeting vector was designed to delete exons 1–5 of the *Lmp7* gene (derived from 129Ola) which encode the first 247 amino acids of the 276 amino acid protein. A neomycin-resistance cassette was used to replace approx. 2.6 kb of the *Lmp7* genomic sequence, from the *Sal*I site 30 bp upstream of the translational start site to the *Hin*cII site at the 3′ end of exon 5. The HSV-*tk* gene was included at the 3′ end of the genomic sequences. The targeting construct was electroporated into E14.1 ES cells and positive clones were used to generate chimeric males that were mated with C57BL/6 or 129Ola females. Heterozygous progeny were intercrossed to obtain LMP-7$^{-/-}$ mutant mice.

Southern blotting confirmed the absence of exons 1–5 in mutant mice. Northern blotting of spleen and thymic mRNA showed that the expression of the closely linked TAP-1 and TAP-2 genes was unaffected.

Phenotype

LMP-7$^{-/-}$ mice were born at the expected frequency, appeared healthy and bred normally under conventional animal housing conditions. Mutant mice were physically indistinguishable from wild-type littermates. Peripheral lymphoid organs and thymi were of normal size and cellularity, with normal numbers and subsets of T and B cells. A wide range of cell surface markers on T and B cells was normal, including MHC class II. However, cell surface expression of MHC class I was reduced by 25–45% on all major lymphoid subpopulations and on macrophages of LMP-7$^{-/-}$ mice. H-2K and H-2D molecules were affected equally. LMP-7$^{-/-}$ splenocytes or lymph node cells were inefficient in presenting the H-Y antigen to transgenic H-Y TCR-bearing $CD8^+$ cells *in vitro*. The addition of synthetic peptides to LMP-$7^{-/-}$ splenocytes was able to restore wild-type levels of MHC class I expression on mutant cells.

Comments

LMP-7 plays an important role in the intracellular supply of endogenous peptides required for MHC class I antigen presentation.

Reference

[1] Fehling, H.J. et al. (1994) Science 265, 1234–1237.

Other names
Lipoprotein lipase

Gene symbol
Lpl

Accession number
MGI: 96820

Area of impact
Metabolism

General description

LPL is a 52 kDa enzyme associated with heparan sulfate proteoglycans on endothelial cells. It is expressed in most tissues and high expression can be found in cardiac and skeletal muscles and adipose tissue. LPL hydrolyzes triglycerides in chylomicrons and very low density lipoproteins (VLDL). LPL-regulated lipolysis controls the size of lipoprotein particles and the exchange of apolipoproteins and lipids among lipoprotein complexes. Deficiencies of human LPL (type I hyperlipoproteinemia) cause high plasma chylomicron and VLDL levels and low levels of LDL and HDL.

KO strain 1 construction[1]

An 8 kb fragment of the murine *Lpl* gene (including the translational start site in exon 1, intron 1, exon 2, and parts of introns) was replaced with a neomycin-resistance gene. HSV-*tk* was placed at the 5′ end of the construct. Targeted ES cells were injected into C57BL/6 blastocysts.

KO strain 2 construction[2]

A neomycin-resistance gene was inserted into exon 8 (aa 380). HSV-*tk* was placed at the 5′ end of the construct. E14 ES cells were electroporated and targeted cells were injected into C57BL/6 blastocysts.

Phenotype

Already at birth, LPL-null mice had a 3-fold increase in triglycerides and a 7-fold increase in VLDL-cholesterol levels. After suckling, LDL$^{-/-}$ pups became cyanotic, died very rapidly after birth (within 48 hours), and displayed nearly 100 times higher levels of triglycerides. Capillaries were filled with chylomicrons. Intracellular fat droplets and adipose tissues were reduced. Overexpression of human LPL under the control of a muscle specific promoter rescued the LPL-null phenotype and normalized the levels of lipoproteins in the serum. Mice heterozygous for the LPL mutation had reduced LPL activity, slightly elevated levels of triglycerides (1.5–3-fold), and survived into adulthood. LPL$^{+/-}$ mice also had defects in VLDL clearance.

Comments

Mice carrying the *cld* mutation are deficient in both hepatic and lipoprotein lipases.

References

1 Weinstock, P.H. et al. (1995) J. Clin. Invest. 96, 2555–2568.
2 Coleman, T. et al. (1995) J. Biol. Chem. 270, 12518–12525.

Other names
Low-density lipoprotein receptor-related protein

Gene symbol
Lrp

Accession number
MGI: 96828

Area of impact
Metabolism

General description

The multifunctional low-density lipoprotein (LDL) receptor-related protein (LRP) is a member of the LDL receptor family. It is ubiquitously expressed and has very high expression in the liver. It is produced as a precursor of 600 kDa (LRP600) that is intracellularly cleaved into LRP85 and LRP515. LRP515 confers ligand binding whereas LRP85 confers cell anchorage and the cytoplasmic domain. LRP is a clearance receptor and binds various ligands including lipoprotein lipase, apoE-rich remnant lipoproteins, α_2-macroglobin/protease complexes, plasminogen activator inhibitor 1 complexes (PAI-1), the tissue type plasminogen activator (t-PA), urokinase type plasminogen activator (u-PA), bacterial exotoxins, viruses and lactoferrin. The main function of LRP may be the clearance of remnant chyomicrons from the plasma. LRP also mediates the uptake and degradation of u-PA/PAI-1 complexes.

KO strain construction

A neomycin resistance cassette was inserted into the murine *Lrp* gene to interrupt the coding sequence between amino acids 803 and 804 (based on the human LRP sequence). Two copies of HSV-*tk* were placed at the 5′ end of the construct. AB-1 EES cells were targeted and injected into C57BL/6 blastocysts.

Phenotype

LRP$^{-/-}$ mice died in early embryonic development at the stage of implantation. LRP$^{+/-}$ mice were phenotypically normal. LRP$^{-/-}$ fibroblasts failed to bind α_2-macroglobulin. LRP$^{-/-}$ mice were impaired in their degradation of u-PA/PAI-1 complexes, which could interfere with implantation of embryos in the uterus. The uptake of the receptor-associated protein (RAP) was also impaired in LRP-null fibroblasts.

References
1 Herz, J. et al. (1992) Cell 71, 411–421.
2 Willnow, T.E. and Herz, J. (1995) J. Cell Sci. 107, 719–726.
3 Willnow, T.E. et al. (1995) Proc. Natl Acad. Sci. USA 90, 4537–4541.

Other names
Lymphotoxin-beta

Gene symbol
Ltb

Accession number
MGI: 104796

Area of impact
Immunity and inflammation, cytokines

General description

The LT-β gene lies in the class III region of the MHC very close to the TNF and lymphotoxin-α genes. LT-β is a ligand in the TNF family and is a type II transmembrane protein that forms a heteromeric complex with LT-α. The complex binds to a specific TNF-type receptor called the LT-β receptor but not to the two TNF receptors. LT-α is also found in a homomeric secreted form without LT-β. In general, the lymphotoxin proteins are involved in the development of the lymph nodes and the organization of the secondary lymphoid system. The LT-α/β complex is detected only on the surface of activated T, B and NK cells, whereas the receptor is generally found on non-hematopoietic cells.

KO strain construction

From a 129SvJ genomic clone, portions of exons 3 and 4 which comprise the receptor-binding domain were deleted and replaced with a neomycin-resistance gene. This approach removed the entire β sheet extracellular domain conserved among LT-α, LT-β and various other TNF family members. The arms of the targeting construct were 2.3 and 2.5 kb and did not extend into either of the flanking TNF or B144 genes. The insert with two copies of *tk* attached was transfected into W9.5 ES cells and homologous recombinants were injected into C57BL/6 blastocytes. Chimeric males were bred to C57BL/6 females.

Phenotype

In the absence of LT-β expression, no surface LT-α was observed; nor could a soluble form of the LT-β receptor bind to activated T and B cells. The secretion of LT-α was assumed to be normal in these mice. LT-$\beta^{-/-}$ mice lacked peripheral lymph nodes (LN) (except for the cervical and mesenteric nodes) and Peyer's patches. Like the LT-$\alpha^{-/-}$ knockout mouse, the organization of the spleen was disrupted, with some scrambling of the red and white pulp, loss of the marginal zone, and loss of the normally tight segregation of T and B cells into separate regions in the white pulp. Within the mesenteric LN, there was some alteration in the expression of subcapsular MOMA-1. These observations indicate that the LT-α/β pathway is involved in the development of some, but not all, LN as well as the organization of both the spleen and aspects of the LN.

Circulating IgG and IgM levels were normal; however, both serum and fetal IgA titers were greatly reduced. The ability of mutant mice to mount an anti-SRBC response was defective and splenic germinal centers did not form. Immunization with NP-CGG led to normal Ig titers although the level of affinity maturation was reduced, possibly reflecting the lack of organized splenic germinal centers. $CR1^+$ follicular dendritic cells (FDC) were absent in both the spleen and LN but germinal centers were present in the LN. This observation suggests that germinal centers do not require a FDC network to develop, at least in the LN.

Comments

The disparity between the LT-α and LT-β knockout mice (i.e. the clear presence of mesenteric LN in the LT-β KO mouse) suggests that all LT-related function is not mediated by the heteromeric complex alone, and that there may exist a novel signaling path for homomeric LT-α.

Acknowledgements
Richard Flavell
Howard Hughes Medical Institute and Section of Immunobiology, Yale University School of Medicine, New Haven, CT, USA

References

1 Koni, P.A. et al. (1997) Immunity 6, 491–500.
2 Browning, J.L. et al. (1993) Cell 72, 847–856.
3 De Togni, P. et al. (1994) Science 264, 703–707.
4 Rennert, P.D. et al. (1996) J. Exp. Med. 184, 1999–2006.
5 Ettinger, R. et al. (1996) Proc. Natl Acad. Sci. USA 93, 13102–13107.

Other names
Leukotriene A$_4$ hydrolase

Gene symbol
Lta4h

Accession number
MGI: 96836

Area of impact
Immunity and inflammation

General description

The initial steps in the metabolism of arachidonic acid (AA) to leukotrienes (LT) are catalyzed by arachidonate 5-lipoxygenase (5-LO) and result in the intermediate compound LTA$_4$. LTA$_4$ can then be converted to LTB$_4$ by LTA$_4$ hydrolase, or conjugated with glutathione by LTC$_4$ synthase to form LTC$_4$. LTB$_4$ is a powerful chemotactic agent for and activator of neutrophils, while LTC$_4$ (and the peptidyl LT) cause mucus secretion and possess strong bronchoconstrictive and vasodilative properties.

KO strain construction

Fragments identical to the corresponding 129 strain genomic regions were inserted 5′ and 3′ of the neomycin-resistance gene (*neo*) such that bases 739–853 of the hydrolase cDNA were replaced by *neo* upon homologous recombination. An HSV-*tk* gene was inserted outside the areas of homology to allow negative selection against random integrants. Targeting was performed in 129Ola ES cells. Positively targeted cell lines were microinjected into C57BL/6 blastocysts. Resultant chimeric offspring were mated with 129SvEv mice to maintain the 129 background.

Phenotype

LTA$_4$ hydrolase-deficient mice developed normally and were healthy. These animals were unable to produce detectable LTB$_4$ in response to intraperitoneal treatment with Zymosan A, but retained their capacity to make LTC$_4$ and its peptidyl LT derivatives.

Comments

LTA$_4$ hydrolase$^{-/-}$ mice, when compared to wild-type and 5-LO-deficient animals, provide an excellent model system to study the relative contributions of LTB$_4$ and the peptidyl LT to complex inflammatory responses.

Acknowledgements
Beverly Koller
Department of Medicine, University of North Carolina at Chapel Hill, Chapel Hill, NC, USA

References

[1] The LTA_4 hydrolase$^{-/-}$ mouse is not yet published.
[2] Medina, J.F. et al. (1991) Biochem. Biophys. Res. Commun. 176, 1516–1524.

Ly-6A

Other names
Sca-1, TAP

Gene symbol
Ly6a

Accession number
MGI: 107527

Area of impact
Immunity and inflammation, T cell

General description

Ly-6A is a glycosyl phosphatidylinositol-anchored molecule expressed on most peripheral lymphocytes, thymocytes, osteoblasts, hematopoietic precursors (including stem cells) as well as non-hematopoietic fibroblasts and kidney epithelial cells. In the peripheral lymphoid organs, Ly-6A expression is upregulated on activated lymphocytes. Although a ligand of Ly-6A has not yet been identified, crosslinking of Ly-6A by monoclonal antibodies activates T and B lymphocytes in the presence of the appropriate secondary signals.

KO strain construction

A 1.7 kb fragment containing exons 1–3 of the *Ly6a* gene was replaced with the 1.1 kb *Xho*I/*Sal*I fragment of pMC1*neo* containing the *neo*^{r} gene driven by the HSV-*tk* promoter and a polyoma enhancer. An HSV-*tk* cassette was inserted into the targeting plasmid to allow for negative selection. The targeting construct was introduced into E14TG2a (129Ola) ES cells and positive clones were injected into C57BL/6J blastocysts. The Ly-6A mutation has been backcrossed to C57BL/6 (Ly-6A.2 allele) for eight generations. The line is currently being backcrossed to BALB/c (Ly-6A.1 allele).

Phenotype

Ly-6A$^{-/-}$ mice were healthy and had normal numbers and percentages of all hematopoietic lineages. However, T lymphocytes from Ly-6A-deficient animals proliferated at a significantly higher rate in response to antigens and mitogens than did T cells from wild-type littermates. In addition, Ly-6A mutant splenocytes generated more CTLs compared to wild-type splenocytes when co-cultured with alloantigen. This enhanced proliferation was not due to alterations in the kinetics of response, sensitivity to stimulant concentration, or cytokine production by the T cell population, and was manifest in both *in vivo* and *in vitro* T cell responses. Moreover, T cells from Ly-6A-deficient animals exhibited a prolonged proliferative response to antigenic stimulation.

Comments

Ly-6A is not required for normal hematopoietic development. It is involved in signaling via the TCR but its mode of action remains unclear. The data suggest that Ly-6A acts to downmodulate T lymphocyte responses to antigen.

Acknowledgements
William L. Stanford
Samuel Lunenfeld Research Institute, Mount Sinai Hospital, Toronto, ON, Canada

Reference

[1] Stanford, W.L. et. al. (1997) J. Exp. Med. 106, 705–717.

Gene symbol
Lyl1

Accession number
MGI: 96891

Area of impact
Hematopoiesis, transcription factors

General description

The *Lyl1* gene encodes a member of the bHLH transcription factor protein family. lyl-1 was originally isolated from a T cell acute lymphoblastic leukemia (T-ALL) cell line due to its involvement in a chromosomal translocation t(7;19). It is expressed in cells of the hematopoietic lineage (c-*kit*$^+$ pro-myelocytes and megakaryocytes) as well as in endothelial cells.

KO strain construction

The targeting vector was designed to replace the 3′ end of the *Lyl1* gene (including its bHLH region) with a *lacZ/neo* cassette, creating an in-frame *Lyl1/lacZ* fusion. The targeting vector was introduced into E14.1 ES cells (129Ola) and positive clones were injected into C57BL/6 blastocysts. Mutant mice were backcrossed onto 129Ola and C57BL/6.

Phenotype

lyl-1$^{-/-}$ mice had no obvious phenotype. The development of neither endothelial nor hematopoietic cells was impaired.

Comments

The absence of the T cell leukemia oncoprotein lyl-1 did not affect hematopoiesis in the mouse. This result stands in contrast to the absence of blood cell formation in mice lacking the T cell leukemia oncoprotein Tal1/SCL, a bHLH gene related to lyl-1[3].

Acknowledgements
Fred Sablitzky
University College London, London, UK

References

1 The lyl-1$^{-/-}$ mouse is not yet published.
2 Mellentin J.D. et al. (1989) Cell 58, 77–83.
3 Shivdasani, R.A. et al. (1995) Nature 373, 432–434.

Other names
p56/58^{lyn}, p53/56^{lyn}

Gene symbol
Lyn

Accession number
MGI: 96892

Area of impact
Hematopoiesis, signal transduction

General description

The Src-related kinase Lyn is a non-receptor protein tyrosine kinase expressed preferentially in myeloid cells, B lymphocytes and platelets, and in endothelial cells in the brain. Two species of the Lyn protein, p53^{lyn} and p56^{lyn}, are produced due to alternative splicing. Lyn is thought to participate in signal transduction from cell surface receptors that lack intrinsic tyrosine kinase activity. Biochemical studies have shown that Lyn is physically associated with various cell surface receptors, such as the B cell antigen receptor, the high-affinity FcεRI complex, CD40 and the IL-2 receptor, among others. Lyn is also a strong candidate for a primary signal transducer of responses to LPS, since it is rapidly activated following LPS treatment and associates with CD14. A significant portion of the functional genomic *Lyn* gene (*Lyn*-L) is duplicated to form a pseudogene (*Lyn*-H), which is distinguished by its lack of 3′ coding exons as well as intron deletions and point mutations within the 5′ coding exons. The *Lyn*-L locus is transcribed, whereas the *Lyn*-H locus is not.

KO strain 1 construction[1]

A targeting vector was designed with a unique specificity for the active *Lyn* gene. The *Lyn* promoter and associated regulatory sequences (approx. 11.5 kb of genomic sequence) were replaced with a PGK*neo* expression cassette. The construct contained a long arm of homology of 5.3 kb and a short arm of 1.1 kb. The targeting vector was electroporated into E14 ES cells (129Ola-derived) and positive clones were injected into C57BL/6 blastocysts. Chimeras were mated with C57BL/6 mice and offspring were intercrossed to generate Lyn$^{-/-}$ mutants. The null mutation was confirmed by Southern analysis, RT-PCR and assay of Lyn autokinase activity in spleen and liver extracts.

KO strain 2 construction[2]

The functional *Lyn* gene derived from a C57BL/6 mouse genomic library was disrupted by inserting the neomycin-resistance gene cassette into the fourth exon encoding the N-terminal half of the SH3 domain. The PGK1*neo* cassette was used for positive selection and the PGK1/HSV-*tk* cassette for negative selection. The targeting vector was electroporated into E14 ES cells and positive clones were injected into C57BL/6 blastocysts. Chimeras were crossed with C57BL/6 mice and progeny heterozygotes were intercrossed to generate homozygous Lyn$^{-/-}$ mutants with genetic background C57BL/6J × 129. These Lyn$^{-/-}$ mice have since been backcrossed to C57BL/6J. The null mutation was confirmed by Southern analysis and immunoblotting of cell lysates from mutant spleen and cerebellum. Truncated forms of the Lyn protein were not detected.

KO strain 3 construction[3]

The functional *Lyn* gene was disrupted by the replacement of exons 3–7 with a neomycin-resistance cassette. The HSV-*tk* selection cassette was included at the 3′ end of the *Lyn* sequences. The targeting vector was electroporated into AB-1 ES cells and positive clones were injected into C57BL/6 blastocysts. Targeting into the active *Lyn* gene was confirmed by the appropriate Southern/blotting PCR analyses. Germ line chimeras were crossed with C57BL/6 females to generate heterozygotes which were intercrossed to establish homozygous animals. The mutation was confirmed to be a complete loss of function mutation by the absence in Lyn$^{-/-}$ bone marrow of Lyn RNA and both the p53 and p56 isoforms of the Lyn protein.

Phenotype

Lyn$^{-/-}$ mice were born at the expected Mendelian ratio, and were viable, fertile, and apparently healthy when young. The number of B cells in Lyn$^{-/-}$ mice was decreased by approximately half in the peripheral tissues. By 12–20 weeks of age, the relative decrease in B cell numbers had increased. In addition, these B cells did not respond normally to a number of stimuli, including B cell antigen receptor (BCR) crosslinking, LPS, and CD40 ligand. Normal germinal centers failed to develop, suggesting defects in some aspects of T-dependent immune responses. Induction of tyrosine phosphorylation on a variety of cellular proteins, such as Vav, Cbl and HS1, upon BCR cross-linking was also impaired in these B cells. In KO strain 3 Lyn$^{-/-}$ mice, although slight impairments of the initial phases of signaling through the BCR were present, downstream signaling through the MAPK kinase cascade was enhanced 2–4-fold over wild type.

Despite impaired BCR-mediated signaling, concentrations of IgM and IgA in sera were remarkably elevated in Lyn$^{-/-}$ mice, and production of anti-nuclear and anti-dsDNA autoantibodies was detected. Within the myeloid compartment, there was progressive development of severe extramedullary hematopoiesis with the accumulation of large numbers of myeloid cells and hematopoietic precursors in the spleen. Enhanced responses to cytokine stimulation were also seen in Lyn$^{-/-}$ myeloid cells. Histological study showed progressive splenomegaly and enlargement of lymph nodes that became evident with age. The spleen contained a significant number of plasma cells as well as unusual lymphoblast-like cells carrying Mac1 antigen and cytoplasmic immunoglobulin M. These cells spontaneously secreted a large amount of IgM *in vitro*. By age 8–10 months, Lyn$^{-/-}$ mice showed signs of anemia and increased mortality. Significant numbers of Lyn$^{-/-}$ mice showed glomerulonephritis caused by immune complex deposition in the kidney, an indication of autoimmune disease. Nevertheless, despite the low B cell numbers and high IgM levels with autoimmunity, immune responses to foreign antigens were largely normal in $Lyn^{-/-}$ mice.

$Lyn^{-/-}$ B cells were impaired in the induction of Fas expression after CD40 ligation and exhibited a reduced susceptibility to Fas-mediated apoptosis. Moreover, BCR cross-linking in $Lyn^{-/-}$ B cells suppressed Fas expression induced by costimulation with CD40 ligand and IL-4. The accumulation of lymphoblast-like and plasma cells in $Lyn^{-/-}$ mice may be caused by impaired Fas-mediated apoptosis after activation.

Unlike $Lyn^{+/+}$ mast cells, crosslinking of FcεRI in $Lyn^{-/-}$ mast cells failed to induce protein tyrosine phosphorylation of various substrates and evoked a delayed, slow Ca^{2+} mobilization. However, degranulation, adhesion, and the production of cytokines occurred normally in $Lyn^{-/-}$ mast cells. Passive cutaneous anaphylaxis reaction occurred normally in KO strain 2 $Lyn^{-/-}$ mice but not in KO strain 1 $Lyn^{-/-}$ mice. Comparing the proliferative response of splenic B cells to intact anti-IgM antibodies with that to F(ab′)2 anti-IgM antibodies, FcγRIIB-mediated downregulation of BCR signaling was not properly functional in the $Lyn^{-/-}$ B cells. It has been shown that Lyn is involved in B cell triggering by CD38 ligation and IL-5 for proliferation and isotype switching.

Triple mutant $Hck^{-/-}Fgr^{-/-}Lyn^{-/-}$ mice[3]

These animals lack all of the predominant Src family kinases in myeloid hematopoietic cells. Despite this, myeloid cell development was largely normal. Triple mutant mice suffered the same expansion of the myeloid cell compartment, with extramedullary hematopoiesis and splenomegaly, as seen in $Lyn^{-/-}$ single mutant animals. FcR phagocytosis in macrophages from these mice was only modestly affected, while LPS-mediated signal transduction was completely normal. Signaling through β1 integrins was impaired in cells of triple mutants, much as described for neutrophils of $Hck^{-/-}Fgr^{-/-}$double mutant mice.

Comments

The Lyn kinase serves both positive and negative regulatory roles in hematopoietic signal transduction. It appears to be involved not only in clonal expansion and terminal differentiation of peripheral B cells but also in the elimination of autoreactive B cells. The negative regulatory role of this kinase is non-redundant with other Src family kinases and its absence in $Lyn^{-/-}$ mice may explain the development of autoimmunity in these animals. It is possible that the importance of Lyn as a signaling component of cell surface receptors other than the B cell receptor is masked in $Lyn^{-/-}$ mice by the compensatory actions of other Src family kinases. A unique function for Lyn in myeloid cell signaling has not been revealed and the role of Lyn in signaling by the FcεRI remains unclear.

Acknowledgements
Margaret Hibbs
Ludwig Institute for Cancer Research, Melbourne Tumour Biology Branch, Royal Melbourne Hospital, Victoria, Australia
Hirofumi Nishizumi
The Institute of Medical Science, The University of Tokyo, Tokyo, Japan
Clifford Lowell
Department of Laboratory Medicine, UCSF, San Francisco, CA, USA

References

1 Hibbs, M.L. et al. (1995) Cell 83, 301–311.
2 Nishizumi, H. et al. (1995) Immunity 3, 549–560
3 Meng, M. and Lowell, C.A. (1997) J. Exp. Med. 185, 1661–1670.
4 Chan, V.W.F. et al. (1997) Immunity 7, 69–71.
5 Yamanashi, Y. et al. (1991) Science 251, 192–194.
6 Wang, J. et al. (1996) J. Exp. Med. 184, 831–838.
7 Nishizumi, H. et al. (1997) J. Immunol. 158, 2350–2355.
8 Crowley, M.T. et al. (1997) J. Exp. Med. 186, 1027–1039.

α_2-Macroglobulin

Other names
MAM

Gene symbol
A2m

Accession number
MGI: 87854

Area of impact
Cardiovascular

General description

Mouse α_2-macroglobulin (MAM) is a member of the proteinase inhibitors of the α_2-macroglobulin (A2M) family. The inhibitors block proteinases from all classes by a steric trapping mechanism. In mouse plasma, two different types of A2M are observed: the tetrameric mouse A2M and the monomeric murinoglobulin. MAM mRNA is expressed predominantly in the liver and during embryogenesis from 13 days pc onwards. The α_2-macroglobulins are thought to function protectively, trapping and eliminating unwanted proteinases and possibly cytokines and growth factors of different types and origins. The *in vivo* target proteinases and cytokines are unknown.

KO strain construction

The targeting construct (insertion vector) contained a genomic 7.5 kb *Sst* I/*Eco*RI fragment (129J genomic library) comprising exons 16–19 of the *A2m* gene. A 1.8 kb *Xho*I/*Cla*I fragment encoding the hygromycin B phosphotransferase gene driven by the PGK promoter was placed into intron 17, thereby replacing 0.7 kb of intronic sequences. The E14 ES cell line (129Ola) was injected into C57BL/6 blastocysts and these were implanted into F1 (C57BL female × CBA/J male) foster mice. Chimeric mice were mated to C57BL/6 mice to obtain heterozygous agouti pups. Homozygous mice used in the experiments were a mixture of the 129Ola and C57BL/6J mouse strains.

Phenotype

Liver MAM mRNA and plasma protein were absent in homozygous MAM-deficient mice and reduced to 50% in heterozygotes. MAM$^{-/-}$ mice were viable and produced normal-sized litters with normal sex ratios. Characterization of adult MAM$^{-/-}$ mice included feeding them diets with different fat contents and a choline- and methionine-free diet supplemented with ethionine to induce pancreatitis. To test the immune response and lung and liver toxicity, mice were treated with endotoxin, bleomycin and carbon tetrachloride, respectively. Knockout mice were more resistant to endotoxin but more sensitive to choline-free diet supplemented with ethionine. Regulation of murinoglobulin mRNA expression during pregnancy was analyzed as a possible back-up mechanism for the deficiency in α_2-macroglobulin. In

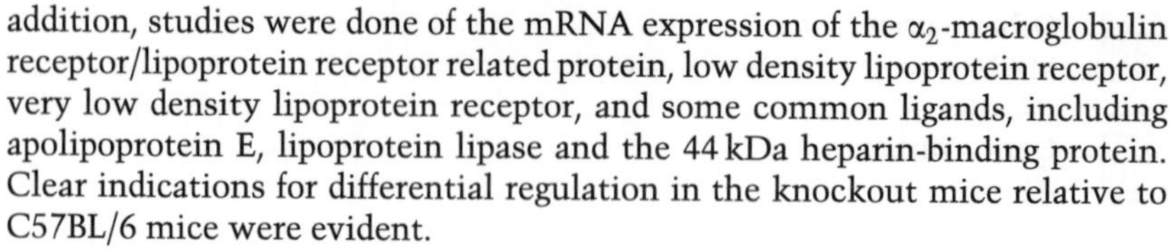

addition, studies were done of the mRNA expression of the α_2-macroglobulin receptor/lipoprotein receptor related protein, low density lipoprotein receptor, very low density lipoprotein receptor, and some common ligands, including apolipoprotein E, lipoprotein lipase and the 44 kDa heparin-binding protein. Clear indications for differential regulation in the knockout mice relative to C57BL/6 mice were evident.

Comments

Some caution is needed in the interpretation of the diet results. The variation in susceptibility among inbred strains of mice to such treatments is well known. The $MAM^{-/-}$ mice used in these experiments were a genetic mixture of the 129Ola and C57BL/6J mouse strains. Further experiments will be carried out on mice with the silenced MAM gene in a C57BL/6 background, on murinoglobulin-deficient mice as well as on doubly deficient mice ($MAM^{-/-}$ $MUG^{-/-}$).

Acknowledgements
Fred Van Leuven
Experimental Genetics Group, Center for Human Genetics, Leuven, Belgium

References

[1] Umans, L. et al. (1995) J. Biol. Chem. 270, 19778–19785.
[2] Van Leuven, F. (1992) Trends Biochem. Sci. 7, 185–187
[3] Van Leuven, F. et al. (1992) Eur. J. Biochem. 210, 319–327.
[4] Umans, L. et al. (1994) Genomics 22, 519–529.

Mad1

Other names
Mad

Gene symbol
Madh1

Accession number
MGI: 109452

Area of impact
Hematopoiesis

General description

Mad1 is encoded by one of four members of the *Mad* gene family. The *Mad* genes encode bHLH-Zip transcriptional repressors which bind to CACGTG sites as heterodimers with Max and antagonize the function of the Myc oncoprotein family. Members of the Myc family are thought to be key regulators of proliferation and differentiation. Heterodimers between Myc and Max transcriptionally activate genes required for the G_1-to-S cell cycle transition. When deregulated, Myc inhibits differentiation, promotes apoptotic cell death, and cooperates with other oncogenes in tumorigenesis. In contrast, transcriptional repression by Mad1 : Max heterodimers has been proposed to initiate cell cycle withdrawal and terminal differentiation. Notably, Mad1 expression is upregulated during terminal differentiation of most cell types (concomitant with downregulation of Myc), often as an immediate early response to differentiation-inducing stimuli.

KO strain construction

Two gene replacement vectors were employed that deleted overlapping stretches of the *Madh1* promoter and first exon, and created null alleles at the RNA expression level. One of these constructs drove the expression of *lacZ* off the endogenous *Madh1* regulatory elements. Analysis was performed predominantly on the inbred 129Sv genetic background. Similar phenotypes were observed for animals on hybrid 129B6F2 and inbred N4-C57BL/6 backgrounds.

Phenotype

Targeted disruption of the murine *Madh1* gene inhibited cell cycle exit of granulocytes following the colony-forming cell stage, resulting in increased proliferation and delayed differentiation of low proliferative-potential cluster-forming cells. Although the number of terminally differentiated peripheral blood granulocytes was unaffected due to an apparent compensatory increase in cluster-forming cell apoptosis, recovery of the granulocytic compartment following bone marrow ablation was significantly enhanced. Ectopic expression of other Mad family members was also observed, suggesting cross-regulation at the transcriptional level.

Comments

These results demonstrate that Mad1 regulates cell cycle withdrawal during granulocyte differentiation, and suggest that the relative levels of Myc versus Mad1 mediate a balance between cell proliferation and terminal differentiation or apoptosis. Functional reduncancy and cross-regulation between Mad family members may allow for apparently normal differentiation of most tissues in the absence of Mad1. The delayed differentiation in Mad1$^{-/-}$ mice is reminiscent of the phenotypes observed due to null mutations in some cyclin-dependent kinase inhibitors. The similarity of these phenotypes argues that multiple pathways cooperate to mediate cell cycle withdrawal during differentiation, or that there may be previously unrecognized epistatic relationships between these factors.

Acknowledgements
Kevin Foley, Grant McArthur and Robert Eisenman
Fred Hutchison Cancer Research Center, Division of Basic Sciences, Seattle, WA, USA

References

[1] Foley, K.P. et al. (1998) EMBO J. 17, 774–785.
[2] Ayer, D.E. et al. (1993) Cell 72, 211–222.

MAOA

Other names
Monoamine oxidase A

Gene symbol
Maoa

Accession number
MGI: 96915

Area of impact
Neurology, metabolism

General description

MAOA is a mitochondrial enzyme that inactivates neurotransmitters of the monoamine family such as dopamine, serotonin and norepinephrine. These neurotransmitters control movement, mood and excitation. MAOA and MAOB are encoded by separate genes on the X chromosome and share 70% amino acid similarity.

KO strain construction

Mice lacking the *Maoa* gene were generated accidentally by the replacement of about 17 kb of the *Maoa* gene with an IFN-β transgene. The IFN-β minigene consisted of a 1588 bp *Acc* I/*Apa* I fragment derived from plasmid pKb-IFN-β. Transgene integration caused the deletion of exons 2 and 3 of the *Maoa* gene. C3H/HeJ eggs were injected.

Phenotype

A considerable increase (9-fold at day 1 and 6-fold at day 12) was observed in the amount of serotonin in MAOA$^{-/-}$ pups, with a return to close to normal levels in older mice. Brain and liver MAOA activity was abolished in these mice. Mutant mice lacked the characteristic barrel-like clustering of layer IV neurons in the primary somatosensory cortex. MAOA$^{-/-}$ mice displayed a wide array of behavioral abnormalities from birth through maturity (hyperactive startle response, violent movements during sleep, tremulousness, abnormal posture). Increased male aggessiveness and increased male–male wounding were also observed. The major behavioral alterations in adult males were manifested as:

1. Bite wounds present on males housed in groups from time of weaning.
2. Short latency to the first appearance of biting attacks in resident intruder tests after a long period of breeding or after a period of isolation.
3. Prolongation of time spent in the center during open-field tests.
4. Reduced immobility during the Porsolt swim test.

MAOA$^{-/-}$ mice also failed to develop cortical whisker barrels. All of these effects could be suppressed through application of a serotonin synthesis inhibitor, whereas catecholamine synthesis suppression had no effect.

Comments

MAOA controls the amount of serotonin and norepinephrine in neurons, and may be linked to aggression.

There is an X-linked human mutation in MAOA in which affected individuals show borderline mental retardation and abnormal behavior, including impulsive aggression, arson, attempted rape and exhibitionism. Affected males exhibit markedly disturbed monoamine metabolism and an absence of functional MAOA in cultured fibroblasts. A non-conservative point mutation is found in all affected males and all female carriers which introduces a stop codon at position 296 of the deduced amino acid sequence. Note, however, that inhibition of MAOA in normal adult males is not consistently associated with aggressive behavior. Mutations in MAOA and MAOB have been implicated in the mental retardation in some patients with Norrie disease.

Acknowledgements
Edward De Maeyer
Institute Curie, Orsay, France

References

[1] Cases, O. et al. (1995) Science 268, 1763–1766.
[2] Brunner, H.G. et al. (1993) Science 262, 578–580.
[3] Mark, J.S. et al. (1995) Curr. Biol. 5(9), 997–999.

MAP1B

Other names
Microtubule-associated protein 1b

Gene symbol
Mtap1

Accession number
MGI: 97174

Area of impact
Neurology

General description

MAP1B belongs to a family of neuronal microtubule-associated proteins. This 255 kDa protein is expressed broadly in neurons and glia in various areas of the brain. It is expressed most highly in regions of the late embryo, where extensive axogenesis is occurring. MAP1B forms a complex with a heavy and light chain (LCI) component of two other proteins. It is thought that MAP1B has roles in brain development and the maintenance of neuronal plasticity.

KO strain construction

neo^r was inserted into exons coding for the microtubule-binding motif of MAP1B. *tk* was spliced to the 3′ end. The vector was electroporated into E14.1 ES cells (129). Clones selected for homologous recombination were injected into C57BL/6 blastocysts. Chimeric offspring were mated with C57BL/6.

Phenotype

MAP1B$^{-/-}$ mice died during embryogenesis. Mice heterozygous for the mutation displayed a variety of defects, including motor system abnormalities, decreased growth rates, and a lack of visual acuity in one or both eyes. Histochemical analysis showed abnormal morphology of the Purkinje cell dendritic processes in severely affected mice. The Purkinje cells failed to react with MAP1B antibodies and showed reduced staining with MAP1A antibodies. Similar immunochemical and histological changes were observed in the retina, olfactory bulb and hippocampus.

Comments

MAP1B plays a role in the development of dendritic processes.

Reference

[1] Edelmann, W. et al. (1996) Proc. Natl Acad. Sci. USA 93, 1270–1275.

Mash-1

Other names
Mammalian achaete-scute homolog 1

Gene symbol
Mash1

Accession number
MGI: 96919

Area of impact
Development

General description

Mash1 is one of two mammalian genes encoding basic helix-loop-helix transcription factors related to the *Drosophila achaete-scute* genes. It is transiently expressed in precursor cells of the CNS (basal telecephalon, hypothalamus, ventral thalamus, retina, midbrain, hindbrain, alar plate of the spinal cord), olfactory epithelium and PNS (sympathetic ganglia, parasympathetic ganglia, enteric ganglia).

KO strain construction

A replacement vector deleted a 1.5 kb *Not* I/*Hpa*I genomic fragment containing the entire protein-coding sequence of the *Mash1* gene and replaced it with a PGK*neo* selectable cassette in the same transcriptional orientation as *Mash1*. The mutation was introduced in the R1 ES line (129Sv × 129Sv-CP) and has been back-crossed with the CD1 outbred strain.

Phenotype

Animals heterozygous for the mutation appear normal and are fertile. Animals homozygous for the mutation die less than 24 hours after birth, with apparent breathing or feeding difficulties. Development of the olfactory epithelium, sympathetic ganglia, parasympathetic ganglia, enteric ganglia and retina, is affected. Most olfactory neuron progenitor cells fail to express the progenitor cell markers neurogenin 1 and NeuroD, and the differentiation markers SCG10 and neuron-specific β-butulin, and they die several days after their differentiation arrest. Sympathetic neuron precursors expressing NCAM and neurofilaments 68 and 160 kDa fail to differentiate into β-tubulin, peripherin and neuron-specific enolase-positive neurons. A subset of enteric neurons, which are serotoninergic and transiently express sympathoadrenal markers, is missing. The differentiation of late retinal cell types rod and horizontal and bipolar cells is delayed and the final number of bipolar cells is reduced.

Acknowledgements
Francois Guillemot, IGBMC, Illkirch, France

References

1 Guillemot, F. et al. (1993) Cell 75, 463–476.
2 Blaugrund, E. et al. (1996) Development 122, 309–320.
3 Cau, E. et al. (1997) Development 124, 1611–1621.
4 Sommer, L. et al. (1995) Neuron 15, 1245–1258.
5 Tomita, K. et al. (1996) Genes to Cells 1, 765–774.

Mash-2

Other names
Mammalian achaete-scute homolog 2

Gene symbol
Mash2

Accession number
MGI: 96920

Area of impact
Development

General description

Mash2 is one of two mammalian genes encoding basic helix-loop-helix transcription factors related to the *Drosophila achaete-scute* genes. The gene is expressed in oocytes, in preimplantation embryos, at high levels in the ectoplacental cone and chorion and at low levels in head mesenchyme of postimplantation embryos, and in the labyrinthine and spongiotrophoblast layers of the placenta. *Mash2* is genomically imprinted, with the paternally-inherited allele initially expressed by groups of trophoblast cells (E6.5–E7.5) but completely repressed at E8.5.

KO strain construction

The vector deletes a 510 bp *Mlu*I/*Hpa*I genomic fragment coding for most of the Mash-2 protein (amino acids 20–189), and replaces it with an in-frame *lacZ* sequence and a PGK*neo* cassette. The mutation was introduced in the R1 ES line (129Sv × 129Sv-CP) and has been back-crossed with the CD1 outbred strain.

Phenotype

Heterozygous animals inheriting the mutant allele from their father are normal and fertile. Heterozygous animals inheriting the mutant allele from their mother present the same phenotype as homozygous mutant animals. Homozygous mutant animals die at E10.5 from placental failure. Spongiotrophoblast cells are missing from mutant placenta and their precursors are absent in mutant ectoplacental cones. The chorionic ectoderm and labyrinthine trophoblast are reduced and the giant cell layer of the placenta is enlarged. The mutant placental phenotype can be rescued in chimeras with tetraploid wild-type cells, allowing development of viable homozygous mutant adult mice. This demonstrates that Mash-2 is not required outside of the trophoblast lineage.

Comments

Mash2 maps to the distal region of chromosome 7, within a cluster of imprinted genes including insulin-2, insulin-like growth factor-2 and H19.

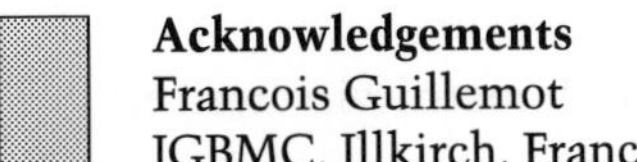

Acknowledgements
Francois Guillemot
IGBMC, Illkirch, France

References
[1] Guillemot, F. et al. (1994) Nature 371, 333–336.
[2] Guillemot, F. et al. (1995) Nature Genet. 9, 235–241.

Matrilysin

Other names
Matrix metalloproteinase 7 (Mmp7), MAT, pump-1, matrilysin uterine, EC 3.4.24.23

Gene symbol
Mmp7

Accession number
MGI: 103189

Area of impact
Oncogenes

General description

Matrix metalloproteinases (MMP) are implicated in the basement membrane destruction associated with late-stage tumor cell invasion and metastasis. However, matrilysin, an MMP family member, is expressed in a high percentage of early-stage human colorectal tumors.

KO strain construction

Exons 3 and 4 of the matrilysin gene were replaced with a *neo* cassette. The vector was electroporated into R1 ES cells (129Sv). Clones showing homologous recombination were injected into C57BL/6 blastocysts. Chimeras were backcrossed four generations to C57BL/6 and then crossed with $Min^{+/-}$ mice.

Phenotype

Matrilysin expression was examined in benign intestinal tumors from wild-type mice heterozygous for the Apc^{Min} allele (Min^{+}). Matrilysin was detected in the tumor cells, where, surprisingly, it was predominantly immunolocalized to the luminal surface of dysplastic glands, rather than the basement membrane or extracellular matrix. Matrilysin-deficient Min^{+} mice had a reduction in mean tumor multiplicity of approximately 60%, and a significant decrease in the average tumor diameter.

Comments

Matrilysin is a suppressor of the Min phenotype, perhaps functioning by a mechanism which is independent of matrix degradation. These results argue for the use of MMP inhibitors in the treatment and prevention of early-stage colon cancer.

Reference
[1] Wilson, C.L. et al. (1997) Proc. Natl Acad. Sci. USA 94, 1402–1407.

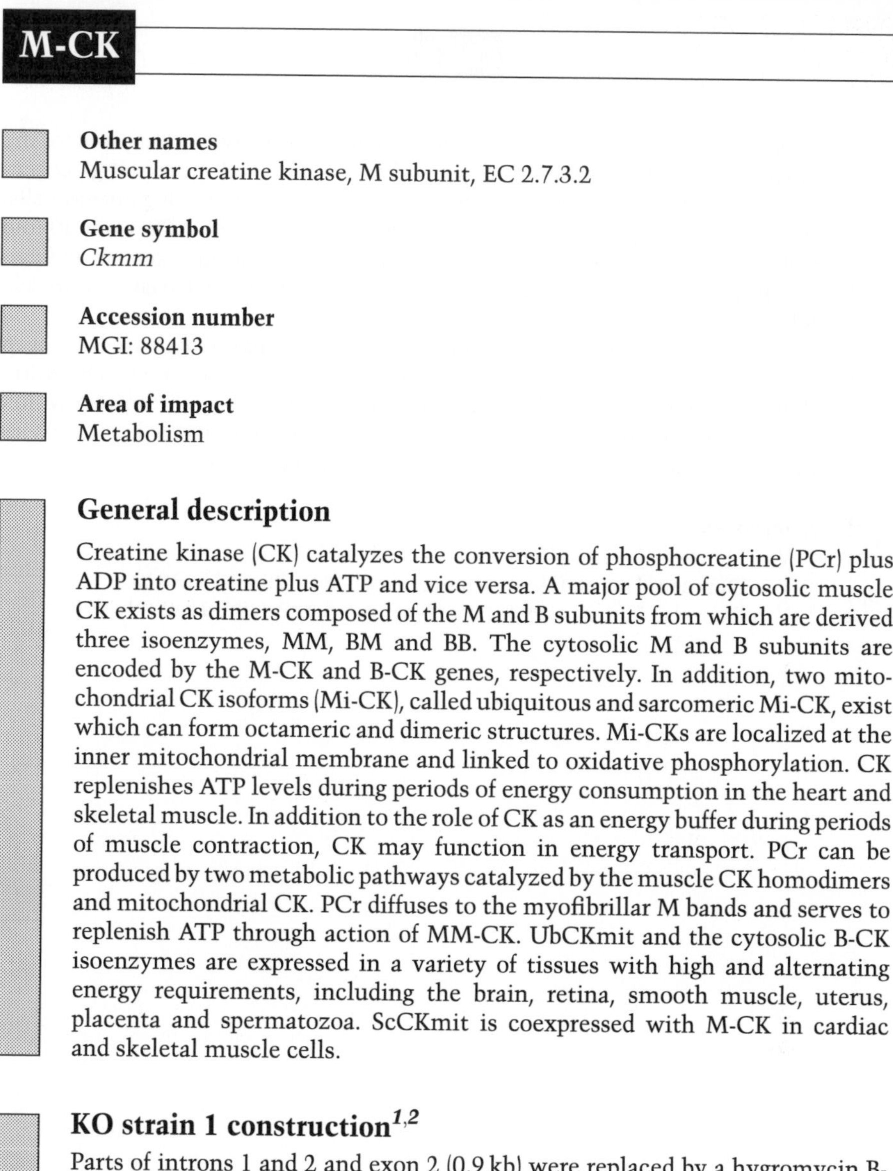

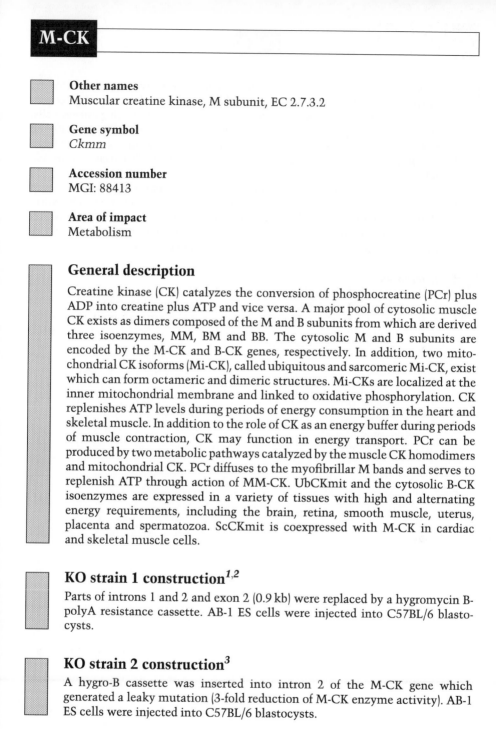

Other names
Muscular creatine kinase, M subunit, EC 2.7.3.2

Gene symbol
Ckmm

Accession number
MGI: 88413

Area of impact
Metabolism

General description

Creatine kinase (CK) catalyzes the conversion of phosphocreatine (PCr) plus ADP into creatine plus ATP and vice versa. A major pool of cytosolic muscle CK exists as dimers composed of the M and B subunits from which are derived three isoenzymes, MM, BM and BB. The cytosolic M and B subunits are encoded by the M-CK and B-CK genes, respectively. In addition, two mitochondrial CK isoforms (Mi-CK), called ubiquitous and sarcomeric Mi-CK, exist which can form octameric and dimeric structures. Mi-CKs are localized at the inner mitochondrial membrane and linked to oxidative phosphorylation. CK replenishes ATP levels during periods of energy consumption in the heart and skeletal muscle. In addition to the role of CK as an energy buffer during periods of muscle contraction, CK may function in energy transport. PCr can be produced by two metabolic pathways catalyzed by the muscle CK homodimers and mitochondrial CK. PCr diffuses to the myofibrillar M bands and serves to replenish ATP through action of MM-CK. UbCKmit and the cytosolic B-CK isoenzymes are expressed in a variety of tissues with high and alternating energy requirements, including the brain, retina, smooth muscle, uterus, placenta and spermatozoa. ScCKmit is coexpressed with M-CK in cardiac and skeletal muscle cells.

KO strain 1 construction[1,2]

Parts of introns 1 and 2 and exon 2 (0.9 kb) were replaced by a hygromycin B-polyA resistance cassette. AB-1 ES cells were injected into C57BL/6 blastocysts.

KO strain 2 construction[3]

A hygro-B cassette was inserted into intron 2 of the M-CK gene which generated a leaky mutation (3-fold reduction of M-CK enzyme activity). AB-1 ES cells were injected into C57BL/6 blastocysts.

Phenotype

KO strain 1 M-CK$^{-/-}$ mice were viable and fertile and had no apparent abnormalities. Absolute muscle force was normal but the null mice had a defect in burst activities. Fast-twitching fibers selectively compensated for the M-CK deletion by increased mitochondrial volume and increased glycogenolysis/glycolysis. ATP and PCr levels were not changed in resting muscle cells. However, phosphate energy exchange between PCr and ATP were 20-fold reduced. Moreover, PCr levels were normally reduced during exercise, indicating that other pathways than M-CK can contribute to PCr usage in muscle cells.

The leaky KO strain 2 M-CK mutants did not exhibit the increased glycogen consumption or glycogen contents observed in M-CK-null mice. The mitochondrial volume of muscle fibers was also normal. However, the energy flux through the CK enzyme reaction was still reduced, and muscle burst activity was decreased, correlating with levels of M-CK expression.

References

[1] Van Deursen, J. et al. (1993) Cell 74, 621–631.
[2] Van Deursen, J. et al. (1992) Nucleic Acids Res. 20, 3815–3820.
[3] Van Deursen, J. et al. (1994) Proc. Natl Acad. Sci. USA 91, 9091–9095.

mdr1a P-glycoprotein

Other names
mdr3 P-glycoprotein

Gene symbol
Pgy3

Accession number
MGI: 97570

Area of impact
Metabolism

General description

mdr1a P-glycoprotein is a plasma membrane protein that can actively extrude from the cell many different amphipathic compounds, including important anti-cancer and other drugs, xenobiotic toxins, and some steroid hormones. mdr1b P-glycoprotein can cause multidrug resistance in tumor cells by lowering intracellular drug concentration. mdr1a is primarily found in the apical membrane of polarized epithelia in the intestine and renal proximal tubules, in the canalicular membrane of hepatocytes and in the blood–brain barrier. In the blood–brain barrier it can effectively limit the accumulation in the brain of many drugs (e.g. ivermectin, vinblastine, digoxin, cyclosporin A, loperamide, and many others) by transporting these compounds back into the bloodstream. mdr1a P-glycoprotein in the intestinal epithelium can actively contribute to the direct excretion of many drugs from the bloodstream into the intestinal lumen as has been demonstrated for the drugs digoxin and paclitaxel (Taxol). Moreover, mdr1a P-glycoprotein in the intestinal epithelium limits the uptake of orally administered drugs. mdr1a P-glycoprotein in the liver and kidney also contributes to the excretion of several compounds into bile and urine, respectively.

KO strain construction

A hygromycin-selectable cassette was inserted in a 12.5 kb *Xba*I-*Sal*I *Pgy3* genomic fragment, replacing a 1.6 kb *Nhe*I fragment containing exons 6 and 7 of *Pgy3* which encode the putative transmembrane segments 2, 3 and 4, and the first cytoplasmic and second intracellular loop of the protein. E14 ES cells were targeted. The homozygous mouse stocks are maintained in a 129Ola or in an FVB genetic background.

Phenotype

mdr1a$^{-/-}$ mice were viable and fertile under normal animal facility conditions. Extensive histological analysis revealed no substantial differences between mutant and wild-type mice. No physiological abnormalities or differences in lifespan were observed. The mdr1a$^{-/-}$ mice were markedly hypersensitive to a range of drugs and pesticides such as ivermectin (50–100-fold), vinblastine (3-fold), and doxorubicin (2-fold). This hypersensitivity resulted from effects on several pharmacological processes in which the mdr1a P-glycoprotein plays an important role. mdr1a$^{-/-}$ mice exhibited a markedly (3-fold) increased bioavailability of orally administered drugs such as paclitaxel. mdr1a$^{-/-}$ mice also display delayed elimination of many drugs from the bloodstream and the whole body. The phenotype of mdr1a$^{-/-}$ mice is consistent with an important role of mdr1a P-glycoprotein in the protection of organisms against xenobiotic toxins, primarily those ingested with food.

Comments

The phenotype suggests that pharmacological blocking of mdr1-type P-glycoproteins with effective inhibitors may lead to drastically increased brain penetration, increased oral bioavailability, decreased elimination, and increased toxicity of many drugs. This implies certain risks, but also therapeutic opportunities for the use of such blockers in patients.

Acknowledgements
A.H. Schinkel
Netherlands Cancer Institute, Amsterdam, The Netherlands

References

1 Schinkel, A.H. et al. (1994) Cell 77, 491–502.
2 Schinkel, A.H. et al. (1995) J. Clin. Invest. 96, 1698–1705.
3 Schinkel, A.H. et al. (1996) J. Clin. Invest. 97, 2517–2524.
4 Mayer, U. et al. (1996) Br. J. Pharmacol 119, 1038–1044.
5 Sparreboom, A. et al. (1997) Proc. Natl Acad. Sci. USA 94, 2031–2035.

mdr1b P-glycoprotein

Other names
mdr1 P-glycoprotein (P-gps)

Gene symbol
Pgy1

Accession number
MGI: 97568

Area of impact
Metabolism

General description

mdr1b P-glycoprotein is a plasma membrane protein that can actively extrude from the cell many different amphipathic compounds, including important anti-cancer and other drugs, xenobiotic toxins, and some steroid hormones. It can cause multidrug resistance in tumor cells by lowering intracellular drug concentrations. mdrb1 is primarily found in bone marrow cells and the hematological compartment, kidney, liver, secretory epithelium of pregnant uterus, placental trophoblasts, adrenal gland and ovary.

KO strain construction

A neomycin-selectable cassette was inserted in a 9.6 kb *Apa*I *Pgy1* genomic fragment, replacing a small *Nco*I fragment containing exons 3 and 4 of *Pgy1* which encode the first putative transmembrane segment and some flanking sequences of the protein. The insertion also causes a frameshift. E14 ES cells were targeted. The homozygous mouse stocks are maintained in a 129Ola or in an FVB genetic background.

Phenotype

mdr1b $^{-/-}$ mice displayed normal viability, fertility and lifespan under normal animal facility conditions. Extensive histological and hematological analyses and analysis of serum clinical chemistry revealed no substantial differences between mutant mice and wild-type mice. No physiological abnormalities or differences in lifespan were observed. The size of the adrenal gland was not altered in mdr1b $^{-/-}$ mice. The mdr1b $^{-/-}$ mice were somewhat hypersensitive to anthracyclines like daunorubicin and doxorubicin (2-fold). This hypersensitivity may have resulted from increased sensitivity of bone marrow cells. Partially purified hematopoietic progenitor cells of mdr1b$^{-/-}$ mice had a decreased capacity to extrude the P-glycoprotein substrate and diagnostic dye rhodamine 123. Analysis of the disposition of the cardiac glycoside digoxin showed little effect of the mdr1b knockout on the tissue distribution and elimination of this drug, although the adrenal gland and ovaries accumulate somewhat (2-fold) more digoxin. The penetration of digoxin in fetuses of (17-day) pregnant mdr1b $^{-/-}$ mice was not detectably increased. The phenotype of mdr1b $^{-/-}$ mice suggests some contribution of the mdr1b P-glycoprotein in the protection of organisms against xenobiotic toxins, but its role in this respect appears to be secondary to that of the mdr1a P-glycoprotein.

Comments

The phenotype of mdr1b$^{-/-}$ mice indicates that the very high expression of mdr1b in the adrenal gland, secretory epithelium of pregnant uterus and placental trophoblasts, and the clear expression in ovary and hematopoietic progenitor cells are not essential for basic vital functions.

Double mdr1a$^{-/-}$mdr1b$^{-/-}$ KO strain construction

The neomycin-selectable targeting construct for the *Pgy1* gene was targeted next to the hygromycin-disruption of the *Pgy3* gene. The homozygous mouse stocks are maintained in a 129Ola or in an FVB genetic background.

Phenotype of double KO mutant

mdr1a$^{-/-}$/1b$^{-/-}$ mice displayed normal viability, fertility and lifespan under normal animal facility conditions (analyzed up to 1 year of age). Extensive histological and hematological analyses and analysis of serum clinical chemistry revealed no substantial differences between mutant mice and wild-type mice. No physiological abnormalities were observed. The size of the adrenal gland was not altered in mdr1a$^{-/-}$/1b$^{-/-}$ mice. Representation of several lymphocyte subclasses was not abnormal in these mice. The cytotoxic activity of NK cells was not changed. mdr1a$^{-/-}$/1b$^{-/-}$ mice displayed very extensive alterations in drug distribution and elimination similar to mdr1a$^{-/-}$ mice, with highly increased brain penetration and delayed body elimination of drugs. In addition, drug accumulation in adrenal glands and ovaries was increased. Hepatobiliary excretion of several drugs (digoxin) was significantly decreased in mdr1a$^{-/-}$/1b$^{-/-}$ mice, whereas the direct intestinal excretion of drugs (digoxin, paclitaxel) was nearly abrogated. Partially purified hematopoietic progenitor cells of mdr1a$^{-/-}$/1b$^{-/-}$ mice had a strongly reduced capacity to extrude the P-glycoprotein substrate and diagnostic dye rhodamine 123, even when compared to hematopoietic progenitor cells from the single mdr1a$^{-/-}$ or mdr1b$^{-/-}$ mice. This suggests a potential role for the endogenous Mdr1-type P-glycoproteins in protection of bone marrow against cytotoxic anti-cancer agents. mdr1a$^{-/-}$/1b$^{-/-}$ mice should provide a useful model system to further test the pharmacological roles of the drug-transporting P-glycoproteins and to analyze the specificity and efficacy of P-glycoprotein-blocking drugs.

Comments

The phenotype of mdr1a$^{-/-}$/1b$^{-/-}$ mice indicates that the mdr1a and mdr1b P-glycoproteins are not vital to the normal physiology of mice. This suggests that it may be feasible to completely block mdr1 P-glycoprotein activity in humans by pharmacological means without severe physiological side-effects.

Acknowledgements
A.H. Schinkel
Netherlands Cancer Institute, Amsterdam, The Netherlands

References

1 Schinkel, A.H. et al. (1997) Proc. Natl Acad. Sci. USA 94, 4028–4033.
2 Borst, P. and Schinkel, A.H. (1997) Trends Genet. 13, 217–222.

Other names
Mdr2 P-glycoprotein, phosphatidylcholine (PC) translocator

Gene symbol
Pgy2

Accession number
MGI: 97569

Area of impact
Metabolism

General description

Mdr2 is a plasma membrane protein of the ATP-binding cassette family. It carries out transmembrane translocation of phosphatidylcholine (PC) and some PC analogs. Mdr2 is present primarily in the bile canalicular membrane of hepatocytes, although some RNA expression has also been observed in spleen, skeletal and heart muscle, lung, and adrenal gland.

KO strain construction

A neomycin-selectable cassette was inserted in an 8.5 kb *Eco*RI *Pgy2* genomic fragment, replacing a *Xho*I/*Cla*I fragment containing exons 1 and 2 of *Pgy2* and the putative promoter region. E14 ES cells were targeted. Homozygous mutant mice were originally created on a 129Ola genetic background but are now maintained in the FVB genetic background to improve breeding efficiency.

Phenotype

Mdr2$^{-/-}$ mice were viable and fertile but, depending on their genetic background, some displayed decreased fertility and a shortened lifespan. Mdr2$^{-/-}$ mice developed a severe liver pathology, resembling a non-suppurative inflammatory cholangitis with portal inflammation and ductular proliferation[1]. Depending on genetic background, older Mdr2 $^{-/-}$ mice developed pre-neoplastic lesions in the liver which could progress to metastatic liver cancer[2]. The primary cause of this pathology was the absence of phosphatidylcholine from the bile of Mdr2$^{-/-}$ mice. The bile salt output was not significantly changed. Secondary to the absence of PC and the resulting damage to the hepatocytes, the bilary excretion of cholesterol and glutathione was also diminished, whereas the bulk bile flow was increased. In heterozygous Mdr2$^{+/-}$ mice, the bile flow, cholesterol, and glutathione output were not significantly altered but the biliary excretion of PC was intermediate between that of wild-type and Mdr2$^{-/-}$ mice. Phospholipid excretion into bile was normalized and all liver pathology was absent in Mdr2$^{-/-}$ mice carrying a MDR3 (the human homolog of Mdr2) transgene expressed in the liver[3]. It is likely that the detergent action of the high concentrations of hydrophobic bile salts in the absence of PC damages the canalicular membranes. In the presence of normal concentrations of PC in bile of wild-type mice (and even with intermediate PC levels in bile of

$Mdr2^{+/-}$ mice), bile salts are taken up in mixed micelles with PC, resulting in a drastically reduced detergent action of the bile salts.

Comments

The phenotype of the $Mdr2^{-/-}$ mice shows that Mdr2 P-glycoprotein in the bile canalicular membrane is necessary for the translocation of PC from the hepatocyte into the bile, and thus essential for maintaining the structural integrity of the biological membranes that come in contact with bile. The human counterpart of the $Mdr2^{-/-}$ mouse has recently been described[4,5]. Absence of the MDR3 P-glycoprotein results in severe liver disease, requiring liver transplantation.

Acknowledgements
P. Borst
Netherlands Cancer Institute, Amsterdam, The Netherlands

References

1 Smit, J.J.M. et al. (1993) Cell 75, 451–462.
2 Mauad, T.H. et al. (1994) Am. J. Pathol. 145, 1237–1245.
3 Crawford, A.R. et al. (1997) J. Clin. Invest. 100, 2562–2567.
4 Deleuze, J.F. et al. (1996) Hepatology 23, 904–908.
5 De Vree, J.M.L. et al. (1998) Proc. Natl Acad. Sci. USA 95, 282–287.
6 Smith, A.J. et al. (1994) FEBS Lett. 354, 263–266.
7 Oude Elferink, R.P.J. et al. (1995) J. Clin. Invest. 95, 31–38.

Megalin

Other names
gp330, LDL receptor-related protein, Heyman nephritis antigen, LRP

Gene symbol
Gp330

Accession number
MGI: 95794

Area of impact
Neurology

General description

Megalin is a member of the low-density lipoprotein (LDL) receptor gene family. It mediates the endocytic uptake of various macromolecules, including cholesterol-carrying lipoproteins, proteases and anti-proteinases. Megalin is expressed on the apical surfaces of epithelial tissues such as the neuro-epithelium.

KO strain construction

A 1.5 kb fragment of genomic sequence corresponding to regions 3′ of amino acid Asn4204 were selected for homologous recombination. ES clones were used to generate chimeras. Null mutants produced no detectable protein.

Phenotype

Megalin$^{-/-}$ mice died perinatally from respiratory insufficiency. Epithelial tissues that normally express megalin showed abnormalities, including lung and kidney. In brain, a holopresencephalic syndrome characterized by a lack of olfactory bulbs, forebrain fusion, and a common ventricular system resulted from impaired proliferation of neuroepithelium. An insufficient supply of cholesterol during development causes similar syndromes in humans and animals.

Comments

Megalin is important for forebrain development. The ability of megalin to bind lipoproteins suggests that it may be part of the maternal–fetal lipoprotein transport system, mediating the endocytic uptake of essential nutrients during post-gastrulation development.

Reference

[1] Willnow, T.E. et al. (1996) Proc. Natl Acad. Sci. USA 93, 8460–8464.

Mel-18

Other names
Zfp144

Gene symbol
Zfp144

Accession number
MGI: 99161

Area of impact
Development

General description

This gene was first identified in a melanoma cell line (hence Mel-18). It is a nuclear protein characterized by possession of a RING finger domain, a variant zinc-finger domain. It is related to the *Polycomb* group of genes in *Drosophila*, which control the expression of homeotic genes. The gene is widely expressed during mouse embryogenesis.

KO strain construction

A 1.0 kb *Eco*RI fragment containing the fifth exon, which encodes an ATG start codon and the RING finger domain, which is necessary for DNA-binding activity, was deleted and replaced by an MC1*neo* cassette in the reverse transcriptional orientation. The vector was introduced into R1 129 ES cells, which were aggregated with morulae to generate chimeras. The mutant phenotype was examined on a mixed C57BL/6 × 129 background.

Phenotype

Homozygous mutant mice are born but fail to thrive and most die between 3 and 6 weeks after birth. Cause of death is unclear but may relate to intestinal obstruction. Examination of skeletal preparations revealed subtle posterior homeotic transformations of the vertebrae. Expression boundaries of a number of *Hox* genes were moved anterior in the developing scleretomes at E11.5.

Comments

The homeotic phenotype observed in Mel-18 mutants is similar to that seen in *Bmi1* mutants, another polycomb-related gene in mammals, suggesting that these genes are involved in maintenance of repression of *Hox* gene expression during mesoderm development.

Reference

[1] Akasaka, T. et al. (1996) Development 122, 1513–1522.

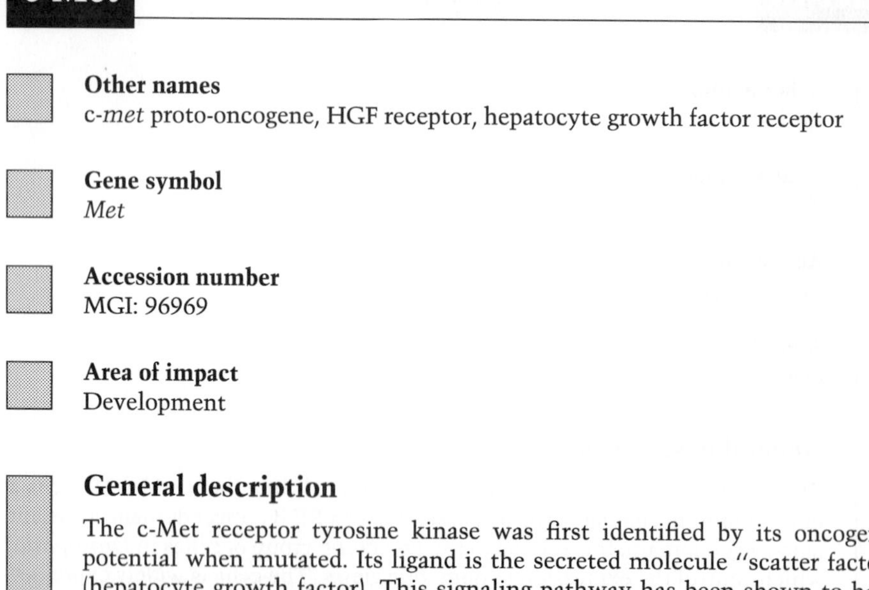

c-Met

Other names
c-*met* proto-oncogene, HGF receptor, hepatocyte growth factor receptor

Gene symbol
Met

Accession number
MGI: 96969

Area of impact
Development

General description

The c-Met receptor tyrosine kinase was first identified by its oncogenic potential when mutated. Its ligand is the secreted molecule "scatter factor" (hepatocyte growth factor). This signaling pathway has been shown to have multiple effects, including induction of cell motility, growth, invasion and migration, and morphogenesis. c-Met is expressed during development in the myogenic lineage (ventro-lateral dermomyotome and migrating myogenic precursor cells). Expression is also found in all epithelia and in subsets of neurons (motoneurons, structures in the telencephalon).

KO strain construction

A fragment of 2 kb that included the codon for an invariant lysine essential for kinase activity was deleted and replaced by a neo^r cassette. The vector was introduced into E14.1 129 ES cells, which were injected into C57BL/6 blastocysts. The phenotype was analyzed on a hybrid B6/129 background.

Phenotype

Homozygous embryos die between E14.5 and 16.5. The embryonic lethality is probably caused by a defect in placental development which affects the generation of the labryrinth layer. The liver of homozygous mutant animals is smaller and shows various degrees of damage: abnormal morphology of hepatocytes, increased sinusoids, and apoptosis which in extreme cases can lead to a loss of the majority of the liver parenchyma. c-Met$^{-/-}$ ES cells cannot contribute to the liver.

In addition, it was shown that myogenic cells are absent in the limbs, shoulders, diaphragm and tip of the tongue. In contrast, axial muscles are normal. The affected muscles are formed by migration from the somites. c-Met is expressed in the migrating muscle cells and the mesenchyme of the limb bud expresses HGF strongly. Migrating myogenic precursor cells are not formed in c-Met$^{-/-}$ embryos, and as a result, the skeletal muscles derived from migrating cells do not form. The observed phenotype in skeletal muscle is cell-autonomous, i.e. c-Met$^{-/-}$ ES cells cannot contribute to the skeletal muscles.

Comments

HGF/Met interaction may be a primary signal for muscle migration. The phenotype of mice with the SF/HGF$^{-/-}$ mutation is identical to that of mice with the c-Met$^{-/-}$ null mutation.

Acknowledgements
Carmen Birchmeier
Max-Delbrück-Centrum für Molekulare Medizin,Berlin, Germany

References

[1] Bladt, F. et al. (1995) Nature 376, 768–771.
[2] Schmidt, C. et al. (1995) Nature 373, 699–702.
[3] Brand-Saberi, B. et al. (1996) Dev. Biol. 179, 303–308.

Other names
GlcNAc-TI, *N*-acetylglucosaminyltransferase I

Gene symbol
Mgat1

Accession number
MGI: 96973

Area of impact
Development

General description

The *Mgat1* gene encodes GlcNAc-TI (UDP-*N*-acetyl-D-glucosamine:glycoprotein (*N*-acetyl-D-glucosamine to α-D-mannosyl-1,3-(R1)-β-D-mannosyl-(R2) β-1,2-*N*-acetyl-D-glucosaminyltransferase; EC 2.4.1.101) that catalyzes the addition of a GlcNAc residue in β1,2 linkage to $Man_5GlcNAc_2Asn$ to initiate the synthesis of complex or hybrid *N*-linked glycans. Two *Mgat1* gene transcripts of ~2.9 kb and ~3.3 kb have different 5′ UTRs and are differentially expressed in tissues. The *MGAT1* gene resides on human chromosome 5q31.2–31.3; a syntenic region occurs on mouse chromosome 11.

KO strain 1 construction[1]

A 3.8 kb *Sac*I fragment in pGEM7Zf(+) containing the *Mgat1* coding region was disrupted at the *Xma*III site by a PGK/neomycin-resistance gene in reverse orientation. A PGK/HSV-*tk* gene was inserted 5′ of the *Mgat1* gene fragment. The vector was linearized with *Hin*dIII. The two vectors obtained were electroporated into WW6 ES cells (~75% 129Sv, ~20% C57BL/6, ~5% SJL), and positive clones were injected into C57BL/6 blastocysts. The chimeric males were bred to CD1 females and heterozygotes were intercrossed to create homozygous mutants.

KO strain 2 construction[2]

The targeting construct inserted a neomycin-resistance cassette (pMC1*neo*-polyA) into the single protein encoding exon of the *Mgat1* gene. The *neo* cassette was flanked by 2.4 kb and 0.8 kb of the *Mgat1* gene. The vector was linearized with *Sac*I and electroporated into D3 ES cell lines. Homologous recombination was confirmed by Southern blotting. The ES cells were injected into C57BL/6J blastocysts, and chimeric males were mated with C57BL/6J females to generate heterozygous mice. Heterozygotes were intercrossed to produce homozygous mutants.

Phenotype

KO strain 1 mice carrying two disrupted *Mgat1* alleles died at mid-gestation. No stable *Mgat1* transcripts were detected and the embryos had no GlcNAc-TI activity or complex *N*-glycans, showing that the homozygous mutation was null. Mgat1$^{-/-}$ mice at E9.5 were retarded in development and had fewer somites, reduced neural tissue, and an open neural tube. Generally they had only one prominent branchial arch and a primitive heart. Small hemorrhages were visible. Histological examination revealed cellular disorganization but no particular defect that could be a specific cause of death. The heterozygous embryos and mice were indistinguishable from wild type littermates.

KO strain 2 mice also died at midgestation. At E9.5, they were one-half to two-thirds the size of wild-type embryos, and also had fewer somites. Some embryos had inverted heart loops and impaired vascularization.

The fact that Mgat1$^{-/-}$ embryos survived to midgestation suggests that complex or hybrid *N*-glycans are not required for blastocyst development and implantation. However, pre-implantation mutant blastocysts were shown by lectin binding to possess complex *N*-glycans, and by RT-PCR to contain apparently equivalent amounts of *Mgat1* RNA as Mgat1$^{+/+}$ blastocysts[3,4]. This RNA was a product of the *Mgat1* gene and thus must have been due to persistence of maternal *Mgat1* gene transcripts in Mgat1$^{-/-}$ blastocysts. By E5.5, complex *N*-glycans had disappeared from Mgat1$^{-/-}$ embryos and development continued to E9.5 in the absence of this subset of *N*-glycans.

A cell type-specific defect due to the lack of GlcNAc-TI was identified in chimeras formed with Mgat1$^{-/-}$ WW6 cells (KO strain 1). The cells were able to form beating embryoid bodies of differentiated morphology which contained endothelial tubes with red blood cells, but they exhibited a difference when tracked in chimeras by virtue of the β-globin transgene present at the telomere of chromosome 3 in WW6 ES cells[5]. The transgene is transcriptionally inert and can be readily detected by DNA:DNA *in situ* hybridization. Mgat1$^{-/-}$ WW6 cells contributed like Mgat1$^{+/+}$ and Mgat1$^{+/-}$ WW6 cells to essentially all tissues of E10.5 to E16.5 chimeras. However, they essentially failed to contribute to the organized layer of bronchial epithelium. In Mgat1$^{-/-}$ embryos at E9.5, a lung bud was present with a space where the bronchus should be, but there was no evidence of an organized layer of bronchial epithelium. Therefore, complex or hybrid *N*-glycans are essential for the generation or maintenance of this cell type[6].

While Mgat1$^{+/-}$ heterozygous mice had no obvious phenotype, Mgat1$^{+/-}$ WW6 cells contributed only about half as well as Mgat1$^{+/+}$ WW6 cells to organized epithelium in bronchi[6]. This suggests that mice with a single functional *Mgat1* allele may be compromised somehow in the lung. Cell type-specific and functional changes are currently being investigated in lungs of Mgat1$^{+/-}$ mice.

Comments

Mgat1$^{-/-}$ embryos show generalized cellular disorganization, suggesting that complex or hybrid *N*-glycans are involved in cell–cell interactions. Although a null mutation at the *MGAT1* locus should be lethal in humans, it is possible that some type of lung pathology may arise in people with one inactive *MGAT1* allele.

Acknowledgements
Pamela Stanley
Albert Einstein College Medicine, New York, NY, USA

References

1. Ioffe, E. et al. (1994) Proc. Natl Acad. Sci. USA 91, 728–732.
2. Metzler, M. et al. (1994) EMBO J. 13, 2056–2065.
3. Campbell, R. et al. (1995) Glycobiology 5, 535–543.
4. Ioffe, E. et al. (1997) Glycobiology 7, 913–919.
5. Ioffe, E. et al. (1995) Proc. Natl Acad. Sci. USA 92, 7357–7361.
6. Ioffe, E. et al. (1996) Proc. Natl Acad. Sci. USA 93, 11041–11046.
7. Kumar, R. et al. (1992) Glycobiology 2, 383–393.
8. Yang, J. et al. (1994) Glycobiology 4, 703–712.

Other names
N-Acetylglucosaminyltransferase III, GlcNAc-TIII, EC2.4.1.144

Gene symbol
Mgat3

Accession number
MGI: 104532

Area of impact
Unclear

General description

The *Mgat3* gene encodes GlcNAc-TIII which transfers the bisecting GlcNAc residue to the core β-linked Man residue in a subset of *N*-glycans. The gene encodes an RNA of ~4.7 kb that is differentially expressed in adult tissues, most highly in brain and kidney. The gene is located on chromosome 15 in mouse in a region syntenic to chromosome 22q in humans.

KO strain 1 construction[1]

A 1.8 kb *Bam*HI/*Sma*I fragment derived from a 129SvJ mouse genomic library containing the single *Mgat3* coding exon was cloned into the *Bam*HI site of the pflox vector (which contains sequences for neo^r and *tk*). The flanking 2.0 kb *Bam*HI fragment and an 8.5 kb *Sma*I fragment were also added to the pflox vector. R1 ES cells were electroporated with targeting vector. The null mutant allele was created in ES cells by Cre-*lox* site-directed recombination. Chimeric mice were generated by microinjection of ES cells into C57BL/6 blastocysts.

KO strain 2 construction[2]

An ~3 kb *Nsi*I genomic DNA fragment containing the *Mgat3* gene coding region cloned in pGEM7Zf(+) was disrupted by insertion at an *Nco*I site of a PGK/neomycin-resistance gene in reverse orientation. A PGK/HSV-*tk* gene was inserted at *Hin*dIII/*Xho*I sites 3′ of the *Nsi*I fragment. This targeting vector was linearized with *Xho*I. WW6 ES cells (~75% 129Sv, ~20% C57BL/6 , ~5% SJL[3]) gave male chimeras that transmitted agouti but neither the β-globin transgene of WW6 cells[3], nor the disrupted *Mgat3* allele. A female chimera mated to CD1 transmitted both the transgene and the disrupted allele. The strain is being maintained by intergenerational matings. Backcrossing to 129SvJ is in progress and is at F10.

Phenotype

KO strain 1 Mgat3$^{-/-}$ mice had no apparent phenotype in a C57BL/6 background other than a lack of GlcNAc-TIII activity and a deficiency of GlcNAc-bisected *N*-linked oligosaccharides. These mice were viable and healthy and reproduced normally. Internal organs were of normal cellularity and morphology, including brain and kidney. No abnormalities were seen in circulating leukocytes, erythrocytes or in serum metabolites reflective of kidney function.

KO strain 2 Mgat3$^{-/-}$ mice were viable and fertile but were obtained in higher numbers than predicted. Transmission distortion of the disrupted *Mgat3* allele also occurred from both male and female Mgat3$^{-/-}$ mice. An mRNA capable of producing a truncated protein was transcribed from the targeted *Mgat3* allele but no active product was formed. Mgat3$^{-/-}$ mice lacked GlcNAc-TIII activity in kidney extract, and glycoproteins from kidney or brain did not bind the lectin E-PHA under conditions optimized for recognition of *N*-glycans with a bisecting GlcNAc.

KO strain 2 Mgat3$^{-/-}$ mice were a little smaller than their littermates but appeared healthy and had normal lifespans. Histological examination of multiple organs revealed no significant lesions. Mice from this background and Mgat3$^{-/-}$ mice from heterozygotes of a tenth generation backcross to strain 129SvJ exhibited an altered leg clasp response when suspended by the tail, and an altered gait. However, these characteristics were not observed in KO strain 1 Mgat3$^{-/-}$ mice. This discrepancy may reflect differences in genetic background.

Comments

The phenotype of KO strain 1 Mgat3$^{-/-}$ mice suggests that GlcNAc-TIII is not essential for normal development, homeostasis or reproduction. The complex phenotype of KO strain 2 Mgat3$^{-/-}$ mice suggests that, in some genetic backgrounds, *N*-glycans with a bisecting GlcNAc may be involved in diverse cellular functions including neuronal function.

Acknowledgements
Pamela Stanley
Albert Einstein College of Medicine, New York, NY, USA

References

[1] Priatel, J.J. et al. (1997) Glycobiology 7, 45–56.
[2] The KO strain 2 Mgat3$^{-/-}$ mouse is not yet published.
[3] Ioffe, E. et al. (1995) Proc. Natl Acad. Sci. USA 92, 7357–7361
[4] Bhaumik, M. et al. (1995) Gene 164, 295–300.

β_2-Microglobulin

Other names

β2m

Gene symbol

B2m

Accession number

MGI: 88127

Area of impact

Immunity and inflammation, T cell

General description

β_2-Microglobulin (β2m) is a 12 kDa polypeptide which associates with the heavy chain of MHC class I molecules to form, along with an antigenic peptide, the mature class I MHC complex. In addition, β2m associates with some non-classical MHC class I proteins, as well as the neonatal intestinal Fc receptor. β2m is also found in relatively high concentration as a soluble protein in serum. The association of β2m with the MHC class I heavy chain is necessary for the class I complex to be transported from the ER to the cell surface. Thus, the lack of β2m protein results in the functional deletion of MHC class I molecules on the cell surface. MHC class I molecules are integral membrane proteins present on virtually all vertebrate cells and consist of a heterodimer between the highly polymorphic α-chain and the β2m protein. These cell surface molecules play a pivotal part in the recognition of antigens, the cytotoxic response of T cells, and the induction of self-tolerance. It is possible, however, that the function of MHC class I molecules is not restricted to the immune system, but extends to a wide variety of biological reactions including cell–cell interactions. For example, MHC class I molecules seem to be associated with various cell surface proteins, including the receptors for insulin, epidermal growth factor, luteinizing hormone and the β-adrenergic receptor. In mice, MHC class I molecules are secreted in the urine and act as highly specific olfactory cues which influence mating preference.

KO strain 1 construction[1]

A targeting vector was constructed which contained the neomycin-resistance vector inserted into exon 2 of the *B2m* gene. The vector was electroporated into D3 ES cells, and the correctly mutated ES cells were injected into C57BL/6J blastocysts. Chimeric heterozygous mice (129 × C57BL/6) were crossed to obtain homozygous $\beta 2m^{-/-}$ mice. The inactivation of the *B2m* gene was confirmed by Northern and Southern blotting, as well as by immunoprecipitation from metabolically labeled mutant cells.

KO strain 2 construction[2,3]

The targeting vector was designed such that the *B2m* gene was disrupted by the insertion of a neomycin-resistance cassette into exon 2. The construct contained approx. 5 kb of *B2m* sequences, from the 5′ flanking region to the second intron. The neomycin-resistance cassette was positioned in the same transcriptional orientation as the *B2m* sequences. The targeting vector was electroporated into E14 TG2a ES cells and positive clones were injected into C57BL/6 blastocysts. Chimeras were mated to C57BL/6 or B6/D2 animals. Heterozygous offspring were intercrossed to generate $\beta 2m^{-/-}$ mice. The null mutation was confirmed by immunostaining of mutant thymocytes and lymph node cells.

Phenotype

β2m-deficient mice were born at normal Mendelian frequencies and were grossly normal. The mutant mice did not have any MHC class I molecules on the surfaces of their cells, although they did express low amounts of immature heavy chain in the ER. The mice also failed to express functional Fcγ receptor on the intestinal epithelium of neonates. The number of $TCR\alpha\beta^+$ $CD4^-8^+$ T cells was drastically reduced (100–150-fold reduction), while the number of $TCR\alpha\beta^+$ $CD4^+8^-$ T cells and sIg^+ B cells was unchanged. The T cell precursor populations in the mutant mouse thymus were normal, as were the $\gamma\delta^+$ T cells. The mutant mice lacked CTL precursors and were unable to stimulate CTL responses when mixed with allogeneic cells in a mixed lymphocyte reaction (MLR). Cells from the β2m-deficient mice could be killed by activated, allogeneic CTLs, but only when effector cells were at a 9-fold higher concentration than was normally required. Proliferative responses to allogeneic cells were normal. When $\beta 2m^{-/-}$ mice were infected with lymphocytic choriomeningitis virus (LCMV), $CD4^+$ class II-restricted CTLs were induced, indicating that $CD4^+$ T cells can compensate, at least partially, for the lack of $CD8^+$ T cells in the $\beta 2m^{-/-}$ mice.

Double MHC class $I^{-/-}$MHC class $II^{-/-}$ mice

Mice doubly mutated so as to lack both MHC class I and MHC class II were created by crossing $\beta 2m^{-/-}$ mice (heterozygous for a disrupted $A_\beta{}^b$ allele) to $A_\beta{}^b$-deficient mice (heterozygous for the disrupted *B2m* allele)[8]. Double mutant mice were healthy and bred normally under sterile conditions. Both $CD4^+$ and $CD8^+$ T cells were depleted in double mutant mice but the B cell compartment was normal. Double mutant mice were able to mount specific antibody responses to T-independent antigen. Mutant spleen cells were poor responders and stimulators in MLR but mutant mice showed surprisingly normal rejection of allogeneic skin grafts.

Comments

The data demonstrate that the expression of β2m is necessary for the normal formation and surface expression of the MHC class I complex. In the absence of functional class I molecules, $CD8^+$ T cell maturation is severely impaired, indicating the crucial role that MHC class I plays in the selection of $CD8^+$ T cells. The results also confirm that the β2m molecule is a necessary component of the neonatal Fcγ receptor (FcRn). β2m-deficient mice have impaired CTL activity, but this appears to be at least partially compensated for by the development of $CD4^+$, class II-restricted cytotoxic T cells in these mutant mice. Doubly-mutated mice lacking both MHC class I and MHC class II show an unexpected degree of immunocompetence, evidence of compensatory mechanisms in the immune system.

Acknowledgements
Rudolf Jaenisch
Department of Biology, Massachusetts Institute of Technology, Cambridge, MA, USA

References

1. Zijlstra, M. et al. (1989) Nature 342, 435–438.
2. Koller, B.H. and Smithies, O. (1989) Proc. Natl Acad. Sci. USA 86, 8932–8935.
3. Koller, B.H. et al. (1990) Science 248, 1227–1230.
4. Zijlstra, M. et al. (1990) Nature 344, 742–746.
5. Bix, M. et al. (1991) Nature 349, 329–331.
6. Zijlstra, M. et al. (1992) J. Exp. Med. 175, 885–657.
7. Muller, D. et al. (1992) Science 255, 1576–1578.
8. Grusby, M.J. et al. (1993) Proc. Natl Acad. Sci. USA 90, 3913–3917.

Other names
Glutamate receptor, metabotropic 1; metabotropic glutamate receptor 1

Gene symbol
Grm1

Accession number
MGI: 95825

Area of impact
Neurology

General description

Glutamate is a major excitatory neurotransmitter in the brain. It binds to several kinds of glutamate receptors. Some receptors (AMPA, NMDA and kainate) flux ions across the membrane, whereas members of the other class (metabotropic GluRs) are coupled to G proteins. There are eight metabotropic GluRs (mGluR1–8). mGluR1 activates PLC, rather than cAMP, and has been postulated to play a role in synaptic plasticity.

KO strain 1 construction[1]

neo^r was inserted into the first coding exon (exon 1) of the *Grm1* gene. This vector was electroporated into D3 (129Sv) ES cells. Homologous recombinants were injected into C57BL/6 blastocysts. Chimeric mice were mated with C57BL/6, and +/− mice were intercrossed to produce progeny for study (129 × B6). No mGluR1 protein was detected.

KO strain 2 construction[2]

*neo-lac*Z was inserted near the second intracellular loop. An unstable fusion protein was made which was not inserted into the dendrites.

Phenotype

KO strain 1 mGluR1$^{-/-}$ mice were viable but exhibited characteristic cerebellar deficiencies resulting in ataxic gait and intention tremors. The anatomy of the cerebellum was not overtly perturbed. However, Purkinje cells appeared to receive multiple innervations from climbing fibers (rather than the one-to-one pattern typically expected). Excitatory synaptic transmission from parallel fibers (PFs) to climbing fibers appeared functional, and voltage-gated calcium channels of Purkinje cells appeared normal. Short-term plasticity to paired stimuli appeared normal for both parallel and climbing fibers. However, long-term depression (LTD) was clearly deficient (480 single PF stimuli in 2 seconds). The conditioned eyeblink response was also impaired. The results also suggested that the activation of both AMPA and mGluR1 receptors at parallel fiber synapses leads to LTD when paired with depolarization of Purkinje cells. Long-term potentiation (LTP) in the hippocampus was also substantially reduced in these animals, and there was a moderate reduction in context-specific associative learning. The gross anatomy of this region appeared normal.

Both KO strain 1 and 2 showed similar findings, but the affected region of the hippocampus differed[1,2].

Comments

mGluR1 is not "in line" in the generation of LTP, but rather modulates synaptic plasticity in the cerebellum.

References

1 Aiba, A. et al. (1994) Cell 79, 377–385 and 365–375.

2 Conquet, F. et al. (1994) Nature 372, 237–243.

mGluR2

Other names
Glutamate receptor, metabotropic 2; metabotropic glutamate receptor subtype 2

Gene symbol
Grm2

Accession number
MGI: 95826

Area of impact
Neurology

General description

mGluR2 is the glutamate receptor which is coupled to inhibition of the cAMP cascade through G protein. mGluR2 is expressed on the neurons in the CNS, particularly in the hippocampus, cerebellum and accessory olfactory bulb.

KO strain construction

The targeting vector consisted of 15.3 kb genomic sequences in which the downstream sequence of the translation initiation site in exon 2 was replaced with the neomycin-resistance gene. A HSV-*tk* gene fragment was attached to the 3′ end for negative selection. Chimeric mice generated from mGluR2-disrupted CCE ES cells (from an inbred mouse line 129SvJ) were mated with BDF1 ((C57BL/6 × DBA/2) F1).

Phenotype

The mGluR2-deficient mice grew and mated normally. They showed no behavioral abnormalities, nor any gross anatomical changes in the brain. They also showed no alterations in basal synaptic transmission, paired-pulse facilitation, or tetanus-induced long-term potentiation (LTP) at the hippocampal mossy fiber–CA3 synapses. Long-term depression (LTD) induced by low-frequency stimulation at the hippocampal mossy fiber–CA3 synapses, however, was almost fully abolished. The mutant mice performed water maze learning tasks normally.

Comments

The pre-synaptic mGluR2 is essential for inducing LTD at the mossy fiber-CA3 synapses, but this hippocampal LTD does not seem to be required for spatial learning.

Acknowledgements
Shigetada Nakanishi
Kyoto University, Faculty of Medicine, Kyoto, Japan

References
1. Yokoi, M. et al. (1996) Science 273, 645–647.
2. Tanabe, Y. et al. (1992) Neuron 8, 169–179.
3. Yokoi, M. and Nakanishi, S. (1996) In: Gene Targeting and New Developments in Neurobiology (S. Nakanishi et al., eds). Tokyo: Japan Scientific Societies Press, pp. 115–123.

Other names
Glutamate receptor, metabotropic 4; metabotropic glutamate receptor 4

Gene symbol
Grm4

Accession number
MGI: 95828

Area of impact
Neurology

General description

mGluR4 is a member of the G protein-linked family of receptors for glutamate that are negatively coupled to cAMP. Its expression is highest in granule cells of the cerebellum, which send parallel fibers that synapse with Purkinje cell dendrites. Coactivation with climbing fibers induces a form of long-term depression (LTD) that may be involved in learning motor tasks.

KO strain construction

neo^r was inserted into amino acids 246–291 in the N-terminal domain of mGluR4. It was flanked by *tk* on the 3′ end. The vector was electroporated into R1 ES cells (129). Homologous recombinants were aggregated with CD1 embryos to derive chimeric mice. Germ line offspring were mated with CD1 to create 129 × CD1 hybrids for analysis.

Phenotype

The mGluR4$^{-/-}$ mice developed normally and showed normal brain anatomy and behavior in the open field. Short-term plasticity in the Purkinje cells was impaired, although LTD was normal. This corresponded with an impairment in motor learning on a rotating rod. The glutamate analog L-2-amino-4-phosphonobutyric acid (LAP4) had no effect on Purkinje cell synaptic responses from mutant mice, but caused synaptic depression in wild-type controls.

Comments

These results suggest that mGluR4 is required for optimal motor function.

Acknowledgements
John Roder
Samuel Lunenfeld Research Institute, Mount Sinai Hospital, Toronto, ON, Canada

Reference

[1] Pekhletski, R. et al. (1996) J. Neurosci. 16, 6364–6373.

Other names
Glutamate receptor metabotropic 5; metabotropic glutamate receptor 5, class I mGluR

Gene symbol
Grm5

Accession number
MGI: 95829

Area of impact
Neurology

General description

Metabotropic glutamate receptors are G-coupled to G proteins. mGluR5 is expressed widely in the CNS, including the hippocampus. Broadly acting agonists (ACPD) and antagonists (MCPG) have suggested a role for class I mGluRs (mGluR1, mGluR5) in synaptic plasticity involving long-term potentiation (LTP) in the hippocampus.

KO strain construction

neo^r was inserted into exon 1 of the *Grm5* gene and flanked by *tk* on the 3′ end. The vector was electroporated into R1 ES cells (129). Clones showing homologous recombination were aggregated with CD1 embryos. Germ line chimeras were intercrossed to CD1 to produce (129 × CD1) offspring for analysis.

Phenotype

The mGluR5$^{-/-}$ mice showed normal survival and gross behavior. Sections through the CNS revealed normal development. Synaptic transmission in the hippocampus was also normal, but LTP in CA1 cells and dendate gyrus was low. Performance in the spatial component of the Morris water maze was deficient, as was context-dependent fear conditioning. LTP in CA3 neurons was normal. LTP in CA2 at the level of AMPA receptors was expressed normally, but patch clamp analysis showed a selective loss of $NMDA_{EPSC}$. This LTP_{NMDA} could be restored by augmenting PKC and was mimicked in normal cells by blocking PKC in wild-type neurons.

Comments

mGluR5 is coupled to PKC, which acts on NMDARs to selectively enhance their function during LTP. This step is important for spatial learning and memory.

Acknowledgements
Zhengping Jia and John Roder
Samuel Lunenfeld Research Institute, Mount Sinai Hospital, Toronto, ON, Canada

Reference

[1] Lu, Y.M. et al. (1997) J. Neurosci. 17, 5196–5205.

mGluR6

Other names
Glutamate receptor, metabotropic 6; metabotropic glutamate receptor subtype 6

Gene symbol
Grm6

Accession number
MGI: 95830

Area of impact
Neurology

General description

mGluR6 is the glutamate receptor which is coupled to the inhibition of the cAMP cascade through G protein in transfected CHO cells. It responds selectively to L-2-amino-4-phosphonobutyrate (LAP4), and is localized restrictedly at the post-synaptic site of the retinal ON-bipolar cells.

KO strain construction

The targeting vector consisted of a 13.6 kb genomic sequence in which the 1.2 kb fragment encoding a part of the transmembrane region of mGluR6 was replaced with the neomycin-resistance gene. A HSV-*tk* gene fragment was attached to the 5′ end for negative selection. Chimeric mice generated from mGluR6-disrupted CCE ES cells (from an inbred mouse line 129SvJ) were mated with BDF1 ((C57BL/g × DBA/2) F1)).

Phenotype

The mGluR6-deficient mice developed normally and showed no apparent behavioral abnormalities. They showed a loss of ON responses but unchanged OFF responses to light. They displayed no obvious changes in retinal cell organization, nor in the projection of optic fibers to the brain. They showed visual behavioral responses to light stimulation, as examined by shuttle box avoidance behavior experiments using light exposure as a conditioned stimulus. Additionally, they showed unaltered locomotor activity in a daily light–dark cycle, and exhibited light-stimulated induction of Fos immunoreactivity in the suprachiasmatic nucleus. However, mutant mice showed reduction of the sensitivity of pupillary responses to light stimulation and impairment of the ability to drive optokinetic nystagmus in response to visual contrasts.

Comments

mGluR6 is essential in synaptic transmission from photoreceptors to the ON bipolar cell and contributes to discrimination of visual contrasts.

Acknowledgements
Shigetada Nakanishi
Department of Biological Sciences, Kyoto University Faculty of Medicine, Kyoto, Japan

References

1 Masu, M. et al. (1995) Cell 80, 757–765.
2 Nakajima, Y. et al. (1993) J. Biol. Chem. 268, 11868–11873.
3 Nomura, A. et al. (1994) Cell 77, 361–369.
4 Iwakabe, H. et al. (1997) Neuropharmacology 36, 135–143.

MHox

Other names
Paired mesoderm homeobox 1

Gene symbol
Pmx1

Accession number
MGI: 97712

Area of impact
Development

General description

MHox is a homeodomain protein, most closely related to the *paired* class of homeodomain proteins. It is expressed in lateral mesoderm and in the mesenchyme of the limb buds and facial primordia during embryogenesis.

KO strain construction

A PGK*neo* cassette was introduced into the 5′ end of the homeodomain. The targeting vector was introduced into AB-1 129Sv ES cells, which were injected into C57BL/6 blastocysts. The mutant phenotype was examined on a B6 × 129 hybrid background.

Phenotype

Homozygous mice died neonatally, with respiratory compromise. They showed cranial abnormalities, including microcephaly, low-set ears and a pointed snout. Cleft palate was observed. Examination of the skeletons revealed multiple skeletal abnormalities, in both membranous and endochondral bone. In the skull, many neural crest derivatives were missing or hypoplastic. Defects were also observed in a subset of long bones and vertebrae.

Comments

MHox seems to be required in the process of formation of pre-skeletal condensations from undifferentiated mesenchyme.

Reference

[1] Martin, J.F. et al. (1995) Genes Dev. 9, 1237–1249.

Other names
Macrophage inflammatory protein 1α, stem cell inhibitor

Gene symbol
Scya3

Accession number
MGI: 98260

Area of impact
Immunity and inflammation, cytokines

General description

MIP-1α is a member of the CC subfamily of chemokines that has pro-inflammatory and stem cell inhibitory activities *in vitro*. MIP-1α expression can be induced in macrophages by LPS treatment and in monocytes by binding to plates coated with ICAM-1 or endothelial cell monolayers. MIP-1α can also be induced in mast cells, Langerhans cells, fibroblasts and T cells *in vitro*. MIP-1α is chemotactic for monocytes, T cells and neutrophils *in vitro*, and activates mast cells and basophils. MIP-1α can inhibit the proliferation of hematopoietic early progenitor cells *in vitro* and *in vivo*. The importance of MIP-1α *in vivo* has been difficult to determine because some other β chemokines have similar activities and bind to the same receptors.

KO strain construction

In the replacement-type targeting vector, 300 nucleotides of genomic DNA upstream of the mRNA start site, exon 1 and half of exon 2 of the MIP-1α gene were deleted and replaced with the PGK*neo* cassette. The *neo* gene was flanked by two regions of homology of 9 kb and 0.8 kb, respectively. The HSV-*tk* gene was also included to permit negative selection of random integrants. BK4 ES cells (a derivative of 129-derived E14TG2A) were transfected with the targeting vector by electroporation. Positive clones were injected into C57BL/6 blastocysts to generate chimeras. Offspring of chimeras included MIP-1$\alpha^{+/-}$ heterozygotes which were interbred to yield MIP-1$\alpha^{-/-}$ homozygotes

The null mutation was confirmed by Northern analysis of RNA from bone marrow macrophages stimulated with L929 cell-conditioned media. Heterozygotes showed reduced levels of MIP-1α RNA, while essentially no MIP-1α RNA was detected in cells of MIP-1$\alpha^{-/-}$ mice. Any mRNAs transcribed from the mutant gene were considered unlikely to form a functional protein because the deletion included almost half of the MIP-1α coding region.

Phenotype

MIP-1$\alpha^{-/-}$ mice were born at the expected Mendelian ratio and had no gross anomalies in either development or major organ histology. No overt hematologic abnormalities were noted in the mutant mice. Antigen processing, T cell activation and B cell differentiation were normal. MIP-1$\alpha^{-/-}$ mice were resistant to the coxsackievirus B3/20-induced myocarditis seen in wild-type mice. The absence of myocarditis in mutant mice was not due to any natural, strain-specific differences between 129 and C57BL/6J mice. The ability of mutant mice to recruit or retain mononuclear cells in the infected heart may be the source of the resistance. Influenza virus-infected MIP-1$\alpha^{-/-}$ mice had reduced pneumonitis and delayed clearance of virus compared to infected wild-type mice. T cells derived from *Listeria monocytogenes*-infected MIP-1$\alpha^{-/-}$ mice were significantly impaired in their ability to function in adoptive transfer-mediated protection of naive recipients.

Comments

MIP-1α is required for an inflammatory response to viral infection and cannot be replaced by other chemokines. MIP-1α may be required for the recruitment of immunocompetent T cells to the sites of viral infection. MIP-1α is not necessary for normal hematopoiesis and does not appear to be required to maintain hematopoietic stem cells in a quiescent state *in vivo*.

Acknowledgements
Don Cook
Schering Plough Research Institute, Kenilworth, NJ, USA

References

[1] Cook, D.N. et al. (1995) Science 269, 1583–1585.
[2] Cook, D.N. (1996) J. Leukocyte Biol. 59, 61–66.

Other names
Müllerian-inhibiting substance, anti-Müllerian hormone

Gene symbol
Amh

Accession number
MGI: 88006

Area of impact
Development

General description

During male development in mammals, the Sertoli cells of the testis produce a substance that actively induces the regression of the Müllerian ducts, thus preventing the formation of female reproductive organs. The female ovaries do not produce this substance, MIS, creating a permissive environment for female reproductive tract development. Testosterone produced by the Leydig cells of the testis promotes differentiation of the Wolffian ducts in males.

MIS is a homodimeric glycoprotein, a member of the large TGFβ growth factor gene family.

KO strain construction

A 0.6 kb fragment of MIS sequence encoding a portion of exon 1, intron 1 and exon 2, was replaced with a PGK*neo* cassette in the opposite transcriptional orientation. The vector was introduced into 129Sv AB-1 ES cells, which were injected into C57BL/6 blastocysts. The mutant phenotype was examined on a hybrid 129 × B6 and an inbred 129 background.

Phenotype

MIS homozygous mutants were viable. Females were morphologically normal but males developed normal male reproductive tracts as well as additional Müllerian duct derivatives, including uteri, oviducts and ovaries. MIS mutant males had poor fertility, which was a result of physical blockage of sperm transfer by the presence of the female reproductive tracts. MIS-deficient females were fully fertile. Leydig cell hyperplasia and occasional neoplasia was also observed in mutant males, suggesting that MIS is also a negative regulator of Leydig cell proliferation.

In mice doubly mutant for MIS and *Tfm* (a mutation in the androgen receptor), the male reproductive tract is absent and the female tract is fully developed. These mice develop as overt females with vaginal openings.

Comments

This study confirms the importance of MIS in suppressing female genital tract development in males. Mutations in MIS in humans are associated with pseudo-hermaphroditism.

Reference
[1] Behringer, R.R. et al. (1994) Cell 79, 415–425.

Other names
MutL (*E. coli*) homolog 1

Gene symbol
Mlh1

Accession number
MGI: 101938

Area of impact
DNA repair, cancer, meiosis

General description

MLH1 functions in the repair of DNA replication errors in all tissues examined and in processes involved in meiotic recombination. It is a homolog of the yeast MLH1 protein. Germ line mutations of the *Mlh1* gene are frequently found in hereditary non-polyposis colon cancer (HNPCC).

KO strain 1 construction[1]

A PGK-*hprt* mini-gene was used to replace exon 4 of the *Mlh1* gene. This exon encodes a highly conserved portion present in all MutL-like proteins. The construct was introduced into AB2.2 ES cells (129SvEv). The deletion of exon 4 produced a null allele as demonstrated by immunoblot analysis. Mice were backcrossed into the C57BL/6 strain.

KO strain 2 construction[2]

Exon 2 was deleted and replaced by a PGK*neo* cassette. PMC1*tk* was placed at the 3′ end of the construct. E14-1 ES cells were targeted and injected into C57BL/6 blastocysts. Chimeric males were mated with C57BL/6 females.

Phenotype

KO strain 1 MLH1-deficient animals were viable but prone to the spontaneous development of lymphomas, skin tumors and most notably intestinal adenomas and adenocarcinomas[3]. Microsatellite instability was observed in all tissues analyzed. Both male and female MLH1-deficient mice of both KO strains were infertile. Mutant testes were approximately half the normal size. Spermatids and spermatozoa were absent and spermatocytes failed to progress beyond the pachytene stage of meiosis. Mutant females underwent a normal estrous cycle and had normal mating and reproductive behavior but were sterile due to impaired meiosis. Spermatocytes from MLH1-deficient mice exhibited high levels of prematurely separated chromosomes and arrest in the first division of meiosis. MLH1 foci appeared to localize to sites of crossing-over on meiotic chromosomes in KO strain 1 mice.

Comments

The analysis of MLH1-deficient mice suggests a link between DNA mismatch repair and meiotic crossing-over. These mice will be useful as a model for hereditary non-polyposis colorectal cancer.

Acknowledgements
R. Michael Liskay
Oregon Health Sciences University, Portland, OR, USA

References

1 Baker, S.M. et al. (1996) Nature Genet. 13, 336–342.
2 Edelmann, W. (1996) Cell 85, 1125–1134.
3 Prolla, T.A. et al. (1998) Nature Genet. 18, 276–279.

Other names
HRX, ALL-1

Gene symbol
Mll

Accession number
MGI: 96995

Area of impact
Development

General description

Mll was first identified by its disruption by chromosomal translocation in human acute lymphoid-myeloid leukemias. It possesses a conserved SET domain found in *Drosophila trithorax* and *Polycomb* group genes, which regulate homeotic gene expression in a positive or negative manner, respectively. Expression is widespread throughout development with strong expression in the nervous system and somites.

KO strain construction

A promoterless *lacZ*-RSV-*neo* cassette was inserted in frame in exon 3b, proximal to the conserved SET motifs. The vector was inserted in D3 129Sv ES cells, which were injected into C57BL/6 blastocysts.

Phenotype

Heterozygotes were small at birth and had retarded growth. They were anemic and B cell populations were reduced. Skeletal abnormalities were observed and included bidirectional homeotic transformations of vertebral identities (C7 → C6, T3 → T2 and T13 → L1, L6 → S1). Expression boundaries of *Hox* genes were altered. Heterozygotes were hypofertile but some successful heterozygous crosses were performed. Homozygous embryos were non-viable beyond day 10.5 and showed loss of expression of *Hox* genes.

Comments

Mll appears to be required for positive regulation of mammalian *Hox* genes in a dose-dependent manner.

Acknowledgements
R. Hanson
Washington University School of Medicine, Washington, DC, USA

Reference

[1] Yu, B.D. et al. (1995) Nature 378, 505–508.

Other names
Thrombopoietin receptor

Gene symbol
Mpl

Accession number
MGI: 97076

Area of impact
Hematopoiesis

General description

The proto-oncogene c-*Mpl* encodes the receptor for thrombopoietin (TPO), the primary regulator of megakaryocyte and platelet production. c-Mpl is a member of the cytokine receptor superfamily with sequence similarity to the erythropoietin receptor and the G-CSF receptor. Expression of c-Mpl in normal mice appears to be restricted primarily to hematopoietic tissue, primitive hematopoietic stem cells, megakaryocytes and platelets.

KO strain 1 construction[1]

The targeting vector consisted of a 6.6 kb c-*Mpl* mouse genomic fragment (originally derived from strain 129) with a neomycin-resistance cassette inserted into the third exon of the c-*Mpl* gene, disrupting the cytokine receptor domain 1. The targeting construct was electroporated into ES-D3 C-12 cells, a subclone of D3 ES cells. Positive clones were injected into C57BL/6J blastocysts and chimeric males were mated with C57BL/6J females. Heterozygotes were crossed to obtain homozygous mutants. The null mutation was confirmed by RT-PCR of spleen RNA from mutant mice.

KO strain 2 construction[2]

Exons 1–5 of the c-*Mpl* gene (originally derived from a 129Sv genomic clone) were removed and replaced with a β-galPGK*neo* resistance cassette. A 1.8 kb region of the 5′ flanking sequence and 9.2 kb of the 3′ flanking sequence were included in the construct. The targeting vector was electroporated into W9.5 129Sv ES cells. Chimeric males were mated with C57BL/6 females and heterozygotes were interbred to generate homozygous mutants. The null mutation was confirmed by ligand-binding experiments using radioiodinated TPO, by Northern analysis of bone marrow cell RNA, and by the inability of mutant cells to respond to TPO in culture.

Phenotype

Heterozygotes (c-Mpl$^{+/-}$) showed no significant alterations in either platelet counts, platelet volume or megakaryocyte numbers. c-Mpl$^{-/-}$ mice were viable, healthy and showed no gross abnormalities. There were no significant differences in c-Mpl$^{-/-}$ red blood cells, total white blood cells, neutrophils, bands or eosinophils, as determined by differential cell counts. Size and cellularity of lymphoid organs were normal and no alteration was detected in bone marrow, spleen and thymus when cell type ratios or cell maturation markers of B and T cells were measured.

However, complete blood cell counts performed on c-Mpl$^{-/-}$ and c-Mpl$^{+/+}$ mice revealed a dramatic 10–20-fold drop in platelet counts in the gene-targeted animals which occurred with 100% penetrance. Histopathology demonstrated a similar 6-fold reduction in megakaryocytes in spleen and bone marrow. The remaining megakaryocytes were of lower ploidy than control megakaryocytes. Analysis of CFU-megakaryocyte from c-Mpl$^{-/-}$ and normal mice showed that the number of megakaryocyte progenitors was reduced in mutants compared to controls, indicating that the c-Mpl ligand has megakaryocyte-CSF activity. Furthermore, analysis of progenitor cells from the other hematopoietic lineages revealed a dramatic decrease in CFU-GM, BFU-E and CFU-MIX in c-Mpl$^{-/-}$ mice.

As is the case in other thrombocytopenic animals, circulating TPO levels were dramatically elevated in c-Mpl$^{-/-}$ mice but no increase in TPO mRNA level could be detected in any mutant organs. In contrast to normal platelets, c-Mpl$^{-/-}$ platelets were incapable of binding, internalizing or degrading TPO; the absence of receptors for TPO dramatically affected the clearance of TPO from the plasma. Reconstitution of normal platelet levels in c-Mpl$^{-/-}$ mice by injection of platelets purified from normal mice rapidly decreased plasma levels of TPO to those found in normal mice. It appears that TPO is constitutively released into the circulation at a constant rate, and that its circulating level is directly regulated by the platelets themselves.

Comments

Analysis of c-Mpl-deficient mice indicates that the c-Mpl ligand, thrombopoietin, is the major regulator of platelet production. Signaling through c-Mpl is critical for the maintenance of mature megakaryocyte and platelet numbers through control of progenitor proliferation and megakaryocyte ploidy. Although the absence of TPO seemed to affect only platelets among differentiated cells, the low level of all hematopoietic progenitors in the c-Mpl$^{-/-}$ mice indicates that TPO has a role in a broader spectrum of primitive hematopoietic cells than was first anticipated. This decrease in progenitors from all lineages indicates that TPO probably acts on a very early progenitor cell common to all lineages.

Acknowledgements
Frederic de Sauvage
Genentech Inc., S. San Francisco, CA, USA
Warren Alexander
The Walter and Eliza Hall Institute of Medical Research, Royal Melbourne Hospital, Victoria, Australia

References
[1] Gurney, A.L. et al. (1994) Science 265, 1445–1447.
[2] Alexander, W.S. et al. (1996) Blood 87, 2162–2170.
[3] Carver-Moore, K. et al. (1996) Blood 88, 803–808.
[4] Fielder, P. et al. (1996) Blood 87, 2154–2161.
[5] Vigon, I. et al. (1992) Proc. Natl Acad. Sci. USA 89, 5640–5644.

Other names
Cation-dependent mannose-6-phosphate receptor, CD-MPR, M6pr

Gene symbol
M6pr

Accession number
MGI: 96904

Area of impact
Metabolism

General description

The cation-independent (CI-MPR, MPR300, IGF2R) and the cation-dependent (CD-MPR, MPR46) mannose-6-phosphate receptors sort mannose-6-phosphate tagged molecules to lysosomes. MPRs acts in the *trans*-Golgi network to sort enzymes containing lysosomal targeting structures (MPR300 and MPR46) and bind to ligands on the cell surface and mediate their endocytosis (MPR300). MPR46 does not function in endocytosis of extracellular ligands.

KO strain 1 construction[1]

A neomycin-resistance cassette was inserted into exon 2 of the MPR46 gene. HSV-*tk* was placed at the 3′ end of the construct. Targeted E14 ES cells were injected into C57BL/6 blastocysts.

KO strain 2 construction[2]

A neomycin-resistance cassette was inserted into exon 3 of the MPR46 gene. HSV-*tk* was placed at the 5′ end of the construct. Targeted E14 ES cells were injected into C57BL/6 blastocysts.

Phenotype

MPR46$^{-/-}$ mice appeared normal and were viable. However, these mice partially missort mannose-6-phosphate-tagged proteins and exhibited increased levels of phosphorylated lysosomal enzymes in body fluids and the serum. MPR46$^{-/-}$ thymocytes and cultured fibroblasts secreted high levels of lysosomal enzymes, and fibroblasts had decreased intracellular levels of lysosomal enzymes and accumulated macromolecules within the endosomes/lysosomes due to the lack of lysosomal enzymes.

Comments

IGF-II$^{-/-}$IGF2R$^{-/-}$ MPR46$^{-/-}$ triple mutant mice survive at a low frequency only for first weeks after birth, implying that mannose-6-phosphate-regulated lysosomal molecule trafficking is essential for survival. These mice have a phenotype that resembles the rare human inherited diseases mucolipidosis II (MLII or I-disease) and III (MLIII or pseudo-Hurler polydystrophy)[3].

References

1. Ludwig, T. et al. (1993) EMBO J. 12, 5225–5235.
2. Köster, A. et al. (1993) EMBO J. 12, 5219–5223.
3. Ludwig, T. et al. (1996) Dev. Biol. 177, 517–535.

Other names
BCEI, trefoil factor 1, TFF1

Gene symbol
Bcei

Accession number
MGI: 88135

Area of impact
Metabolism

General description

The human (hpS2) and mouse (mpS2) pS2 proteins belong to the family of trefoil factors (TFF), which are characterized by the presence of one to six cysteine-rich domains called P-domain or TFF domains. Although hpS2 and mpS2 are normally expressed in the gastric mucosa, hpS2 is also abnormally expressed in ulcerative gastrointestinal diseases and in various cancers. In all cases, hpS2 and mpS2 are found in the cytoplasm of epithelial cells. It has been proposed that pS2 functions as a growth factor, protease inhibitor, or a mucin stabilizer to modulate cell growth and protect the integrity of the gastric mucosa.

KO strain construction

A 5′ 4.1 kb pS2 genomic DNA fragment and a 3′ 1.1 kb fragment were generated by PCR and subcloned into the *Kpn*I/*Cla*I sites of the pBS SK$^+$ vector, generating the pRH1-2 and pRH3-4 plasmids, respectively. The 1.1 kb *Cla*I/*Xba*I fragment from pRH3-4 was then subcloned into the *Cla*I/*Xba*I site of pRH1-2, generating plasmid pRH1-4. A 1.3 kb *Bgl*II/*Bam*HI PGK*neo* fragment, which does not contain a polyA signal, was then inserted in the *Bgl*II/*Bam*HI site of pRH1-4, generating the targeting plasmid pRH1-4 *neo*, which was linearized at the pBS *Kpn* site. The pS2 targeting experiment was done in 129SvJ D3 ES cells. The YA13 clone was injected into C57BL/6J blastocysts. Chimeric males were mated with 129SvJ females.

Phenotype

At 3 weeks, the antral and pyloric mucosa were thicker in mpS2$^{-/-}$ mice. Both sexes were equally affected, and the phenotype was fully penetrant. At 5 months, all examined mpS2$^{-/-}$ mice exhibited a circumferential adenoma encompassing the whole antropyloric mucosa. Severe hyperplasia with markedly elongated pits occupied most of the thickness of the mucosa, whereas the glands had a normal appearance. The epithelial cells lining the surface and the elongated pits showed high-grade dysplasia: nuclei were enlarged and hyperchromatic and showed loss of polarity. In addition, the antral and pyloric epithelial cells exhibited a 10-fold increase of the mitotic index and were improperly differentiated and dysfunctional, as they were almost entirely devoid of mucus as shown by PAS staining. In the upper part of the adenoma, the glandular architecture was distorted, with some branching and intraglandular bridging. Foci of carcinoma were observed within 30% of adenomas. Glands were irregular and closely backed together in a "back-to-back" pattern, and sheets of epithelial cells had crossed through the basement membrane.

At 5 months, the villi of the small intestinal mucosa of mpS2$^{-/-}$ mice were enlarged by a thickened lamina propria, whereas the length of the villi was normal. Epithelial cells lining the villi were normally differentiated, exhibited the usual ratio of goblet cells and enterocytes, and stained positively for mucus, which indicates that they were functional. The thickened lamina propria contained inflammatory cells, including lymphocytes, plasmocytes, and a few macrophages.

Comments

mpS2 may function as a gastric-specific tumor suppressor. However, as only 30% of the mpS2$^{-/-}$ mice developed carcinomas, the loss of mpS2 protein on its own is clearly not sufficient for malignancy. Additional genetic alteration may be required, as is the case for human colorectal tumorigenesis. Whereas normal gastric tissues express large amounts of hpS2, about 50% of human gastric carcinomas have lost expression of hpS2. No major alterations in hpS2 have been found in genomic DNA extracted from gastric carcinomas. However, the presence of aberrant hpS2 transcripts has been reported, which suggests that subtle hpS2 gene modifications (such as mutations leading to aberrant splicing events) might exist in some stomach carcinomas.

Acknowledgements
Marie-Christine Rio and Pierre Chambon
CNRS/INSERM, Illkirch, France

References

1 Lefebvre, O. et al. (1996) Science 274, 259–262.
2 Masiakowski, P. et al. (1982) Nucleic Acids Res. 10, 7895–7903.
3 Rio, M.C. et al. (1988) Science 241, 705–708.
4 Luqmani, Y. et al. (1989) Int. J. Cancer 44, 806–812.
5 Tomasetto, C. et al. (1990) EMBO J. 9, 407–414.

Gene symbol
Mpv17

Accession number
MGI: 97138

Area of impact
Development

General description

The Mpv17 strain of mice was generated by insertional mutagenesis using a recombinant virus. These mice carry a retroviral insert in the genome such that the expression of the *Mpv17* gene is abrogated. The *Mpv17* gene appears to be involved in kidney function in post-natal animals.

KO strain construction

CFW pre-implantation mouse embryos were transfected with the recombinant retrovirus MPSV*neo*, which is a replication-defective derivative of the myeloproliferative sarcoma virus altered to carry a *neo* gene. Embryos of 4–16 cells were exposed to virus-producing Rat1 cells (infected with MPSV*neo* and the helper virus F-Mulv) and transferred to pseudopregnant C57BL/6J × CBA foster mothers. Mosaic founder animals carrying a single provirus in the germ line were obtained and bred to homozygosity. The genetic background of the mutant mice was FVB.

Phenotype

Adult mice homozygous for the Mpv17 integration developed nephrotic syndrome and chronic renal failure. Histologically, affected kidneys showed a progressive sclerosis of the glomeruli with deposition of hyaline material in the glomeruli and in renal tubules. Sequences flanking the proviral integration were cloned and shown to be highly conserved during evolution. The flanking probe detected a 1.7 kb RNA which is ubiquitously expressed during embryogenesis and in the adult with high levels in kidney, brain and heart. This RNA was not detected in any tissue of homozygous animals, suggesting that the provirus interferes with expression of stable Mpv17 RNA. Sequence analysis of the cDNA suggested that the gene codes for a peptide of 176 amino acids with a hydrophobic region which suggests a possible membrane association for the putative Mpv17 protein. The Mpv17 protein appears to localize to the peroxisome and to play a role in the metabolism of reactive oxygen species. Reflecting the physiological relationship between the kidney and the inner ear, Mpv17 mutant animals become deaf at an early age, showing pathological changes reminiscent of Alport's syndrome.

Comments

Glomerular lesions similar to those observed in Mpv17 mice are seen in patients with progressive deterioration of renal function due to various renal disorders. The Mpv17 mutant is a potentially useful experimental system for studying mechanisms leading to renal disorders in humans which are not well understood on the genetic or molecular levels. The mouse is presently being used as a model to test anti-oxidant therapy in kidney disease.

Acknowledgements
Hans Weiher
Institute for Diabetes Research, Munich, Germany

References

1 Weiher, H. et al. (1990) Cell 62, 425–434.
2 Karasawa, M. et al. (1993). Hum. Mol. Genet. 2, 1829–1834.
3 Zwacka, R.M. et al. (1994) EMBO J. 13, 5129–5134.
4 Meyer zum Gortesberge, A.M. et al. (1996) Eur. Arch. Otorhinolaryngol. 253, 470–474.

Other names
Mineralocorticoid receptor, glucocorticoid-activated nuclear receptor, aldosterone receptor

Gene symbol
Mlr

Accession number
MGI: 99459

Area of impact
Metabolism, transcription factors

General description

The mineralocorticoid receptor is a ligand-activated nuclear receptor which is closely related to the glucocorticoid receptor. An important function of the mineralocorticoid receptor is stimulation of sodium reabsorption from the distal renal tubule. Glucocorticoids activate both the mineralocorticoid and the glucocorticoid receptor; aldosterone binds only to the mineralocorticoid receptor. Cell-specific responses to aldosterone are achieved by the specific expression of the enzyme 11-βHSD which inactivates glucocorticoids. This protective mechanism is absent in brain and MR can mediate responses to glucocorticoids and aldosterone. In the limbic system of the brain, mineralocorticoids lead to an increase in neuronal activity.

KO strain construction

To inactivate the *Mlr* gene, the major part of exon 3 which encodes the first of the two zinc fingers of the DNA-binding domain was replaced with the β-galactosidase gene inserted in-frame. *Mlr* DNA was cloned from a 129J genomic library. Targeted 129 ES cells were injected into C57BL/6 blastocysts.

Phenotype

Homozygous mutant animals died about 10 days after birth with alterations typical of aldosterone deficiency: hyperkalemia and high urinary sodium loss, and elevated plasma levels of renin, angiotensin II and aldosterone. Aldosterone-mediated sodium reabsorption was strongly reduced. The expression of the mRNAs encoding the subunits of the Na,K-ATPase and of the amilioride-sensitive sodium channel was unaltered.

Acknowledgements
Günther Schütz
German Cancer Research Centre, Heidelberg, Germany

References

[1] Berger, S. et al. (1996) Endocrine Res. 22, 641–652.
[2] Berger, S. et al. (1996) Steroids 61, 236–239.

MRP

Other names
Multidrug resistance (associated) protein, basolateral multispecific organic anion transporter (MOAT), glutathione *S*-conjugate pump (GS-X pump), LTC_4 exporter/transporter

Gene symbol
Mdrap

Accession number
MGI: 102676

Area of impact
Metabolism, inflammation

General description

MRP is a ubiquitously expressed member of the ATP-binding cassette transporter superfamily. This membrane protein mediates the ATP-dependent cellular extrusion of many natural product drugs (e.g. anti-cancer drugs), glutathione *S*-conjugates, glucuronide-conjugates, and cysteinyl leukotrienes. Elevated levels of MRP in tumor cells can cause multidrug resistance by lowering the intracellular drug concentration.

KO strain construction

A targeting vector was constructed by assembling a 3.6 kb *Sac*I/*Xho*I 5′ genomic fragment of the *Mdrap* gene (derived from 129Ola) with a fragment containing a hygromycin resistance gene driven by the mouse PGK promoter, and a 4.4 kb *Bam*HI/*Xho*I 3′ fragment of the *Mdrap* gene. Correct targeting deleted 2.7 kb of *Mdrap* sequences containing exons encoding amino acids 706–822 of the first ATP-binding domain. The targeting vector was introduced into E14 ES cells (129Ola) and positive clones were injected into mouse blastocysts. Chimeric offspring were generated that transmitted the mutation through the germ line. Mouse stocks were maintained as a cross of FVB and 129Ola strains because the 129Ola mice bred poorly.

Phenotype

MRP $^{-/-}$ mice were viable and fertile under normal animal facility conditions. Extensive histological analysis of 4-month-old females and males revealed no substantial differences between mutant and wild-type mice. Experiments using a mix of inside-out outside-in erythrocyte plasma membrane vesicles showed a strongly diminished ATP-dependent uptake of glutathione conjugates (e.g. 2,4-dinitrophenyl *S*-glutathione, ethacrynic acid *S*-glutathione) into the MRP$^{-/-}$ vesicles. Upon stimulation of leukotriene synthesis, *in vitro* cultured bone marrow-derived mast cells (BMMC) secreted similar amounts of leukotriene B_4 (LTB_4) but lower amounts of leukotriene C_4 (LTC_4). *In vivo*, the response to arachidonic acid-mediated ear inflammation was impaired, whereas phorbol ester-mediated ear inflammation and the responses to exogenous LTC_4 and platelet-activating factor were unaltered. The defect in the inflammatory response was probably due to decreased secretion of LTC_4 from leukotriene-synthesizing cells.

Compared to wild-type BMMC, growth of MRP-deficient BMMC cultures was inhibited in the presence of etoposide or vincristine, but not in the presence of cisplatin. The $MRP^{-/-}$ mice were hypersensitive to the anti-cancer drug etoposide (VP-16) but normally sensitive to vincristine. No differences in the tissue distribution of radiolabeled etoposide were detected, and the mutant and wild-type mice showed similar degrees of thrombocytopenia and leukocytopenia upon etoposide administration. The etoposide-sensitive target organ(s), tissue(s), or cells in $MRP^{-/-}$ mice responsible for the lethal effect remain to be identified.

Comments

The phenotype of $MRP^{-/-}$ mice is consistent with a role for MRP as the main LTC_4 exporter in leukotriene-synthesizing cells, as the main high-affinity GS-X pump or MOAT in erythrocytes, and as an important drug exporter in etoposide-sensitive normal cells. The phenotype further suggests that the ubiquitous GS-X pump (MRP) is dispensable in mice (and humans), making the treatment of multidrug resistance with MRP-specific inhibitors (reversal agents) in combination with particular anti-cancer drugs (e.g. vincristine) potentially feasible. In addition, these results suggest the existence of human MRP-deficient patients hypersensitive to particular anti-cancer drugs (e.g. etoposide).

Acknowledgements
P. Borst
Division of Molecular Biology, Netherlands Cancer Institute, Amsterdam, The Netherlands

References

[1] Wijnholds, J. et al. (1997) Nature Med. 3, 1275–1279.
[2] Evers, R. et al. (1997) FEBS Lett. 419, 112–116.

Gene symbol
Msh2

Accession number
MGI: 101816

Area of impact
DNA repair

General description

Msh2 is a homolog of the bacterial MutS DNA mismatch repair gene. The human MSH2 protein was found to bind to insertion/deletion loop-type (IDL) mismatched nucleotides with high affinity and the single base pair G/T mismatch with lower affinity. MutS family proteins have been also implicated in the safeguarding of the genome from increased recombination. MSH2 mutations cosegregate with the majority of hereditary non-polyposis colon cancer (HNPCC). HNPCC is an autosomal dominant syndrome with high penetrance that predisposes to colon cancer and a variety of other tumors, including gastric and endometrial tumors. MSH2 mutations have also been described as a cause of the Muir–Torre syndrome which predisposes to gastrointestinal and genitourital cancers and cancers of the skin and sebaceous glands.

KO strain 1 construction[1]

A neomycin-resistance gene was placed into a genomic *Msh2* fragment such that a stop codon was introduced at amino acid position 640 of the MSH2 protein. E14 ES cells were targeted and injected into C57BL/6 blastocysts.

KO strain 2 construction[2]

A hygromycin-resistance gene was inserted between codons 588 and 589 of the *Msh2* gene, N-terminal to the putative ATP-binding site. E14 ES cells were used and targeted cells injected into C57BL/6 blastocysts. No MSH2 transcripts were observed.

Phenotype

MSH2$^{-/-}$ mice were viable, fertile and had no development defects. However, these mice developed T and B cell lymphomas at an early age. They also developed spontaneous intestinal carcinomas and skin neoplasms. Biochemically, MSH2$^{-/-}$ mice had a general increase in the spontaneous mutation rate and a defect in binding to the mismatch, and acquired microsatellite instability.

Comments

Loss of DNA mismatch repair in MSH2 knockout mice leads to genome instability and a predisposition to cancer.

Acknowledgements
Tak W. Mak
Ontario Cancer Institute and Amgen Institute, Toronto, ON, Canada

References

[1] Reitmair, A.H. et al. (1995) Nature Genet. 11, 64–70.
[2] de Wind, N. et al. (1995) Cell 82, 321–330.
[3] Reitmair, A.H. (1996) Cancer Res. 56, 3842–3849.

Other names
Macrophage scavenger receptor class A, macrophage scavenger receptors type I and type II

Gene symbol
Scvr

Accession number
MGI: 98257

Area of impact
Immunity and inflammation, metabolism

General description

Type I and II (class A) scavenger receptors are expressed in macrophage lineage cells and sinusoidal endothelial cells. They mediate the uptake of a wide range of negatively charged macromolecules, including modified low-density lipoproteins. These receptors also mediate phagocytosis and EDTA-resistant macrophage adhesion.

KO strain construction

Exon 4 of the *Scvr* gene was disrupted by the insertion of a neomycin-resistance gene. HSV-*tk* was placed at the 3′ end for negative selection. A3–1 ES cells derived from 129 mice were electroporated and targeted ES cells were injected into C57L/6J blastocysts.

Phenotype

MSR-A$^{-/-}$ mice were viable, grew normally and were fertile. Peritoneal macrophages from these mice showed decreased EDTA-resistant adhesion to plastic. Degradation of acetyl LDL and oxidized LDL were reduced to 20% and 50% of control values, respectively. However, plasma clearance of modified LDL was normal, suggesting that alternative mechanisms exist for the removal of circulating LDL. MSR-A$^{-/-}$ macrophages showed 20% of the control level of AGE-BSA uptake. MSR-A$^{-/-}$ mice also exhibited increased susceptibility to infection (*Listeria monocytogenes*, herpes simplex virus type 1) and MSR-A$^{-/-}$apoE$^{-/-}$ double knockout mice showed a reduction in atherome size.

Comments

These mice are a model for the study of the involvement of macrophages in atherosclerosis and macrophage-mediated host defense mechanisms.

Acknowledgements
Tatsuhiko Kodama
Research Center for Advanced Science and Technology, University of Tokyo, Tokyo, Japan

References

1 Suzuki, H. et al. (1997) Nature 386, 292–296.
2 Suzuki, H. et al. (1997) Exp. Anim. 46, 17–23.
3 Kodama, T. et al. (1996) Curr. Opin. Lipidol. 7, 287–291.
4 Platt, N. et al. (1996) Proc. Natl Acad. Sci. USA 93, 12456–12460.
5 Sakai, M. et al. (1996) J. Biol. Chem. 271, 27346–27352.

MT-I and MT-II

Other names
Metallothionin I and II

Gene symbol
Mt1 and *Mt2*

Accession number
MGI: 97171
MGI: 97172

Area of impact
Metabolism

General description

Metallothionins (MTs) are low molecular weight molecules having a high cysteine content (30%) which bind to 7–12 metal atoms (zinc, copper, cadmium) per protein via mercaptide linkages. MTs are constitutively expressed at low levels and are highly inducible in response to metal ions, glucocorticoids or lipopolysaccharides. MT-I and MT-II are functionally equivalent enzymes that are localized 6 kb apart in the genome. MT-I and MT-II are expressed in most organs. MT-III is expressed in the brain and MT-IV in stratified squamous epithelia. MTs function in the detoxification of metals, the regulation of copper and zinc homeostasis, the synthesis, assembly, and function of zinc metalloproteins, and protection against reactive oxygen species.

KO strain 1 construction[1]

Pol-*neo* was inserted between the *Mt1* and *Mt2* genes. HSV-*tk* was placed at the 3′ end of the construct. For disruption of *Mt1* and *Mt2*, in-frame stop codons were cloned into the exons of both genes. AB1 ES cells were electroporated and targeted cells were injected into C57BL/6 blastocysts.

KO strain 2 construction[2]

pMC1*neo* was inserted into exon 2 of the *Mt2* gene. For disruption of *Mt1*, a frameshift was introduced into exon 1 using a 20 bp synthetic oligomer. HSV-*tk* was placed at the 5′ end of the construct. E14 ES cells were electroporated and targeted cells were injected into C57BL/6 blastocysts.

Phenotype

MT-I$^{-/-}$MT-II$^{-/-}$ mice were viable and fertile, implying that metallothionins have no role in development and reproduction. These mice were more sensitive to hepatic poisoning by cadmium. Livers from cadmium-injected mutant mice displayed localized necrosis, congestion and hemorrhaging, and focal areas of cell degeneration. Moreover, cells from these mice were more susceptible to oxidative stress affecting the intracellular redox status of cells.

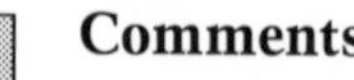

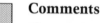

Comments

The main function of MT-I and MT-II appears to be detoxification of metals and the protection of cells from oxidative stress.

References

[1] Masters, B.A. et al. (1994) Proc. Natl Acad. Sci. USA 91, 584–588.
[2] Michalska, A.E. & Choo, K.H.A. (1993) Proc. Natl Acad. Sci. USA 90, 8088–8092.
[3] Lazo, J.S. et al. (1995) J. Biol. Chem. 270, 5506–5510.

Other names
Metal response element binding transcription factor 1 (MREBP), MRP, MRF, MBF-1

Gene symbol
Mtf1

Accession number
MGI: 101786

Area of impact
Transcription factors

General description

MTF-1 (metal responsive transcription factor 1) activates basal and heavy metal induced transcription of metallothionein (MT) genes by binding to MRE (metal responsive elements) sequences in their promotor regions. MTF-1 contains a DNA-binding motif with six C2H2 zinc fingers, and three functional activation domains (AD): an acidic AD, a proline-rich AD and a serine/threonine-rich AD.

KO strain construction

Two-thirds of the first zinc-finger exon were replaced by either the neomycin or hygromycin gene via homologous recombination, eliminating DNA binding of MTF-1. A 146 bp *Kpn*I/*Bam*HI fragment was replaced by either a blunt-ended neomycin phosphotransferase or hygromycin expression cassette in the antisense orientation. 129Sv ES cells from a clone containing one disrupted MTF-1 allele were injected into blastocytes to generate mice heterozygous for the mutated MTF-1 allele. Mice had a heterogeneous 129Sv × C57BL/6 genetic background.

Phenotype

Genotype analysis of embryos from MTF-1$^{+/-}$ intercrosses revealed no viable homozygous mutant embryos older than E14.5 days (0 of 25). Before E13.5, homozygous mutant embryos were obtained at about the expected Mendelian frequency (10 of 41), indicating that implantation and early post-implantation development were not impaired. At E13.5, only 1 of 54 MTF-1 homozygous mutant embryos was dead. Conversely, at E14.5, only 1 of 24 homozygous mutant embryos was found to be still alive. Viable MTF-1$^{-/-}$ embryos at E13.5 were macroscopically indistinguishable from their heterozygous and wild-type littermates. They were uniformly pink in color, well-vascularized and showed no internal bleeding. However, histological analysis of these embryos showed variable degrees of liver damage. Loosened liver structure with enlarged, congested sinusoids and dissociation of the epithelial compartment were observed.

While epithelial cells of the liver appeared morphologically normal by conventional histology, immunohistochemical analysis showed that they had reduced cytokeratin expression. At the same time, α-actin expression in the vascular structure of the liver was not affected. At E14.5, the only null mutant embryo still alive showed necrosis of the liver, diffuse bleeding and edema, areas of necrosis in the mesenchymal tissue and in the vertebral bodies, as well as an almost complete lack of cytokeratin expression, whereas the α-actin expression in the vessel walls again was indistinguishable from heterozygous or wild-type littermates. TUNEL *in situ* staining did not reveal a dramatic increase in apoptosis at E13.5 and E14.5. The cell proliferation rate in the MTF-1 homozygous mutant was not affected, since bromodeoxyuridine incorporation at E13.5 and E14.5 revealed no decrease when compared to wild-type or heterozygous littermates. Living E13.5 homozygous mutants showed not only no macroscopic color change indicative of anemia, but also no change in the number or appearance of erythrocytes or hematopoietic stem cells observed in histological sections. At E13.5, the contribution of enucleated erythrocytes is small so that a deficiency of these cells would not easily be detected. However, the only living E14.5 embryo showed no reduced representation of enucleated erythrocytes, again arguing against a defect in hematopoiesis. Examination of the histology of the placenta in MTF-1 homozygous mutants did not reveal any abnormalities.

Comments

Preliminary data suggest that even fibroblasts obtained from MTF-1$^{-/-}$ embryos before the onset of liver decay have a mutant phenotype, in that they are refractory to oncogene immortalization.

Acknowledgements
Walter W. Schaffner
University of Zurich, Zurich, Switzerland

References

1 The MTF-1$^{-/-}$ mouse is not yet published.
2 Heuchel, R. et al. (1994) EMBO J. 13, 2870–2875.
3 Brugnera, E. et al. (1994) Nucleic Acids Res. 22(15), 3167–3173
4 Radtke, F. et al. (1993) EMBO J. 12, 1355–1362.

MUG1

Other names
Murinoglobulin 1, α-1-inhibitor 3 (A1I3) in rat, hamster and guinea pig

Gene symbol
Mug1

Accession number
MGI: 99837

Area of impact
Cardiovascular

General description

Mouse murinoglobulin is a monomeric 180 kDa plasma glycoprotein and a member of the proteinase inhibitor of the α_2-macroglobulin (MAM) family. Four different murinoglobulin genes have been characterized but only one murinoglobulin protein (MUG1) has been isolated from plasma. The 5 kb MUG1 mRNA is exclusively expressed in mouse liver and only from the third week after birth. Murinoglobulin, but not α_2-macroglobulin (MAM), is subject to regulation of expression during pregnancy, around birth, and in adolescence. The *in vivo* function of the single-chain inhibitors of the murinoglobulin type is not clear.

KO strain construction

The targeting construct was a replacement vector which contained a genomic 7.5 kb fragment *Nhe*I/*Bam*HI (129J library) comprising exons 18–25 of the *Mug1* gene. A mutation and frameshift were created by replacing a 900 bp *Sca*I/*Kpn*I fragment, containing exon 18 and 108 bp of exon 19, by a 1.8 kb *Xho*I/*Cla*I fragment encoding the hygromycin B phosphotransferase gene. The E14 ES cell line (129Ola) was injected into C57BL/6 blastocysts and these were implanted into F1 (C57BL/6 × CBA/J) foster mice. Chimeric mice were mated to C57BL/6 mice to obtain heterozygous agouti pups. Homozygous mice used in the experiments were a mixture of the 129Ola and C57BL/6J strains.

Phenotype

Although only the *Mug1* gene was deleted by targeted inactivation, no MUG mRNA or MUG plasma protein could be detected in homozygous MUG1-deficient mice. This is in accordance with the observation that only one MUG protein (MUG1) was isolated from plasma. The MAM plasma levels of adult mice were not different between C57BL/6 wild-type mice and MUG$^{-/-}$ mice. MUG$^{-/-}$ mice backcrossed to C57BL/6 mice over three generations were fed a choline- and methionine-free diet supplemented with ethionine to induce pancreatitis. MUG knockout mice were more sensitive to the diet than control mice (MUG$^{-/-}$: 44% survival; C57BL/6: 74% survival) but less sensitive than MAM knockout mice (MAM$^{-/-}$: 33% survival). Further experiments will be carried out on MUG-deficient mice backcrossed to C57BL/6.

Comments

Mouse plasma, like that of all rodents but unlike that of mammals, contains two different types of proteinase inhibitors of the A2M family: the tetrameric A2M and the monomeric murinoglobulins.

Double MAM$^{-/-}$MUG$^{-/-}$ KO strain construction

The targeting construct (replacement vector) contained a genomic 7.5 kb *Nhe*I/*Bam*HI fragment (129J library) comprising exons 18–25 of the *Mug1* gene. A 900 bp *Sca*I/*Kpn*I fragment, containing exon 18 and 108 bp of exon 19, was replaced by a 1.1 kb fragment encoding the neomycin gene. This construct was used for electroporation of a hygromycin-resistant heterozygous MAM-deficient cell line with proven germ line transmission of the deficient MAM gene[5]. The E14 ES cell line (129Ola) was injected into C57BL/6 blastocysts and these were implanted into F1 (C57BL/6l × CBA/J) foster mothers. Chimeric mice were mated to C57BL/6 mice to obtain heterozygous agouti pups. Homozygous mice used in the experiments were a mixture of the 129Ola and C57BL/6J mouse strains. Because it was not known if the MAM and MUG genes were closely linked on the same chromosome, two strategies were followed. In the first strategy, a MAM$^{+/-}$ cell line was electroporated with a *Mug1* gene vector to obtain MAM$^{+/-}$MUG$^{+/-}$ cell lines. In the second case MUG-deficient mice were crossed with MAM$^{-/-}$ mice. Both procedures resulted in doubly deficient mice.

Phenotype of double KO mutant

Liver MAM mRNA and MUG mRNAs and MAM and MUG plasma proteins were absent in doubly deficient mice. MAM$^{-/-}$MUG$^{-/-}$ mice were viable, healthy and produced litters of normal size. MAM$^{-/-}$MUG$^{-/-}$ mice were fed a choline- and methionine-free diet supplemented with ethionine to induce pancreatitis. No difference in sensitivity to the diet was found between the control mice (C57BL/6) and the double knockout mice.

Comments

In contrast to MAM$^{-/-}$MUG$^{-/-}$ mice, MAM$^{-/-}$ or MUG$^{-/-}$ single knockout mice were more sensitive to diet-induced pancreatitis than control mice. One explanation could be the difference in genetic background between the single MAM$^{-/-}$ and MUG$^{-/-}$ knockout mice (C57BL/6 background) and double MAM$^{-/-}$MUG$^{-/-}$ knockout mice (129Ola × C57BL/6). Another explanation is that when both proteinase inhibitors are absent, there is an upregulation of a third protein which functionally substitutes for the MAM and MUG deficiencies. Further experiments will have to be carried out on mice with the silenced MAM and MUG genes in a homogeneous C57BL/6 background.

Acknowledgements
Fred Van Leuven
Experimental Genetics Group, Center for Human Genetics, Leuven, Belgium

References

[1] The $MUG^{-/-}$ and $MUG^{-/-}$ $MAM^{-/-}$ mice have not been published.
[2] Van Leuven, (1982) Trends Biochem. Sci. 7, 185–187.
[3] Overbergh, L. et al. (1991) J. Biol. Chem. 266, 16903–16910.
[4] Overbergh, L. et al. (1994) Genomics 22, 530–539.
[5] Overbergh, L. et al. (1995) J. Lipid Res. 36, 1774–1786
[6] Umans, L. et al. (1995) J. Biol. Chem. 270, 19778–19785.

Musk

Other names
Neural fold/somite kinase 2, muscle-specific kinase agrin-receptor, Nsk2

Gene symbol
Nsk2

Accession number
MGI: 103308

Area of impact
Neurology

General description

Musk, also called Nsk2, is a muscle-specific receptor tyrosine kinase which is expressed in patches along skeletal muscle, at sites corresponding to neuromuscular junctions (post-synaptic muscle surface). These Musk patches are thought to be activated by incoming motor nerve fibers that release agrin. Musk is thought to activate a rapsyn-dependent pathway leading to the clustering of acetylcholine receptors and dystroglycan. A separate pathway leads to synapse-specific transcription.

KO strain construction

neo^r was inserted into the third kinase exon encoding part of the catalytic domain. The vector was electroporated into E14.1 ES cells (129) and homologous recombinants were injected into blastocysts. Chimeras were bred to C57BL/6 (+/+, +/−, −/−) to generate 129 × B6 progeny for analysis.

Phenotype

Mice homozygous for the null allele of Musk exhibited a lack of acetylcholine receptor clusters. Neuromuscular synapses did not form in these mice, suggesting a failure in the induction of synapse formation. Nerves innervating skeletal muscle grew all over the skeletal muscle fiber, without forming discrete synapses. These animals died at birth, and exhibited abnormal movement and breathing patterns. There were also effects on the differentiation of pre-synaptic terminals.

Comments

The results obtained suggest that Musk responds to a critical nerve-derived signal (agrin), and in turn activates signaling cascades responsible for all aspects of synapse formation, including organization of the post-synaptic membrane, synapse-specific transcription, and pre-synaptic differentiation.

Agrin$^{-/-}$ mutants also have fewer neuromuscular junctions, but still show some post-synaptic differentiation[4]. This suggests that agrin is not the only organizing signal from the nerve terminal.

References

1 DeChiara, T.M. et al. (1996) Cell 85, 501–512.
2 Valenzuela, D.M. et al. (1995) Neuron 15, 573–589.
3 Glass, D.J. et al. (1996) Cell 85, 513–523.
4 Gautam, M. et al. (1996) Cell 85, 525–535.

Other names
Myeloblastosis oncogene, c-*myb* proto-oncogene

Gene symbol
Myb

Accession number
MGI: 97249

Area of impact
Oncogenes, hematopoiesis

General description

The murine c-myb gene encodes a 636 amino acid product. The protein consists of three functional domains: a DNA-binding domain, a transactivation domain, and a negative regulatory domain. The identification of DNA-binding domains and transcriptional activation domains within its gene product, as well as nucleotide-binding site(s) located within the promoter region of hematopoietic and non-hematopoietic genes, indicates that c-myb is a *trans*-acting transcription factor. Expression patterns of c-myb indicate that the gene product has a primary role in hematopoiesis. High levels of expression are detected in immature hematopoietic cells, with expression decreasing as these cells terminally differentiate. In addition, c-myb expression has been reported in a limited number of human malignancies of neuroectodermal and hematopoietic origin, as well as sporadically in colon, lung and breast carcinomas.

KO strain construction

The 1.2 kb neomycin-resistance gene from pMC1/polyA$^+$ was subcloned into the *Pst*I endonuclease restriction site of exon 6. The c-myb replacement construct contained 8.0 kb of 5′ homology and 1.65 kb of 3′ homology in relationship to the neomycin-resistance gene. Targeted ES-D3 ES cells were injected into C57BL/6J blastocysts. Chimeric males were bred with either C57BL/6J or CD-1 females.

Phenotype

c-myb$^{-/-}$ mutant fetuses died at E15.5 due to anemia. While the hematocrit levels of phenotyically normal fetuses at E15.5 was approximately 40%, the mutant fetuses had hematocrit levels of less than 5%. The mutants appeared unable to switch from primitive (yolk sac-derived) erythropoiesis to definitive (fetal liver-derived) erythropoiesis. Myelopoiesis was also adversely affected in the mutant fetuses in comparison to phenotypically normal littermates.

Comments

The reduction of both erythropoiesis and myelopoiesis in the mutant fetuses in comparison to phenotypically normal littermates strongly suggests that c-myb may function in the proliferation and/or differentiation of early hematopoietic progenitors and possible hematopoietic stem cells.

Acknowledgements
Michael Mucenski
Oak Ridge National Laboratory, Oak Ridge, TN, USA

References

[1] Mucenski, M. et al. (1991) Cell 65, 677–689.
[2] Lin, H. et al. (1996) Curr. Top. Microbiol. Immunol. 211, 79–87.

Gene symbol
Myf5

Accession number
MGI: 97252

Area of impact
Development, transcription factor

General description

The *Myf5* gene is a member of the myogenic basic helix-loop-helix (HLH) factor transcription factor gene family. All induce myogenic differentiation in cultured non-muscle cells, suggesting they might be functionally redundant.

KO strain construction

The vector produced a deletion in the *Myf5* gene and replaced it with a PGK*neo* cassette. The vector was introduced in J1 129 ES cells and injected into C57BL/6 blastocysts. The mutant phenotype was examined on a mixed 129 × B6 background.

Phenotype

Mice lacking Myf5 were unable to breathe and died immediately after birth, owing to the absence of the major distal part of the ribs. Other skeletal abnormalities, except for complete ossification of the sternum, were not apparent. Histological examination of skeletal muscle from newborn mice revealed no morphological abnormalities. Northern blot analysis demonstrated normal levels of muscle-specific mRNAs including *MyoD*, *myogenin* and *Myf6*. However, the appearance of myotomal cells in early somites was delayed by several days.

A myogenin cDNA inserted into the *Myf5* locus by homologous recombination simultaneously disrupted Myf5 function[2]. Mice homozygous for this myogenin "knock-in" mutation developed a normal rib cage and were viable, therefore demonstrating functional redundancy of Myf5 and myogenin for rib formation. Mice doubly homozygous for Myf5 and MyoD mutations show a complete absence of skeletal muscle[3].

Comments

These results suggest that while Myf5 plays a crucial role in the formation of lateral sclerotome derivatives, it is dispensable for the development of skeletal muscle, perhaps because other members of the myogenic HLH family substitute for Myf5 activity.

Acknowledgements
Thomas Braun
Technische Universitat Braunschweig, Braunschweig, Germany
Rudolf Jaenisch
Whitehead Institute, Cambridge, MA, USA

References

[1] Braun, T. et al. (1992) Cell 71, 369–382.
[2] Wang, Y. et al. (1996) Nature 379, 823–825.
[3] Rudnicki, M.A. et al. (1993) Cell 75, 1351–1359.
[4] Rudnicki, M.A. et al. (1992) Cell 71, 383–390.

Other names
Herculin, MRF4

Gene symbol
Myf6

Accession number
MGI: 97253

Area of impact
Development

General description

The *Myf6* gene is a member of the myogenic basic helix-loop-helix (bHLH) family of transcription factors, which have the capacity to transform a variety of cell types in culture into myoblasts. It is expressed after myogenin but before MyoD in the sequence of myotome development *in vitro*. *Myf6* is genetically closely linked to *Myf5*.

KO strain 1 construction[1]

Nucleotides −5 to +207 relative to the transcription start site were replaced by a PGK*neo* cassette in the sense orientation. This deletion removes the initiation codon of *Myf6* and disrupts the gene. The vector was introduced into J1 129 ES cells, which were injected into C57BL/6 or BALB/c blastocysts. The phenotype was identical on a 129 SvJ inbred or mixed 129 × B6 or 129 × BALB/c background.

KO strain 2 construction[2]

The targeting vector deletes 1.9 kb of genomic DNA encompassing all but 69 amino acids of the *Myf6* gene. The deleted region was replaced by a PGK*neo* cassette in the reverse transcriptional orientation. The vector was introduced into AB2-1 129 ES cells, which were injected into C57BL/6 blastocysts. The phenotype was examined on a mixed 129 × B6 background.

KO strain 3 construction[3]

A deletion encompassing the bHLH domain was introduced into the *Myf6* gene.

Phenotype

The three studies on Myf6 mutations produced phenotypes of differing severity. KO strain 2 mice were viable and fertile, with normal muscle differentiation. Slight alterations in expression of muscle-specific genes were observed. *Myogenin* mRNA expression was elevated 4-fold in adult skeletal muscle, suggesting that myogenin may compensate for the absence of Myf6 in the mutants. No change in other myogenic gene expression was seen. Rib anomalies-bifurcations and supernumerary processes were observed. These anomalies are different from those observed in Myf5-deficient mice.

KO strain 3 mice died perinatally. Myotomal development was affected during development, although the late MyoD-driven phase of muscle development did occur, and resulted in extensive trunk myogenesis. Rib anomalies occurred which were more severe than observed in KO strain 2 mice. The KO strain 1 mutation is the most severe, with aberrant and delayed myotome formation, reduction in the size of the axial muscles and loss of distal rib structures. In this mutation, Myf5 expression was shown to be strongly down-regulated, suggesting that the phenotype is caused by essentially loss of Myf5 and Myf6. The two genes are physically linked on the chromosome and the differences in severity of the three mutations probably reflects different *cis*-acting effects of the Myf6 disruption or neighboring Myf5 expression. In all cases, however, muscle development was not grossly impaired, showing that Myf6 is not essential for muscle maturation and adult myogenesis.

References

1 Braun, T. and Arnold, H.-H. (1995) EMBO J. 14, 1176–1186.
2 Zhang, W. et al. (1995) Genes Dev. 9, 1388–1399.
3 Patapoutian, A. et al. (1995) Development 121, 3347–3358.

Gene symbol
Myod1

Accession number
MGI: 97275

Area of impact
Development, transcription factor

General description

MyoD was cloned as a factor capable of converting a variety of cells *in vitro* into myoblasts. It is a member of the myogenic basic helix-loop-helix (bHLH) transcription factor family, which includes MyoD, myogenin, Myf5 and Myf6. These genes are thought to regulate the expression of a number of muscle-specific genes and are activated sequentially during embryonic development. MyoD is expressed later in development than the other myogenic genes.

KO strain construction

The MyoD promoter, exon 1 and half of intron 1 were deleted and replaced by a PGK*neo* cassette. The vector was introduced into J1 129 ES cells, which were injected into C57BL/6 or BALB/c blastocysts. The phenotype was examined on a mixed B6 × 129 or BALB/c × 129 background.

Phenotype

Mice homozygous for the loss of MyoD were viable and fertile. No morphological abnormalities were observed in any muscles examined. Northern analysis with a panel of skeletal muscle-specific probes indicated that MyoD mutant mice accumulate normal levels of muscle-specific mRNAs. Significantly, Myf5 is readily detected in muscle of post-natal MyoD mutant mice. Normally, Myf5 mRNA levels are markedly reduced at day 12 of gestation, about the time MyoD mRNA first appears. These results suggest that Myf5 is repressed by MyoD.

Comments

MyoD is dispensable for muscle development and may have overlapping functions with other myogenic factors. MyoD mutant muscle is deficient in regenerative capacity. Double mutants between Myf5 and MyoD show complete absence of skeletal muscle[2].

Acknowledgements
Michael Rudnicki
McMaster University, Hamilton, ON, Canada

References
1 Rudnicki, M.A. et al. (1992) Cell 71, 383–390.
2 Rudnicki, M.A. et al. (1993) Cell 75, 1351–1359.
3 Megeney, L.A. et al. (1996) Genes Dev. 10, 1173–1183.
4 Wang, Y. et al. (1996) Nature 379, 823–825.

Myogenin

Gene symbol
Myog

Accession number
MGI: 97276

Area of impact
Development

General description

Myogenin is a member of the myogenic family of basic helix-loop-helix (bHLH) transcription factors. In tissue culture, these proteins can act as potent myogenic factors, transforming a number of different cell types into muscle cells. Myogenin is first expressed during embryogenesis in the myotomes at E5.5, following on the earlier expression of Myf5.

KO strain 1 construction[1]

A pMC1*neo* cassette was inserted in the reverse orientation in the first exon, 25 codons downstream of the start of translation. The vector was introduced into AB1 129 ES cells, which were injected into C57BL/6 blastocysts. The phenotype was examined on a mixed 129 × B6 background.

KO strain 2 construction[2]

A neo^r cassette was introduced into the first exon at the *Sac*I site, disrupting the reading frame upstream of the bHLH domain. The vector was introduced into E14 129Ola ES cells, which were injected into B6C3F1 × B6 blastocysts. The chimeras were bred with BALB/c females, so the phenotype was examined on a mixed BALB/c × 129 background.

Phenotype

Homozygous mice were immobile and cyanotic at birth. Skeletal muscle was considerably reduced and skeletal defects were also observed. Histology revealed that myofiber density was much reduced in the mutant muscle areas, and further analysis suggested that myoblasts can form but myotube formation is severely affected. It is proposed that myogenin is not required for initiation of myogenesis, but is required for normal muscle differentiation.

References

1. Hasty, P. et al. (1993) Nature 364, 501–506.
2. Nabeshima, Y. et al. (1993) Nature 364, 532–535.

Other names
Neural cell adhesion molecule, CD56

Gene symbol
Ncam

Accession number
MGI: 97281

Area of impact
Neurology, adhesion

General description

NCAM is a broadly distributed abundant cell surface receptor belonging to the Ig superfamily. Its most clearly defined molecular function is as a homophilic receptor in the formation of cell–cell bonds, with heterophilic properties having been described as well. Although encoded by a single gene, NCAM is expressed in a number of polypeptide variants generated by alternative splicing. These variants differ in the size of their cytoplasmic domains and modes of association with the cell surface membrane. The most prominent variants are NCAM-180, NCAM-140 and NCAM-120.

NCAM is widely expressed in the developing and mature nervous system by both neuronal and glial cells. Roles for NCAM in a variety of processes in the development and function of the nervous system have been proposed, including axonal outgrowth and fasciculation as well as cell migration and learning and memory. NCAM is said to be the major, if not the exclusive, carrier of PSA, a carbohydrate structure thought to be involved in plastic processes in the nervous system.

KO strain 1 construction[1]

The targeting vector was designed to eliminate expression of only the NCAM-180 isoform. An exon 18-targeting vector was created by insertion of a β-actin-driven neomycin phosphotransferase gene (BANF) in place of exon 18. The *neo* gene was transcribed in the opposite direction relative to *Ncam*. A diphtheria toxin A-chain gene transcribed from a β-actin promotor served for negative selection. The genetic background was CD-1.

KO strain 2 construction[2]

The targeting vector was designed to replace most of exons 3 and 4 and the intervening intron with a neomycin-resistance gene, resulting in the elimination of all NCAM isoforms. The mutation was introduced into 129Ola-derived E14-1 ES cells. Chimeras were generated by injection into C57BL/6 blastocysts. Analysis of the phenotype was performed on 129, C57 and mixed backgrounds with comparable results.

KO strain 3 construction[3]

A 500 bp *Bcl*I fragment containing a portion of exon 13 and the downstream intron was replaced with a *neo* cassette. A diphtheria toxin A-chain gene transcribed from a β-actin promoter served for negative selection. The genetic background was 129Sv c+p+. This KO strain secreted the extracellular domain of NCAM.

Phenotype

KO strain 1 NCAM$^{-/-}$ mice were viable, healthy and fertile. The most conspicuous mutant phenotype in KO strain 1 mice was in the olfactory bulb, where granule cells were both reduced in number and disorganized. In addition, precursors of these cells were found to be accumulated at their origin in the subependymal zone at the lateral ventricle. Analysis of the mutant suggested a defect in cell migration through specific loss of the polysialylated form of NCAM-180

KO strain 2 NCAM-deficient mice were born in a Mendelian fashion and appeared healthy and fertile. However, they showed a 10% reduction in total brain size. The absence of NCAM coincided with the almost total disappearance of PSA, indicating that NCAM is in fact the major carrier of this carbohydrate. The animals were also shown to have a much reduced olfactory bulb due to a defect in the migration of olfactory neuron precursors from the subventricular zone of the forebrain into the olfactory bulb. Behavioral testing of the animals in the Morris water maze and an exploratory test revealed deficits in spatial learning and memory as well as exploratory behavior and long-term potentiation (LTP). Increased aggression between males was also observed. A detailed analysis of axonal growth processes in KO strain 2 NCAM-deficient mice (which focused on the dentate gyrus of the hippocampus and its axonal projection, the mossy fibers) demonstrated that quantity, pathfinding and laminar growth of these axons were strongly affected in the absence of NCAM. The basis for these alterations appeared to be deficient fasciculation. Individually growing axons invaded the CA3 pyramidal cell layer, thereby destroying its highly ordered organization. As a consequence, synaptogenesis of the mossy fibers on these target neurons was affected. In addition, the phenotype was quantitatively and qualitatively more severe in adult than in young animals, suggesting an essential role for NCAM in the maintenance of plasticity in the mature nervous system.

Mice heterozygous for the KO strain 3 mutation died *in utero*. Analysis of founder chimeric embryos with high levels of ES cell contribution revealed severe growth retardation and morphological defects by E8.5–9.5. This phenotype was observed for ES cells heterozygous or homozygous for the mutation. These results indicate that the dominant lethality associated with secreted NCAM does not require the presence of membrane-associated NCAM.

Comments

The phenotype of NCAM deficiency is mainly, if not exclusively, linked to the highly polysialylated form, demonstrating the important role of this modification. These results provide the first genetic proof for a vertebrate cell adhesion molecule playing a key role in axonal growth and pathfinding.

Acknowledgements

Brian Key
Department of Anatomy, University of Melbourne, Parkville, Victoria, Australia
Harold Cremer
Laboratoire de Génétique et Physiologie du Développement, Campus de Luminy, Université de le Mediterranée, Marseilles, France
Terry Magnuson
Department Genetics, Case Western Reserve University, Cleveland, OH, USA

References

[1] Tomasiewicz, H. et al. (1993) Neuron 11, 1163–1174.
[2] Cremer, H. et al. (1994) Nature 367, 455–459.
[3] Rabinowitz, J.E. et al. (1996) Proc. Natl Acad. Sci. USA 93, 6421–6424.
[4] Treloar, H.B. et al. (1997) J. Neurobiol. 32, 643–658.
[5] Muller, D. et al. (1996) Neuron 17, 413–422.
[6] Carenini, S. et al. (1996) Cell Tissue Res. 287, 3–9.
[7] Cremer, H. et al. (1997) Mol. Cell. Neurosci. 8, 323–335.
[8] Hu, H. et al. (1996) Neuron 16, 735–743.
[9] Ono, K. et al. (1994) Neuron 13, 595–609.

Other names
Neutral endopeptidase (EC. 3.4.24.11), common acute lymphoblastic leukemia antigen (CALLA), Enkephalinase, CD10

Gene symbol
Mme

Accession number
MGI: 97004

Area of impact
Immunity and inflammation

General description

NEP is a cell surface zinc metallopeptidase that regulates the action of a variety of physiologically active peptides. Identified substances are largely neutral or humoral oligopeptide agonists, such as substance P, bradykinin and atriopeptin. The NEP enzyme is localized in the brush border membrane of the kidney, along mucosal epithelial surfaces, and with peptidergic neurons in the CNS. NEP expression also occurs in inflammatory cells, and in pre-B cells in cases of childhood lymphoblastic leukemia.

KO strain construction

A 6.5 kb genomic fragment (derived from a 129Sv library) containing exons 10–14 of the NEP gene was used to construct the targeting vector. Exon 13 and part of its intron were replaced by PGK*neo*. The targeting fragment was subcloned into pPNT for double selection. The targeting vector was electroporated into J1 ES cells and positive clones were introduced into C57BL/6 blastocysts. Chimeric males were bred with C57BL/6 females to obtain heterozygotes for the mutated allele. Heterozygotes were intercrossed to obtain homozygous $NEP^{-/-}$ mutants. The null mutation was confirmed by Northern analysis of renal $polyA^{+}$ RNA and by assay of renal membranes for phosphoramidon-sensitive esterolytic activity.

Phenotype

Mutant mice were born at the expected Mendelian ratio and were grossly developmentally normal. Coagulation profiles were normal in mutant mice. However, $NEP^{-/-}$ mice showed hypotension, increased sensitivity to endotoxin, and increased microvascular permeability. Mutant mice challenged with endotoxin LPS and D-galactosamine were >100 times more likely to die than wild type littermates, and showed discoloration of their livers due to hemorrhagic necrosis. Mutant mice also showed increased sensitivity to injection with a combination of TNFα and IL-1β, all animals developing irreversible shock. $NEP^{-/-}$ mice could be protected from lethal shock by i.p. injection of human NEP.

Comments

NEP plays a critical role in regulating states of circulatory shock. The site of NEP action must be downstream of the release of TNFα and IL-1.

Acknowledgements
Bao Lu
Children's Hospital, Harvard Medical School, Boston, MA, USA

Reference

[1] Lu, B. et al. (1995) J. Exp. Med. 181, 2271–2275.

Neuregulin

Other names
Heregulin, c-neu receptor, NDF (neu differentiation factor), GGF (glial growth factor), ARIA (acetyl choline receptor inducing activity), SMDF (sensory and motoneuron derived factor)

Gene symbol
Hgl

Accession number
MGI: 96083

Area of impact
Neurology

General description

Neuregulin is a growth and differentiation factor which exists in many variants. These variants are produced by alternatively spliced transcripts, and possibly by the use of different promoters. Common to all variants is an EGF-like domain. The factor interacts directly with the erbB3 and erbB4 receptor tyrosine kinases, but does not bind to erbB2 or erbB1 (EGF receptor). However, rapid phosphorylation of erbB2 in response to neuregulin is observed in cells that coexpress erbB2 and erbB3 or erbB4. This phosphorylation is thought to be the result of receptor heterodimerization and cross-phosphorylation. Mutagenesis in the mouse has demonstrated that erbB2 is an essential coreceptor for transmission of neuregulin signals.

KO strain construction

Two targeting vectors were generated. Exons 7, 8 and 9 of the neuregulin gene (from strain 129) that encode the EGF domain were deleted and replaced either with *neo*, or *lacZ* plus *neo*. The latter generates a neuregulin-*lacZ* fusion allele in which *lacZ* sequences are fused in-frame to neuregulin coding sequences derived from exon 6. In the mouse, both alleles had identical phenotypes when in a homozygous state. The mutations were introduced into E14-1 ES cells derived from 129Ola and the phenotype was analyzed on both pure 129 and mixed 129Ola × C57BL/6 genetic backgrounds.

Phenotype

Homozygous neuregulin$^{-/-}$ animals died on E10.5, probably because of a defect in heart development. Mutant animals displayed an absence of trabeculation in the heart ventricle. In the heart, neuregulin is produced by the endocard; the signal is received by myocardial cells that express erbB4 and erbB2. Targeted mutations of erbB4 or erbB2 result in an identical defect in trabeculation of the heart ventricle. Additional phenotypes were observed in the development of the peripheral nervous system of neuregulin$^{-/-}$ mice. Neural crest-derived neurons were lost and the development of Schwann cells was severely impaired. Neuregulin is produced at sites of cranial ganglia formation and in neurons that project to the periphery (sensory and motoneurons), whereas the erbB3 receptor is found in neural crest cells and in Schwann cells. Targeted mutations of erbB2 or erbB3 result in identical defects in the development of the peripheral nervous system.

Comments

The phenotypes of the neuregulin and erbB2 knockout mutants are identical; this provides genetic evidence for a role for erbB2 as an essential coreceptor for the transmission of neuregulin signals.

Acknowledgements
Carmen Birchmeier
Max-Delbrück-Centrum für Molekulare Medizin, Berlin, Germany

References

[1] Meyer, D. and Birchmeier, C. (1995) Nature 378, 386–390.
[2] Meyer, D. et al. (1997) Development 124, 3575–3586.

Other names
NF-ATc, nuclear factor of activated T cells, cytoplasmic 1

Gene symbol
Nfatc1

Accession number
MGI: 102469

Area of impact
Transcription factor, cardiology

General description

NF-ATc1 belongs to the four-member NF-AT family of transcription factors, which are involved in the activation of early immune response genes such as those for the cytokines. NF-ATc1 encodes the cytoplasmic component of NF-AT. NF-ATc1 is expressed in lymphoid tissues where its expression is augmented by cellular activation. In activated cells, NF-ATc1 is dephosphorylated by calcineurin in a calcium-dependent, cyclosporin-sensitive manner. It then translocates into the nucleus, where it activates the transcription of several genes, particularly IL-2.

KO strain 1 construction[1]

The targeting vector was designed such that the *Apa*I/*Apa*I fragment containing exon 3 of the NF-ATc1 gene was replaced with pMC1*neo*polyA. The targeting vector was transfected into E14 ES cells. Heterozygous mutant clones were injected into C57BL/6J blastocysts. Chimeric mice were bred into the C57BL/6 background and heterozygous offspring were intercrossed to obtain NF-ATc1$^{-/-}$ mice. Heterozygous mutant offspring were also crossed to the CD1 outbred strain.

Heterozygous mutant ES cell lines were subsequently cultured in the presence of an increased concentration of G418 and homozygous mutant ES cell lines were selected. NF-ATc1$^{-/-}$ cells were injected into Rag-1$^{-/-}$ blastocysts (Rag-1-deficient blastocyst complementation assay) to obtain somatic chimeric mice reconstituted with NF-ATc1$^{-/-}$ mature lymphocytes. The null mutation was confirmed by Northern blot analysis.

KO strain 2 construction[2]

The targeting vector was designed such that a 2 kb fragment containing an exon encoding a portion of the Rel similarity domain was replaced with PGK1*neo*-polyA gene. An HSV-*tk* cassette was used for positive–negative selection. The targeting vector was transfected into D3 ES cell clones (129). Positive clones were injected into BALB/c blastocysts to obtain chimeras.

One heterozygous mutant cell clone was transfected with a second targeting construct similar to the first except that the neomycin-resistance gene was replaced with the hygromycin B resistance gene. Clones were selected in the presence of hygromycin B, and homozygous mutant ES cells were injected into Rag-$2^{-/-}$ blastocysts. The absence of DNA-binding activity of the mutant protein encoded by the mutated gene was confirmed by electrophoretic mobility shift assay (EMSA) assay.

Phenotype

Homozygous mutant NF-ATc$1^{-/-}$ mice failed to develop normal aortic and pulmonary cardiac valves and septa and died of circulatory failure before day 14.5 of development. NF-ATc1 mRNA is first expressed in the heart at day 7.5 of development and is restricted to the endocardium, a specialized endothelium that gives rise to the cardiac valves and septum. Activated NF-ATc1 protein is localized to the nucleus only in endocardial cells adjacent to the interface with the cardiac jelly and myocardium, which are thought to provide an inductive stimulus to the valve primordia. Treatment of wild-type embryos with FK506, a specific calcineurin inhibitor, prevents nuclear localization of NF-ATc1, indicating the requirement of the Ca^{2+}/calcineurin/NF-ATc1 signaling pathway for normal cardiac valve and septum morphogenesis.

In KO strain 1 NF-ATc$1^{-/-}$Rag-$1^{-/-}$ chimeric mice, the number of thymocytes was about 50% that of wild-type or heterozygous chimeric mice. The numbers of splenocytes and peripheral lymph node cells were normal. $CD4^-CD8^-$ double negative thymocytes showed evidence of hypoproliferation and a partial G_1 arrest of the cell cycle. In KO strain 2 NF-ATc$1^{-/-}$Rag-$2^{-/-}$ chimeras, some mice had a marked deficiency of thymocytes, peripheral lymph node cells and spleen cells. Despite the reduced numbers of thymocytes and peripheral lymphocytes, the development of thymocytes and B cells was normal.

The proliferation of T and B cells was mildly impaired. IL-2 production by primarily stimulated T cells was either not affected (strain 1) or slightly increased (strain 2). Both KO strains showed impaired production of IL-4 in an *in vitro* Th2-differentiation assay (strain 1) or in response to secondary stimulation (strain 2). Slight increases in serum IgM and IgG2a levels and dramatic decreases in IgG1 and IgE levels in the serum and after *in vitro* stimulation were observed.

Comments

NF-ATc1 is required for the optimal proliferation of peripheral lymphocytes and thymocyte precursors, but is not essential for the production of IL-2. Impaired production of IL-4 by NF-ATc$1^{-/-}$ lymphocytes indicates that NF-ATc1 plays a role in the establishment of the Th2 response in T cells.

The role of NF-ATc1 in cardiac valve and septum development in the mouse suggest that this gene and its regulatory pathways are candidates for genetic defects underlying congenital human cardiac abnormalities.

Acknowledgements
José de la Pompa, Hiroki Yoshida and Tak W. Mak
The Amgen Institute, Ontario Cancer Institute, and University of Toronto, Toronto, ON, Canada
Ann M. Ranger and Laurie H. Glimcher
Department of Cancer Biology, Harvard School of Public Health, Boston, MA, USA

References

[1] dela Pompa, J. et al. (1998) Nature 392, 182–186.
[2] Ranger, A. et al. (1998) Nature 392,186–190.
[3] Yoshida, H. et al. (1998) Immunity 8,115–124.
[4] Ranger, A. et al. (1998) Immunity 8,125–134.
[5] Northrop, J.P. et al. (1994) Nature 369, 497–502.
[6] Tinnerman, L.A. et al. (1996) Nature 383, 837–840.

NF-ATp

Other names
NFAT1, NF-ATc2, nuclear factor of activated T cells, pre-existing

Gene symbol
Nfatc2

Accession number
MGI: 102463

Area of impact
Immunity and inflammation, transcription

General description

The NF-AT (nuclear factor of activated T cells) regulatory complex plays a role in the transcription of many important cytokines and surface receptor molecules. The NF-AT proteins, which cooperate with Fos/Jun transcription factors to mediate the expression of many inducible genes in the immune system, are encoded by at least four distinct genes. NF-ATp is one of the four members which make up the cytoplasmic component of the NF-AT complex. It is expressed in immune system cells, including activated T and B cells, mast cells and NK cells, as well as endothelial cells and certain neuronal cells. Upon activation, NF-ATp is dephosphorylated and translocated to the nucleus. NF-AT proteins are regulated by the phosphatase calcineurin, target of the immunosuppressive drugs cyclosporin A and FK506.

KO strain 1 construction[1]

The targeting vector contained the neomycin-resistance gene inserted in place of an exon encoding 47 amino acids of the highly conserved NF-ATp Rel homology domain. The gene disruption construct was electroporated into D3 ES cells (strain 129), and transfected cells were injected into BALB/c blastocysts. The targeted disruption of the NF-ATp gene was confirmed by Southern blot analysis. Two polypeptides reactive with the NF-ATp antibody were detected by Western blot analysis in cytosolic extracts of the mutant mice; however, gel mobility shift assays showed that these proteins were unable to bind DNA. Western blots demonstrated normal expression of NF-ATc protein.

KO strain 2 construction[2]

A 4.5 kb *Xho*I fragment of the NF-ATp gene, containing most of an essential exon encoding amino acids 390–446 near the N-terminus of the DNA-binding domain, was replaced with a neomycin-resistance cassette. The *tk* gene was inserted at the 3′ end of the targeting construct. The targeting construct was injected into ES cells derived from 129SvJ mice. Heterozygous mice were established from several independent chimeras crossed with C57BL/6J, and bred to homozygosity.

The NF-ATp protein predicted to be encoded by the targeted allele terminated at amino acid 438 because of an altered sequence beyond amino acid 390, and a consequent out-of-frame mutation which resulted in a stop codon. However, protein immunoblot and gel mobility shift assays demonstrated that there was no expression of either the full-length or the predicted truncated protein in cells from homozygous mutant mice. T cells from NF-ATp$^{-/-}$ mice showed a substantial decrease in total NF-AT DNA-binding activity by electrophoretic mobility shift assay. There was no compensatory increase in NF-ATc (NF-AT2) expression in T cells from NF-ATp$^{-/-}$ mice.

Phenotype

NF-ATp$^{-/-}$ mice were born at the expected Mendelian frequency, were healthy and fertile, and had no gross developmental or behavioral defects. An examination of lymphoid organs indicated that the spleens of 11-week-old NF-ATp$^{-/-}$ mice were roughly 18% larger by weight than spleens of their wild-type counterparts; however, histological analysis did not indicate any change in spleen architecture or in the ratio of white to red pulp. A complete blood count and differential leukocyte analysis did not reveal any differences in blood composition between 20-week-old NF-ATp$^{-/-}$ mice and NF-ATp$^{+/+}$ mice. Flow cytometric analysis of splenocytes and thymocytes did not indicate any alteration in the numbers or ratios of T and B cells or CD4^{+} and CD8^{+} populations.

KO strain 1 NF-ATp$^{-/-}$ mice exhibited hyperproliferation of both B and T cells. They showed defects in early transcription of numerous cytokine and cell surface receptor genes, including CD40L, FasL and, particularly, IL-4. The transcription of other cytokines, including IL-13, GM-CSF and TNFα, was also affected but to a lesser degree. The transcription of IL-2 and IFNγ was minimally affected. Despite the early defect in IL-4 transcription, Th2 development as measured by IL-4 production and IgE levels *in vivo* and *in vitro* was enhanced with time under conditions that promoted Th2 differentiation.

When T cells from KO strain 2 NF-ATp$^{-/-}$ mice differentiated with anti-CD3 in the absence of added cytokines followed by restimulation with anti-CD3 *in vitro*, increased levels of mRNA encoding IL-2 and the Th2 cytokines IL-4, IL-5 and IL-13 were transcribed. This observation reflected in part the prolonged kinetics of IL-4 gene transcription in these NF-ATp$^{-/-}$ T cells relative to wild-type T cells. KO strain 2 NF-ATp$^{-/-}$ mice showed increased primary responses to *Leishmania major in vitro*, and mounted increased secondary *in vitro* responses to ovalbumin after primary stimulation *in vivo*. In an *in vivo* model of allergic airway inflammation, the accumulation of eosinophils in the pleural cavity and levels of serum IgE were increased in KO strain 2 NF-ATp$^{-/-}$ mice. Further analysis showed that eosinophil accumulation was a consequence of increased levels of IL-5 production.

Comments

NF-ATp$^{-/-}$ mice do not show an overall immunodeficient phenotype, indicating that other NF-AT proteins can substitute for NF-ATp in regulating the expression of essential immune system genes. However, the unusual tendency toward Th2 cytokine production and enhanced immune responsiveness suggest the existence of target genes uniquely or preferentially negatively regulated by NF-ATp, perhaps IL-4 in particular. This finding does not preclude the possibility that NF-ATp may also exert positive regulatory influences.

Acknowledgements

Laurie H. Glimcher
Department of Cancer Biology, Harvard School of Public Health, Boston, MA, USA
Anjana Rao
Department of Pathology, Harvard Medical School, Boston, MA, USA

References

1 Hodge, M. et al. (1996) Immunity 4, 397–405.
2 Xanthoudakis, S. et al. (1996) Science 272, 892–895.
3 Rao, A. et al. (1997) Annu. Rev. Immunol. 125, 707–747.

NF-κB p50

Other names
NFKB1

Gene symbol
Nfkb1

Accession number
MGI: 97312

Area of impact
Immunity and inflammation, signal transduction

General description

NF-κB is a member of the NF-κB/Rel family of transcription factors, all of which share the conserved N-terminal Rel homology domain. These proteins form dimers that bind to the κB DNA motifs present in many cellular target genes, thus regulating their transcription. The Rel homology domain is required for both dimerization and DNA binding. NF-κB is composed of the p50 and p65 subunits, and was originally identified as an inducible factor able to bind to the κB sequence in the intronic κ light chain enhancer in B cells. NF-κB has since been found to be present in inactive form in the cytoplasm of a wide variety of cell types, and is induced to translocate to the nucleus following stimulation by stress factors, cytokines, LPS, binding to antigen receptors and virus infection. Target genes for NF-κB include those involved in inflammation, acute phase responses, and lymphocyte activation, proliferation, and differentiation. The p50 subunit of NF-κB is generated by proteolysis of a larger precursor called p105 which contains p50 in its N-terminus, and IκBγ in its C-terminus.

KO strain construction

The targeting vector was designed to disrupt the *Nfkb1* gene encoding p105 by inserting the PGK*neo* cassette into the Rel homology domain. The *Nfkb1* gene was isolated from a 129Sv library and a *Not*I site was introduced by PCR mutagenesis into residue 160 in exon 6, which encodes amino acids 134–187 of the Rel homology domain. The targeting vector included 3.7 kb of 5′ homology and 8.0 kb of 3′ homology. The PGK-*tk* cassette was added to the end of the 3′ homology region in the opposite transcriptional orientation. The targeting vector was electroporated into E14 ES cells and positive clones were injected into C57BL/6 blastocysts. Chimeras were crossed to C57BL/6 mice to obtain heterozygotes, which were intercrossed to obtain p50$^{-/-}$ mutant mice.

The disruption of *Nfkb1* in this way was expected to result in a truncated protein unable to either bind DNA or form dimers. The synthesis of IκBγ was expected to proceed independently from an internal promoter. Western blotting of mutant spleen lysates with anti-p50 showed that neither the p105 precursor nor the p50 protein was detectable.

Phenotype

Mutant mice were born at the expected Mendelian frequency and showed no differences from wild-type mice in development, health and reproductive capacity when housed under specific-pathogen-free conditions. However, $p50^{-/-}$ mice were more prone to infection and had an increased incidence of premature death under conventional animal housing conditions. Gross anatomy, including lymphoid organs, was normal. Lymphocyte populations were normal and κ light chain usage in immunoglobulins was unaffected. Expression of MHC class I and II molecules was normal. Purified $p50^{-/-}$ B cells failed to respond to a range of LPS concentrations which induced proliferation in wild-type B cells. Basal and specific antibody production were adversely affected. Total serum Ig was reduced by 4-fold in mutant mice due to a defect in isotype switching that resulted in a significant decrease in all serum Igs except IgM. Mutant mice immunized with a T-dependent antigen failed to produce normal levels of specific antibody. $p50^{-/-}$ mice were able to eliminate extracellular but not intracellular *Listeria monocytogenes*, nor extracellular *Streptococcus pneumoniae*. $p50^{-/-}$ macrophages stimulated with LPS exhibited normal secretion of TNFα and IL-1, but a marked decrease in secretion of IL-6. However, this deficit did not affect control of *Haemophilus influenzae* or *Escherichia coli K1* in $p50^{-/-}$ mice. Surprisingly, $p50^{-/-}$ mice showed increased resistance to infection with encephalomyocarditis virus (EMC) virus. Induced levels of IFN-β transcription were increased several-fold in $p50^{-/-}$ embryonic fibroblasts.

Comments

The p50 subunit of NF-κB is not essential for the development of the immune system but has a critical regulatory role in both non-specific and specific functional immune responses to pathogens. This situation contrasts to that of RelB, which has a primary role in hematopoietic development. p50 is essential for antibody responses to specific immunogens and is required for isotype switching. As well as its positive effects, p50 may exert negative effects by downregulating the transcription of certain genes involved in anti-viral responses.

Reference

1 Sha, W.C. et al. (1995) Cell 80, 321–330.

Other names
Nuclear factor kappa B 2

Gene symbol
Nfkb2

Accession number
MGI: 1099800

General description

The *Nfkb2* gene encodes a mitogen-inducible transcription factor of the Rel/NF-κB family of proteins. This gene is expressed in the thymic medulla, marginal zone and periarterial sheath of the spleen in mice. The NF-κB2 protein is synthesized as a cytoplasmic precursor called p100. Proteolytic processing generates the DNA-binding subunit p52 that activates transcription in conjunction with RelA, RelB, c-Rel and Bcl-3. Aberrant expression of *Nfkb2* is associated with cell transformation.

KO strain construction

The pNF-κB2 targeting vector was generated by introducing a mouse PGK-promoter *neo*-resistance cassette into an *Eco*RI site in exon 4 of the mouse *Nfkb2* gene. This disruption resulted in a truncation of the protein at amino acid 19 and eliminated the NF-κB2 precursor, p100, and the DNA-binding subunit, p52. The targeting vector was introduced into CJ7 ES cells and those carrying a homologous recombination event were injected into C57BL/6 blastocysts. Blastocysts were transferred into pseudopregnant ICR females. Chimeric mice were mated with C57BL/6 mice and further interbred to obtain *Nfkb2*-null animals.

Phenotype

The NF-κB2$^{-/-}$ mice were born at the expected Mendelian ratio and exhibited no gross developmental abnormalities. However, there was a 50–80% reduction in the absolute number of B cells in the spleen and other lymphoid organs. This decrease in the B cell population was accompanied by an increase in splenic T cells. Moreover, NF-κB2$^{-/-}$ mice had an abnormal splenic microarchitecture, characterized by diffuse B cell areas and the absence of a discrete marginal zone. *In vitro*, mutant splenic B cells had only moderate proliferative defects and underwent normal cell maturation to Ig secretion and class switching. B cell precursors in the bone marrow are reduced by 50% between 8 and 12 weeks of age.

Nevertheless, when the NF-κB2$^{-/-}$ mice were challenged with a T cell-dependent antigen (NP-KLH) and -independent antigen (NP-LPS), they displayed impaired immune responses. The deficient T cell-dependent responses in NF-κB2$^{-/-}$ mice correlated with an impaired formation of germinal centers and a disrupted splenic architecture and perifollicular marginal zone. Some of the characteristics displayed by these animals included a strong reduction of marginal metallophilic macrophages (MOMA 1^+) and the lack of follicular dendritic cells (FDC-$M1^+$). Upon challenge with T cell-dependent antigens the spleens of wild-type mice had numerous germinal centers characterized by B cell areas binding peanut agglutinin (PNA) surrounded by IgD^+ cells. In contrast, the spleens of mutant mice had very few cells stained by PNA and a diffuse perifollicular staining of cells expressing surface IgD^+.

NF-κ$B2^{-/-}$ T cells showed normal proliferative responses and a 2–3-fold increase in the production of IL-2 and GM-CSF upon stimulation. However, NF-κ$B2^{-/-}$ mice were defective in clearing the intracellular bacterium *Listeria monocytogenes*, exhibiting a 7-fold higher number of CFU in the spleen compared to wild-type littermates. CTL responses against lymphocytic choriomeningitis virus (LCMV) appeared to be normal.

Comments

NF-κB2 is required for normal splenic architecture, maintenance of the peripheral B cell population, formation of germinal centers and production of normal levels of antigen-specific antibodies in response to T cell-independent and -dependent antigens. The absence of NF-κB2 also results in defective clearance of *L. monocytogenes*.

Several phenotypic defects are common to Bcl-$3^{-/-}$ and NF-κ$B2^{-/-}$ mice, arguing for a synergistic effect of these two proteins in transcription[4,5]. Mice deficient in CD40, CD40L, TNFα, TNFRI, CD19, Lyn, Btk (Xid), c-fos and Oct-2 also share some phenotypic characteristics with NF-κB2-null mice; the implications of these commonalities remain to be elucidated.

Acknowledgements
Jorge H. Caamaño
Department of Pathobiology, School of Veterinary Medicine, University of Pennsylvania, Philadelphia, PA, USA
Rodrigo Bravo
Department of Oncology, Bristol-Myers Squibb Pharmaceutical Research Institute, Princeton, NJ, USA

References

1 Caamaño, J. et al. (1998) J. Exp. Med. 187, 185–196.
2 Weih, F. et al. (1994) Oncogene 9, 3289–3297.
3 Gilmore, T. et al. (1996) Oncogene 13, 1367–1378.
4 Franzoso, G. et al. (1997) Immunity 6, 479–490.
5 Schwarz, E. et al. (1997) Genes Dev. 11, 187–197.

Other names
Nerve growth factor alpha, mGK-4

Gene symbol
Ngfa

Accession number
MGI: 97320

Area of impact
Neurology

General description

NGF was first described as a factor required for both the survival and sprouting of sympathetic and sensory neurons. It was the first member of a growing family of neurotropins. Much of the evidence for the *in vivo* role of NGF comes from antibody-blocking experiments. In culture, NGF also promotes the survival of cholinergic forebrain neurons.

KO strain construction

neo^r was inserted into exon 4 and *tk* was spliced to the 3′ end of the *Ngfa* gene. This was expected to delete the start codon, signal sequence and part of the NGF precursor protein. The vector was electroporated into AB1 ES cells (129). Clones showing homologous recombination were injected into C57BL/6 blastocysts. Chimeric offspring were mated with C57BL/6. The progeny were crossed to generate (129 × C57BL/6) F2 offspring for analysis.

Phenotype

NGF$^{-/-}$ mice showed reduced growth and survival rates (maximum lifespan 4 weeks). This phenotype appeared to be more severe than that observed in TrkA knockouts. NGF$^{-/-}$ animals showed decreased responsiveness to pain resulting from a tail pinch. There was a dramatic loss of sensory and sympathetic neurons. Some animals displayed mild gait abnormalities and developed a tremor, which was most evident during locomotion. Superior cervical ganglia (SCG) dissected from early post-natal mutant animals were markedly smaller than those removed from wild-type littermates and displayed numerous pyknotic nuclei. By post-natal day 14, the SCGs from mutant animals were not visible upon dissection. In contrast, inspection of the nodose ganglia revealed that the size of these ganglia did not differ significantly from those of their wild-type littermates. Dorsal root ganglias of NGF$^{-/-}$ mice exhibited a general reduction in neuronal cell number within cervical and lumbar dorsal root ganglia (particularly in small-diameter CGRP$^+$ neurons). Consistent with this, neuronal (CGRP$^+$) innervation of laminae I and II of the dorsal horn of the spinal cord was substantially reduced. The loss of these cells suggests that cross-reactivity of NT-3 and NT-4 cannot counter the developmental effects of NGF loss.

In the CNS there did not appear to be a significant loss of basal forebrain cholinergic neurons. Cholinesterase projections to the hippocampus and cerebral cortex were not apparently altered.

Comments

The development of sensory and sympathetic neurons, but not forebrain cholinergic neurons, depends on NGF. These neurons respond to mechanical pain.

References

[1] Crowley, C. et al. (1994) Cell 76, 1001–1011.
[2] Klein, R. (1994) FASEB J. July, 738–744.

Other names
Csx

Gene symbol
Nkx2-5

Accession number
MGI: 97350

Area of impact
Development

General description

Nkx2-5, a homeobox gene, is one of the earliest markers of the cardiogenic lineage in mice and its expression is maintained in myocardial cells during embryonic and adult life. Expression also occurs in pharyngeal endoderm, tongue, spleen, thyroid and stomach. It is a homolog of *Drosophila tinman*, essential for formation of cardiac and visceral muscle progenitors in flies.

KO strain construction

The gene was interrupted within helix 3 of the homeodomain (coding exon 2) with a PGK promoter-neomycin resistance cassette (lacking the polyadenylation signal) oriented in the same direction as the gene. The mutation was analyzed whilst outcrossing onto C57BL/6 × C57BL/10 F1 mice. No change in phenotype has been observed over several generations of outcrossing.

Phenotype

Heterozygous mice are normal and viable. Homozygotes are lethal with 100% penetrance. Lethality, beginning around E9, is probably the result of cardiac insufficiency. Homozygous mutants show normal heart development up until the linear heart tube stage, when there is a block to cardiac looping morphogenesis. Mutant hearts remain in a linear conformation and downstream morphogenesis of chambers, valves and trabeculae is highly abnormal, perhaps a secondary consequence of cardiac insufficiency. Linear heart tubes beat with normal pacing, although blood flow appears compromised. Myogenesis, as judged by expression of myofilament genes, is largely normal and robust, although expression of the myosin light chain 2V (*MLC2V*) gene is abolished, and that of *CARP*, encoding a cardiac-restricted nuclear ankyrin repeat protein implicated in regulation of *MLC2V*, is severely downregulated.

The reason for the selective loss of certain myofilament pathways is unknown. Morphogenetically, evidence suggests that mutant hearts retain right and left ventricular chambers and a primitive atrial chamber, although these cannot coordinate the looping process. One possible reason for the block to looping is the lack of expression of the bHLH gene *eHand*. Within the ventricular region of normal hearts, *eHand*'s expression is restricted to the left side of the left ventricle, a pattern that develops dynamically during the early stages of cardiac looping through an inductive interaction with the embryonic left/right patterning system. Left-sided expression of *eHand* may control one aspect of looping morphogenesis that cannot be achieved in Nkx2-5 mutants.

Comments

Ectopic expression of *eHand* has been detected in the caudal pharyngeal endoderm, suggesting that *Nkx2-5* regulates *eHand* both positively and negatively, depending on cellular content. *Nkx2-5* appears also to control pharyngeal patterning.

Acknowledgements
Richard Harvey
The Walter and Eliza Hall Institute of Medical Research, University of Melbourne and Royal Melbourne Hospital, Victoria, Australia

References

[1] Lyons, I. et al. (1995) Genes Dev. 9, 1654–1666.
[2] Zou, Y. et al. (1997) Development 124, 793–804.
[3] Biben, C. and Harvey, R.P. (1997) Genes Dev. 11, 1357–1369.

NMDAR-1

Other names
NMDARδ1, R1, NR1

Gene symbol
Grin1

Accession number
MGI: 95819

Area of impact
Neurology

General description

One type of receptor for glutamate also binds *N*-methyl-D-aspartate (NMDA). The NMDA receptors in the brain are distributed on post-synaptic densities of glutaminergic pathways. The NMDAR is comprised of heteromeric subunits consisting of NMDAR-1, 2A, 2B, 2C and 2D. The NR1 subunit is required for the electrophysiological function of the NMDAR channel. This NMDAR can act as a coincidence detector because it is gated by both voltage and the ligand. Different inputs into a neuron can depolarize it, thus removing the Mg^{2+} block from the pore of the NMDAR channel, allowing Ca^{2+} to flow in and initiate synaptic strengthening in the form of long-term potentiation (LTP).

KO strain 1 construction[1]

The knockout was constructed by the elimination of exons 12–20, a region which encodes 338 amino acids and includes four transmembrane domains. The vector was electroporated into D3 ES cells (129) and selected in G418. Clones showing homologous recombination were injected into blastocysts. Chimeras were mated with C57BL/6 mice and heterozygous offspring were intercrossed to generate F2 hybrids (+/+, +/−, −/−) for analysis.

KO strain 2 construction[2]

The *neo*^{r} gene was inserted to delete the ATG and leader sequence so that no truncated gene product was made. W9 ES cells (129SvJ) were targeted and mice were bred to C57BL/6.

KO strain 3 construction[3]

The targeting vector consisted of a 9 kb genomic sequence in which the *neo*^{r} gene was inserted to delete the ATG and leader sequence so that no truncated gene product was made. A HSV-*tk* gene fragment was attached to the 5′ end for negative selection. CCE ES cells (129SvJ) were targeted and mice were bred to BDF1.

Phenotype

All three strains of NR1$^{-/-}$ mice exhibited lethality at post-natal day 1 (8–15 hours post-birth), possibly due to respiratory failure. Recently, it has been shown that the lifespan of these animals can be artificially prolonged by stimulating breathing through the use of CO_2 and injecting pregnant mothers with the β-adrenergic agonist terbutaline every 4–6 hours at E18.5.

Histological analysis (Nissl, H&E, cytochrome oxidase) of the brains of NR1$^{-/-}$ mice revealed no overt differences compared to heterozygous and wild-type littermates. Cerebellar granule cell layers were present, as were all spinal cord layers which exhibited normal somatotopic organization of the motor columns. In addition, axon pathfinding, initial targeting and crude topographic projections to the brainstem were also observed to be normal in these mice. Using cytochrome oxidase histochemistry on KO strain 2 mice whose lifespan had been prolonged (through the use of terbutalin and CO_2), it was demonstrated that the formation of whisker-related patterns in the brainstem trigeminal nuclei were abnormal[2].

KO strain 1 NR1$^{-/-}$ mice were also impaired in long-term potentiation (LTP) in the hippocampus and learning and memory (LM)[1]. Similar results for LTP and LM were obtained even when this deletion of NR1 was targeted only in the CA1 neurons using a cross to a Cre recombinase line later in development (3 weeks)[4]. The fact that these NR1-null mutants also had aberrant development of whisker barrel fields confirms that NMDA receptors are essential for stabilizing synapses in development. Cortical neuronal cells from KO strain 3 NR1-deficient mice showed no neuronal degeneration to either glutamate or NMDA treatment[3].

Comments

NR1 is essential for activity-dependent neuronal development and rapidly triggered neurotoxicity.

Acknowledgements
Shigetada Nakanishi
Institute for Immunology, Koyota University Faculty of Medicine, Japan

References

[1] Li, Y. et al. (1994) Cell 76, 427–437.
[2] Forrest, D. et al. (1994) Neuron, 13, 325–338.
[3] Tokita, Y. et al. (1996) Eur. J. Neurosci. 8, 69–78.
[4] Tsien, J.Z. et al. (1996) Cell 87, 1327–1338.
[5] Moriyoshi, K. et al. (1991) Nature 354, 31–37.

Other names
Glutamate receptor, ionotrophic, NMDARε1, NR2A

Gene symbol
Grin2a

Accession number
MGI: 95820

Area of impact
Neurology

General description

The NMDAR is one class of receptors that bind glutamate in the CNS. A functional NMDAR channel is heteromeric and is comprised of two subunits, NR1 and NR2. Among the NR2 subunits, 2A is thought to confer Mg^{2+} sensitivity to the NMDAR channel. The role of each subunit in long-term potentiation (LTP) is unknown.

KO strain 1 construction[1]

neo^r was inserted into the exon encoding the second and third transmembrane domains. The vector was flanked by diphtheria toxin (DT) and electroporated into TT2 ES cells (CBA × C57BL/6 Fc). Clones showing homologous recombination were injected into eight-cell ICR embryos. Chimeras were mated with C57BL/6 mice. Heterozygous offspring were intercrossed.

KO strain 2 construction[2]

The targeting vector consisted of a 13.9 kb genomic sequence of the *Grin2a* gene in which a part of transmembrane region was replaced with the neomycin resistance gene. A HSV-*tk* gene fragment was attached to the 3′ end for negative selection. Chimeric mice generated from NR2A-disrupted CCE ES cells (from an inbred mouse line 129SvJ) were mated with BDF1 ((C57BL/6 X DBA/2) F1).

Phenotype

KO strain 1 NR2A$^{-/-}$ mice were born normally, reached adulthood and bred. The neuroanatomy of the brain was normal. Excitatory post-synaptic currents (EPSC) and LTP were reduced. Pre-synaptic function was normal, corresponding with a decrease in the latency to find the hidden platform in the Morris water maze task. Performance on control tasks was normal.

KO strain 2 NR2A$^{-/-}$ mice also developed and mated normally. They showed no histological changes in the cerebellum. They showed normal movements in the motor coordination tasks tested. The NMDA receptor-mediated components of EPSCs in granule cells were reduced.

Comments

The consequences of losing NMDAR-2A are much less severe than those of losing NMDAR-1 or 2B, which are both perinatal lethal.

Acknowledgements
John Roder
Samuel Lunenfeld Research Institute, Mount Sinai Hospital, Toronto, ON, Canada
Shigetada Nakanishi
Kyoto University, Kyoto, Japan

References

[1] Sakimura, K. et al. (1995) Nature 373, 151–155.
[2] Kadotani, H. et al. (1996) J. Neurosci. 16, 7859–7867.
[3] Ishii, T. et al. (1993) J. Biol. Chem. 268, 2836–2843.

Other names
NMDARε2, NR2B

Gene symbol
Grin2b

Accession number
MGI: 95821

Area of impact
Neurology

General description

One type of receptor for glutamate also binds *N*-methyl-D-aspartate (NMDA). This NMDAR can act as a coincidence detector because it is gated by both voltage and the ligand. Different inputs into a neuron can depolarize it, thus removing the Mg^{2+} block from the pore of the NMDAR channel, allowing Ca^{2+} to flow in and initiate synaptic strengthening in the form of long-term potentiation (LTP). The NMDAR is comprised of heteromeric subunits consisting of NMDAR-1, NR2A, 2B, 2C and 2D. The NR2B subunit is widely expressed in the embryo nervous system, but is restricted to the forebrain in adults.

KO strain construction

neo^r was placed in the exon containing the translation initiation site. Diphtheria toxin was added to the 3′ end. The vector was electroporated into TT2 ES cells (B6 × CBA). Selected ES clones were injected into eight-cell embryos of ICR mice. Chimeras were mated with C57BL/6. Heterozygotes were intercrossed to generate mice for analysis.

Phenotype

Homozygous null NR2B$^{-/-}$ mice developed normally to term, but died in the first post-natal day due to an abnormal suckling response. Brain sections showed impaired formation in the whisker-barrel fields in the trigeminal nucleus. In the hippocampus of mice hand-reared for 2–3 days, both long-term depression (LTD) and LTP were missing. Non-excitatory NMDAR post-synaptic potentials (EPSP) were normal in NR2B$^{-/-}$ mice, but NMDAR-mediated EPSPs were absent. These results show that the NR2B subunit is essential for NMDAR channel function. The phenotype was similar to that of NMDAR-1 knockout mice.

Comments

The 2B subunit of the NMDAR is not important for development but is essential for pattern formation and synaptic plasticity.

Acknowledgements
John Roder
Samuel Lunenfeld Research Institute, Mount Sinai Hospital, Toronto, ON, Canada

Reference

1 Kutsuwada, T. et al. (1996) Neuron 16, 333–344.

Other names
Ionotropic glutamate receptor, NMDARε3, NR2C

Gene symbol
Grin2c

Accession number
MGI: 95822

Area of impact
Neurology

General description

The NR2 subunits serve as modulatory subunits that potentiate NMDAR1 receptor activity in a heteromeric formation and confer a functional variability, depending on NR2 subunit composition. NR2A is expressed in many brain regions including the cerebellum, while NR2C is highly expressed in the cerebellar granule cells.

KO strain construction

The targeting vector consisted of a 13.1 kb genomic sequence of the *Grin2c* gene in which a part of transmembrane region was replaced with the neomycin-resistance gene. A HSV-*tk* gene fragment was attached to the 3′ end for negative selection. Chimeric mice generated from NR2C-disrupted CCE ES cells (from an inbred mouse line 129SvJ) were mated with BDF1 ((C57BL/6 × DBA/2) F1). Homozygous NR2A[1] and NR2C mutant mice were mated to generate double mutant mice lacking both NR2A and NR2C.

Phenotype

The $NR2C^{-/-}$ single mutant and $NR2A^{-/-}NR2C^{-/-}$ double mutant mice both developed and mated normally. They showed no histological changes in the cerebellum. The NMDA receptor-mediated components of EPSCs in granule cells were reduced in the NR2C-deficient mice and abolished in mice lacking both NR2A and NR2C. The NR2A- and NR2C-deficient cerebellar granule cells were different in the current–voltage relationship and time course of NMDA receptor responses. The $NR2C^{-/-}$ mice showed normal movements in the motor coordination tasks tested. The $NR2A^{-/-}NR2C^{-/-}$ mice could also manage simple coordinated tasks, but failed more challenging tasks such as staying on a rapidly rotating rod.

Comments

The NR2C receptor plays a role in synaptic transmission in granule cells in the cerebellum. The NMDA receptors 2A and 2C together play an active role in motor coordination.

Acknowledgements
John Roder
Samuel Lunenfeld Research Institute, Mount Sinai Hospital, Toronto, ON, Canada
Shigetada Nakanishi
Kyoto University, Sakyo, Japan

References

1 Kadotani, H. et al. (1996) J. Neurosci. 16, 7859–7867.
2 Ishii, T. et al. (1993) J. Biol. Chem. 268, 2836–2843.

Other names
Myc-oncogene

Gene symbol
Myc

Accession number
MGI: 97250

Area of impact
Neurology, cardiology

General description

N-myc is a proto-oncogene that belongs to the basic helix-loop-helix family of transcription factors. N-myc must bind to max to gain DNA-binding function. N-myc is widely expressed in embryonic brain, lung, heart and other organs.

KO strain 1 construction[1]

neo^r was inserted at the translation initiation site in exon 2 to generate a null mutation. In another construct, insertion of *neo*^{r} into the first intron resulted in a leaky mutation.

KO strain 2 construction[2]

Part of the coding sequence in exon2 was replaced by a *neo* sequence. Translation of the N-myc gene downstream of the disruption was inhibited. Mice were bred to BALB/c.

KO strain 3 construction[3,4]

The first 19 N-myc codons were fused to *neo* resistance coding sequences such that exon 2 was disrupted. The construct was electroporated into D3 ES cells (129 strain) and positive clones were injected into C57BL/6 blastocysts.

Phenotype

Mice which possessed a deletion of the N-myc gene exhibited embryonic lethality, exhibiting defects in the branching morphogenesis of the lung, the neuroepithelium, the sensory ganglia, the gut and heart. These were evident by midgestation (development appeared normal until E10.5). Lethality was observed at E11.5. Death appeared to result from the cardiac failure stemming from hypoplasia of the compact subepicardial layer of the myocardium. Other abnormalities included defects in the limb bud, visceral organs (lung, stomach, liver, heart) and central/peripheral nervous system defects. While fewer neurons were observed in these animals, the basic cytoarchitecture of the nervous system was intact and neuroepithelial stem cells were still present.

A compound heterozygote (leaky N-myc × N-myc null) which had 15% of the normal amount of N-myc protein exhibited midgestational lethality[1,2].

Comments

N-myc is important for the early development of epithelia and organogenesis of the lung and heart.

References

1. Moens, C.B. et al. (1993) Development 119, 485–499.
2. Sawai, S. et al. (1993) Development 117, 1445–1455.
3. Stanton, B.R. et al. (1990) Mol. Cell Biol. 10, 6755–6758.
4. Stanton, B.R. et al. (1992) Genes Dev. 6, 2235–2247.

Other names
Neuronal nitric oxide synthase, type I NOS, ncNOS (neuronal constitutive NOS), bNOS (brain NOS)

Gene symbol
Nos1

Accession number
MGI: 97360

Area of impact
Neurology

General description

Neuronal and endothelial NOS are two of three isoforms of NOS that generate nitric oxide (NO) from L-arginine. nNOS is principally expressed in neurons while eNOS is expressed in vascular endothelial cells, smooth muscle cells and certain neurons. Neuronal NO serves as an intracellular messenger molecule and a retrograde messenger in long-term potentiation (LTP) and depression (LTD), models of learning and memory. Neuronal NO also mediates excitotoxicity and regulates cerebral blood flow and the response to cerebral ischemia. Endothelial NO is responsible for EDRF (endothelium-derived relaxing factor) activity, and induces vasodilation in response to chemical and mechanical stimuli. Endothelial NO production is decreased in atherosclerosis, hypertension, diabetes and aging.

KO strain construction

Genomic regions flanking the first translated exon (second transcribed exon), including the initiation codon ATG, were cloned into the vector pPNT. J1 ES cells of 129SV origin were used. Chimeric mice were mated to C57BL/6 mice.

Phenotype

nNOS knockout mice were viable and fertile. They did not express nNOS as detected by Western blot, NADPH diaphorase staining, immunohistochemical staining, or electron spin resonance techniques. Low levels of NOS catalytic activity (less than 5%) were present in the brain of nNOS knockout mice, due to other isoforms in the brain like endothelial NOS, and to low-level natural splicing variants. These splice variants were present in wild-type mice as well and did not account for the phenotype of the mutant mice.

The nNOS knockout mice are a model of the human disorder pyloric stenosis in which the stomach is enlarged, often to several times the normal size. Electric field stimulation of muscle strips showed abnormal inhibitory junction potentials (IJP). The nNOS knockout mice were resistant to focal and global cerebral ischemia, indicating that the nNOS isoform contributes to neurotoxicity following ischemia[2]. nNOS knockout mice were more aggressive than wild-type littermates, showed decreased latency to flight in an intruder paradigm, and displayed inappropriate sexual behavior. They also showed evidence for non-NO mediated compensation in the cerebrovascular response to hypercarbia and whisker stimulation, nociception, and minimal alveolar concentration for isoflurane anasthesia.

Mice that were doubly mutant for eNOS and nNOS[3] showed diminished LTP, whereas nNOS$^{-/-}$ single mutant and eNOS$^{-/-}$ single mutant mice each showed normal LTP.

Comments

These results provide the first genetic evidence that NOS is involved in LTP in the stratum radiatum, and suggest that the neuronal and endothelial isoforms of NOS can compensate for each other in mice with a single mutation. They further suggest that there is an NOS-independent component of LTP in striatum radiatum and that LTP in stratum oriens is largely NOS-independent.

Acknowledgements
Paul Huang
Cardiovascular Research Center, Massachusetts General Hospital, Charlestown, MA, USA

References

1 Huang, P.L. et al. (1993) Cell 75, 1273–1286
2 Huang, Z. et al. (1994) Science 265, 1883–1885.
3 Son, H. et al. (1996) Cell 87, 1015–1023.
4 O'Dell, T.J. et al. (1994) Science 265, 542–546.
5 Nelson, R.J. et al. (1995) Nature 378, 383–386.
6 Huang, P.L. and Fishman, M.C. (1996) J. Mol. Med. 74, 415–421.

Gene symbol
Notch1

Accession number
MGI: 97363

Area of impact
Development

General description

Notch1 is one of four mammalian homologs of the *Drosophila Notch* gene which is involved in a number of important cell decision events in development. These molecules are transmembrane receptors consisting of tandem EGF-like repeats, NLR (Notch-lin12-related) repeats and six tandem ankyrin repeats in the intracellular domain.

KO strain 1 construction[1]

A PGK*neo* cassette was cloned into a unique restriction site in an exon encoding EGF repeat 32. This disrupts the reading frame of the *Notch1* gene in the extracellular domain. The vector was introduced into CJ7 and CJ9 129 ES cells, which were injected into C57BL/6 blastocysts. The phenotype was studied in a mixed B6 × 129 background.

KO strain 2 construction[2]

A 10.4 kb fragment of the *Notch1* gene, encompassing part of the EGF repeat, the NLR, transmembrane domain and most of the ankyrin repeats, was deleted and replaced with a PGK*neo* cassette. The vector was introduced into R1-S3 129 ES cells which were injected into C57BL/6 blastocysts. The phenotype was examined on a 129 inbred or 129 × B6 or 129 × CD1 mixed background.

Phenotype

Homozygous mutant embryos were resorbed by E11.5. By E9.5, homozygotes were readily distinguishable from their littermates by reduced size, distended pericardia and reduced axis extension. The exact cause of death is not known, but may be associated with circulatory problems. Somites form in the mutants but there is a delay and lack of coordination in segmentation, leading to variations in somite size and apparent fusion of somites. Embryos die too early to fully examine the possible role of Notch1 in the nervous system, but examination of expression of Notch signaling pathway homologs, such as *Hes5, Dll1* and *Mash1*, has revealed altered expression consistent with enhanced neurogenesis. This suggests that Notch1 signaling plays a novel role in somitogenesis but that its role in neurogenesis may be conserved across evolution.

References

[1] Swiatek, P.J. et al. (1994) Genes Dev. 8, 707–719.
[2] Conlon, R.A. et al. (1995) Development 121, 1533–1545.
[3] de la Pompa, J.-L. et al. (1997) Development 124, 1139–1148.

Other names
NFE2-related factor 2, nuclear factor erythroid 2, p45 NRF2

Gene symbol
Nfe212

Accession number
MGI: 108420

Area of impact
Transcription factors

General description

NRF2 is a member of the mammalian CNC ("cap 'n' collar") subfamily of basic-leucine zipper (bZIP) transcription factors. CNC is a homeotic gene involved in the development of the head and neck structure in *Drosophila*. NRF2 shares homology with p45-NFE2, a related transcription factor which is erythroid-specific and implicated in globin gene regulation. NRF2 (p45) forms a heterodimer with a p18 MAF subunit. Unlike NFE2, NRF2 is expressed at varying levels in a wide range of tissues, but most abundantly in the luminal epithelia of the digestive and respiratory organs. Although NFE2 has been implicated in globin gene regulation, and NRF2 has been shown to transactivate reporter genes in cell culture, the function of NRF2 *in vivo* is unknown.

KO strain construction

The targeting vector was based on the pPNT backbone. A 4.2 kb segment of DNA from the 129SVJ genomic library (Stratagene) containing part of exon 4 and all of exon 5 of *Nfe212* was replaced by a 5.5 kb fragment of DNA coding for the bacterial gene *lacZ* followed by the neomycin-resistance cassette. This cassette contained a polyA signal, the *neo* resistance gene for positive selection, and the HSV-*tk* gene for negative selection. Thus, the targeting vector nullified NRF2 function because the CNC bZIP regions were deleted, but was expected to permit monitoring of Nrf2 promoter activity via the activity of the *lacZ* reporter gene.

The targeting construct was introduced into JM1 ES cells derived from 129SVJ mice by electroporation. Positive ES clones were injected into blastocysts and chimeras were bred to produce F1 progeny. F1 heterozygotes were crossed to produce homozygous mutants. The null mutation was confirmed by RNA analysis of sections of adult stomach and intestine.

Phenotype

No phenotype has been detected as yet in NRF2-null mice of up to 15 months of age. Mutant mice were born in the expected Mendelian ratio and were normal in appearance, development, fertility and behavior. No histological, hematological or anatomical anomalies were observed. No β-galactosidase activity was detected, so that the expression of NRF2 through development could not be monitored.

Comments

Unlike p45-NFE2$^{-/-}$ mice (which die due to a lack of platelets), NRF2$^{-/-}$ mice are apparently normal, indicating that NRF2 is dispensable for mouse development, growth and fertility. It cannot be ruled out that the lack of NRF2 results in subtle changes that do not produce a recognizable phenotype under laboratory conditions, but which may be manifested as the mice age or are subjected to specific challenges.

Acknowledgements
Kaimin Chan
Department of Laboratory Medicine, University of California, San Francisco, CA, USA
Yuet Wai Kan
Department of Laboratory Medicine, and Howard Hughes Medical Institute and Cardiovascular Research Institute, University of California, San Francisco, CA, USA

References

1 Moi, P. et al. (1994) Proc. Natl Acad. Sci. USA 91, 9926–9930.
2 Chan, J.Y. et al. (1995) Hum. Genet. 95, 265–269.
3 Chan, K. et al. (1996) Proc. Natl Acad. Sci. USA 93, 13943–13948.
4 Tuan, D. et al. (1985) Proc. Natl Acad. Sci. USA 82, 6384–6389.
5 Moi, P. and Kan, Y.W. (1990) Proc. Natl Acad. Sci. USA 87, 9000–9005.

NT-3

Other names
Neurotropin-3, neuronal survival factor

Gene symbol
Ntf3

Accession number
MGI: 97380

Area of impact
Neurology

General description

Neurotropins play an important role in the survival and differentiation of neurons and the growth of axonal and dendritic projections. NT-3 binds to the Trkc receptor to activate its kinase domain.

KO strain construction

A replacement vector was designed to delete part of the coding region of NT-3 in a 129 strain ES cell line. Chimeric mice were bred onto 129.

Phenotype

NT-3$^{-/-}$ mice displayed severe movement defects of the limbs, and most died shortly after birth. Substantial portions of peripheral sensory and sympathetic neurons were lost, while motor neurons were not affected. Significantly, spinal proprioceptive afferents and their peripheral sense organs (muscle spindles and Golgi tendon organs) were completely absent in homozygous mutant mice. This correlated with a loss of parvalbumin and carbonic anhydrase-positive neurons in the dorsal root ganglion. No gross abnormalities were seen in Pacinian corpuscles, cutaneous afferents containing substance P and calcitonin gene-related peptide, and deep nerve fibers in the joint capsule and tendon. Importantly, the number of muscle spindles in heterozygous mutant mice was half of that in control mice, indicating that NT-3 is present at limiting concentrations in the embryo.

Acknowledgements
Patrik Ernfors
Karolinska Institute, Stockholm, Sweden

References
[1] Ernfors, P. et al. (1994) Cell 77, 503–512.
[2] Ernfors, P. et al. (1995) Neuron 14, 1153–1164.
[3] Kucera, J. et al. (1995) J. Comp. Neurol. 363, 307–320.
[4] el-Shamy, W.M. et al. (1996) Development 122, 491–500.

Other names
Neurotropin 4, neurotropin 5

Gene symbol
Ntf5

Accession number
MGI: 97381

Area of impact
Neurology

General description

Neurotropins play important roles in neuronal survival during vertebrate development. NT-4 alone or in combination with brain-derived neurotrophic factor (BDNF) has been suggested to be crucial for the survival of peripheral sensory and CNS neurons, including motor neurons, in embryogenesis.

KO strain construction

A targeting vector was designed to delete part of the coding region of NT4 in ES cells derived from 129 mice. Chimeras were mated to 129.

Phenotype

$NT4^{-/-}$ mice were viable but exhibited excessive peripheral sensory neuron loss. In contrast, no loss of motor neurons in the facial nucleus, of sympathetic neurons in the superior cervical ganglion, or dopaminergic neurons in the substantia nigra were seen. Furthermore, no facial motor neuron loss was detected in $BDNF^{-/-}NT4^{-/-}$ double knockout mice, whereas the sensory neuron loss was more severe, and was additive compared to that in the single mutants. Our results suggest that NT-4 is required for the survival of a subset of peripheral sensory neurons but not for sympathetic and motor neurons during embryonic development.

Comments

NT-4 is important for the development of sensory neurons.

Acknowledgements
Xin Liu
University of California Los Angeles, Los Angeles, CA, USA

Reference

1 Liu, X. et al. (1995) Nature 375, 238–241.

Nuk

Other names
Neural kinase, Eph receptor B2 (EphB2), Cek5 (chicken)

Gene symbol
Ephb2

Accession number
MGI: 99611

Area of impact
Neurology

General description

Nuk is a member of the Eph family of transmembrane receptor tyrosine kinases. The three ligands that stimulate catalytic activity of Nuk also contain a transmembrane domain and are anchored to the cell membrane. Nuk expression is mainly confined to the developing and adult nervous system. In the embryo, Nuk is most concentrated in the ventral neural tube and within the axons of the PNS.

KO strain construction

Two mutations were generated in the Nuk locus: (a) a protein-null mutation (Nuk-1) which deleted a 5′ coding exon, and (b) a second mutation (Nuk-*lac*Z) which deleted the exons coding for the tyrosine kinase domain and replaced them with an "in-frame" fusion with *lacZ*. This mutation led to the synthesis of a Nuk–β-gal fusion receptor containing the entire extracellular, transmembrane and juxtamembrane domains of Nuk fused to β-gal.

The R1 ES cell line (129) was used to generate chimeras which were inbred to 129. Two hybrid backgrounds, 129 × CD1 and 129 × C57BL6, were also used.

Phenotype

Animals homozygous for either the Nuk-1 or Nuk-*lacZ* mutation were viable, long-lived and fertile. The forebrain of Nuk-$1^{-/-}$ mice exhibited a specific axon pathfinding defect in the anterior commissure, which is a major communication linkage between neurons in the left and right lobes of the temporal cortex. The Nuk receptor is thus required for the guidance of these commissural axons in the embryo. The ability of Nuk to control axon guidance in the embryonic brain appears to involve a non-cell autonomous function as the ventral hypothalamus expresses Nuk protein and the cortical anterior commissural axons, which are migrating over the ventral hypothalamus, express Nuk ligands, which are themselves anchored to the surface of the axon by a hydrophobic transmembrane domain. Furthermore, expression of the Nuk-*lacZ*-encoded Nuk-β-gal fusion receptor lacking the tyrosine kinase domain was able to support the pathfinding of these axons, indicating that a reverse signal, emitted by the Nuk extracellular domain is transmitted into the Nuk ligands on the migrating anterior commissure axons[1]. Biochemical data support this notion and indicate that the reverse signal involves tyrosine phosphorylation of the intracellular domain of the transmembrane ligands and subsequent activation of a novel signal transduction cascade[3]. This reverse signal transduced into the acP (pars posterior) axons by the transmembrane ligands does not preclude a requirement for Nuk tyrosine kinase activity in the ventral hypothalamus or elsewhere. Thus, we have hypothesized that the interaction of Nuk with its transmembrane ligands results in bidirectional tyrosine kinase-mediated signal transduction into both the receptor- and ligand-expressing cells.

Nuk is one of four Eph family receptors (the Elk/EphB subclass) whose tyrosine kinase domains are activated following exposure to any of the three known transmembrane ligands[2]. As three of the Elk/EphB subclass receptors (Elk/EphB1, Nuk/EphB2 and Sek4/EphB3) are highly expressed in the nervous system, it is possible that there may be some level of redundancy between these molecules. Consistent with this idea, Nuk and Sek4 are co-expressed in specific regions of the developing nervous system, including hindbrain rhombomeres r3 and r5 and in defined areas of the forebrain and midbrain. Like Nuk mutants, Sek4 homozygotes were viable and long-lived. Analysis of the Sek4 mutant brains indicated a fairly normal morphology in the major axon tracts. $Nuk^{-/-}Sek4^{-/-}$ double homozygote mice exhibited much more exacerbated neural phenotypes affecting additional axon pathways in the brain, including two other commissural tracts, the corpus callosum and the habenular commissure[4]. In addition, a number of the double mutants died shortly after birth, and this was associated with the presence of a cleft palate.

Comments

Nuk is an axon guidance molecule and interacts genetically with Sek4.

Acknowledgements
Mark Henkemeyer
University of Texas Southwestern Medical Center, Dallas, TX, USA

References

1 Henkemeyer, M. et al. (1996) Cell 86, 35–46.
2 Gale, N.W. et al. (1996) Neuron 17, 9–19.
3 Holland, S. et al. (1996) Nature 383, 722–725.
4 Orioli, D. et al. (1996) EMBO J. 15, 6035–6049.

Nurr1

Other names
RNR-1, NOT

Gene symbol
Nurr1

Accession number
MGI: 108416

Area of impact
Development, neurology

General description

Nurr1 is an orphan nuclear receptor belonging to the large family of proteins which includes the receptors for retinoic acid, thyroid hormone and steroid hormones. Nurr1 is expressed almost exclusively in the CNS. Sites of expression include the cortex, hippocampus, brainstem and the substantia nigra, where Nurr1 is confined to the dopamine (DA) neurons. During development, Nurr1 is expressed very early in the ventral midbrain, appearing before other markers for DA neurons such as tyrosine hydroxylase.

KO strain construction

Nurr1 was targeted by a replacement strategy in which exons 2, 3 and part of exon 4 (derived from strain 129) were replaced by a neomycin-resistance gene. The targeted gene lacked the translational start site as well as the exons encoding the N-terminal domain and the central DNA-binding domain. The targeting vector was transfected into E14 ES cells and positive clones were injected into C57BL/6 blastocysts.

Phenotype

Homozygous Nurr1$^{-/-}$ mice were hypoactive and died within 24 hours of birth. The cause of death was unknown but was probably due to an inability to suckle. Strikingly, homozygous mutant mice lacked known midbrain DA markers such as tyrosine hydroxylase, aldehyde dehydrogenase 2, DA D2 receptor, and glial cell line-derived neurotropic factor signal transducing receptor component c-ret. In addition, cresyl violet-stained sections revealed a disorganization of cells in the ventral midbrain of Nurr1$^{-/-}$ mice. Thus, Nurr1 is absolutely required for the generation of DA neurons.

The striatum is the major innervation target of DA midbrain neurons. mRNAs encoding enkephalin, substance P and choline acetyltransferase were expressed in normal patterns in striatum of Nurr1$^{-/-}$ brains, although the level of substance P mRNA appeared somewhat reduced. Nurr1 expression continued into adulthood in the substantia nigra and ventral tegmental area, suggesting that Nurr1 may also have a functional role in mature DA neurons.

HPLC was used to measure levels of catecholamines in the brains of heterozygous newborn and adult mice. As expected, DA was virtually absent in these regions in Nurr1$^{-/-}$ newborn animals. Interestingly, both newborn and adult heterozygotes showed significantly reduced levels of DA, indicating that nigrostriatal DA levels are critically affected by gene dosage. The reduction of DA indicates that Nurr1 also plays a critical role in the maintenance of the differentiated phenotype in mature DA neurons.

Comments

The finding that Nurr1 may also play a role in mature DA cells raises the intriguing possibility that Nurr1 ligands, when identified, may become useful in the treatment of Parkinson's disease and other disorders of midbrain DA circuitry. As Nurr1 can also form heterodimers with one of the retinoid receptors, RXR, it can be speculated that retinoids may also be involved in the development of the DA cells and that retinoids could have therapeutic importance for DA disorders. As indicated above, Nurr1 is expressed in the CNS at sites other than in the DA cells. Thus, it is likely that Nurr1 knockout mice will display additional phenotypes as they are further characterized.

Acknowledgements
Thomas Perlmann
Ludwig Institute for Cancer Research, Stockholm, Sweden

References

[1] Zetterström, R.H. et al. (1997) Science, 276, 248–250.
[2] Perlmann, T. and Jansson, L. (1995) Genes Dev. 9, 769–782.
[3] Zetterström, R.H. et al. (1996) Mol. Brain Res. 41, 111–120.
[4] Zetterström, R. et al. (1996) Mol. Endocrinol. 10, 1656–1666.

Other names
Ornithine-δ-aminotransferase

Gene symbol
Oat

Accession number
MGI: 97394

Area of impact
Metabolism

General description

OAT catalyzes the reversible reaction of ornithine into Δ1-pyrroline-5-carboxylate (P5C). This enzymatic reaction is required for the synthesis of arginine from glutamate. Ornithine is converted to citrulline and arginine necessary for the urea cycle in the liver. OAT is widely expressed in intestinal mucosal cells, pericentral vein hepatocytes, renal tubular cells, brain, neural retina and retinal pigment epithelium. Autosomal recessive OAT deficiency in humans causes slowly progressing blinding chorioretinal degeneration (gyrate atrophy, GA) and hyperornithinemia.

KO strain construction

A neomycin-resistance cassette was inserted into exon 3 at codon 40. AB1 ES cells were electroporated and targeted cells were injected into C57BL/6 blastocysts.

Phenotype

Neonatal $OAT^{-/-}$ mice were viable and normal in appearance. However, within hours after birth, $OAT^{-/-}$ pups stopped feeding, became lethargic, exhibited tremors, and died after 24–48 hours. Intraperitoneal injections of L-arginine increased the survival of OAT-null mice. Injections of L-arginine were not required for survival after 14 days of birth. OAT-null mice displayed neonatal hypo-ornithinemia (reduced plasma levels of ornithine, arginine and citrulline, and enhanced plasma ammonium levels). Only after weaning did $OAT^{-/-}$ mice become hyperornithinemic (10–15-fold above normal mice), implying that OAT has distinct roles in neonatal and adult animals. Over a period of several months, OAT-null mice developed progressive retinal degeneration. In particular, photoreceptors (disorganization and loss of photoreceptors) and the retinal pigment epithelium (crystalloid inclusion, accumulation of phagosomes) were affected.

Comments

In stark contrast to OAT-null mice, human GA patients do not have neonatal symptoms.

References

1 Wang, T. et al. (1994) Nature Genet. 11, 185–190.
2 Wang, T. et al. (1996) J. Clin. Invest. 97, 2753–2762.

OBF-1/OCA-B/Bob1

Other names

Oct-binding factor 1, POU domain class 2 associating factor 1

Gene symbol

Pou2af1

Accession number

MGI: 105086

Area of impact

Immunity and inflammation, transcription factors

General description

OBF-1/OCA-B/Bob1 is a 256 amino acid proline-rich transcription coactivator without obvious homologies. It is expressed in a highly B cell-specific manner (expression is found in B lymphocytes of all differentiation stages) and is also transiently inducible in T cells following T cell activation. *In vitro* transcription and transfection studies have shown that OBF-1 is required for B cell-specific transcriptional activity of immunoglobulin promoters. OBF-1 biochemically interacts with the DNA-binding domain (the POU domain) of Oct-1 or Oct-2 and is thereby recruited to some octamer sites in the promoters of various genes. The interaction of OBF-1 with either Oct-1 or Oct-2 leads to transcriptional co-activation.

KO strain 1 construction[1]

The targeting construct was designed to replace a 4.5 kb genomic fragment (129Sv) encompassing murine OCA-B exons 2, 3 and 4 with a neomycin-resistance cassette. An HSV-*tk* cassette was placed downstream of the long homology region and served as a negative selection marker. Two independent ES cell clones (E14 and CJ-7) were injected into C57BL/6 blastocysts. Heterozygous offspring of the germ line-transmitting chimeras were interbred to obtain homozygotes. Backcrosses into BALB/c and C57BL/6 background are currently underway. The null mutation was confirmed by Northern blotting of spleen polyA$^+$ RNA.

KO strain 2 construction[2]

The targeting construct was based on pBluescript (Stratagene) and contained two fragments from the OBF-1 gene as well as the PGK-neomycin and thymidine kinase genes for selection and counterselection, respectively. In the targeted allele, 250 out of the 256 amino acids of the OBF-1 protein were replaced by the PGK*neo* cassette. The targeting vector was electroporated into E14 ES cells (129SV strain). Chimeric mice were generated by aggregation, followed by breeding to C57BL/6 mice. The null mutation was confirmed by Northern blotting of spleen RNA.

KO strain 3 construction[3]

The targeting vector was designed such that a 5.5 kb fragment containing exons 2, 3 and 4 was deleted and replaced by the *neo*-resistance cassette from pMC1*neo*polyA. The short homology arm was a 2.3 kb *Hin*dIII genomic fragment derived from the first intron of the mouse Bob1 gene. The long homology arm was a 4.9 kb *Eco*RI fragment commencing in intron 4 and extending 3′ of the gene. The HSV-*tk* cassette was appended to the outer end of the long arm. The Bob-1 sequences used for the targeting vector were derived from a C57BL/6 genomic bank. The targeting vector (lacking bacterial vector sequences) was electroporated into the C57BL/6 ES cell line N1. Positive clones were injected into BALB/c blastocysts. Chimeric offspring were crossed with C57BL/6 mice. The null mutation was confirmed by RT-PCR analysis of spleen RNA.

Phenotype

Heterozygous mice showed no obvious phenotype. Homozygous mutant mice were viable, healthy, fertile and grossly normal, other than exhibiting a slight reduction in body weight. Surprisingly, the rearrangement and transcription of immunoglobulin genes was largely unaffected. No defect in the antigen-independent phase of B cell differentiation was observed, and no reductions in the numbers of thymocytes, peripheral T cells or peritoneal B1 ($CD5^+$) B cells were noted. However, antigen-dependent maturation of B cells was severely impaired. Mice deficient in OBF-1 had reduced numbers of the more mature IgD^{hi}/IgM^+ B cells in spleen, blood, lymph nodes and bone marrow, and a severe reduction in the number of recirculating B cells, but otherwise showed normal B cell differentiation. Although the serum IgM level was normal or higher in unimmunized animals, there was a severe deficiency in secondary immunoglobulin isotypes IgG, IgA and IgE. In addition, OBF-$1^{-/-}$ mice were incapable of mounting both T cell-independent (Ti) and T cell-dependent (Td) immune responses. Furthermore, there was a striking absence of germinal center formation in the spleen, lymph nodes and Peyer's patches after immunization with Td antigen, although primary follicles were visible.

Cell transfer experiments implicated an intrinsic B cell defect that is observed only *in vivo*, since OBF-$1^{-/-}$ B cells were stimulated essentially normally *in vitro*. Isotype switching experiments using cultured splenic B cells indicated that OBF-$1^{-/-}$ B cells were inherently capable of undergoing correct switching when stimulated with LPS, and LPS plus IL-4.

Comments

The mechanism(s) and direct target(s) of OBF-1 responsible for the observed deficiencies in mutant mice are as yet unknown. The antigen-independent stages of B-cell development are normal in OBF-$1^{-/-}$ mice but the antigen-driven maturation of B cells is strongly inhibited. Considering the multiple defects observed in the knockout mice, it is likely that OBF-1 regulates as yet unidentified factor(s) involved in B cell activation/maturation. While OBF-1 does not appear to be directly involved in isotype switch recombination, the reduction in the level of secondary isotype expression observed in OBF-$1^{-/-}$ B cells indicates that either OBF-1 functions directly at the switched immunoglobulin loci, or that it regulates other factors involved in immunoglobulin transcription in activated B cells that have undergone isotype switching. The proliferative response to surface IgM crosslinking is impaired in resting B cells from OBF-$1^{-/-}$ mice, suggesting that the signaling process through the antigen receptor might be affected. Thus, OBF-1 may be involved either directly or indirectly in the regulation of factors involved in cell–cell interactions that provide critical signals for B cell maturation, including the response of B cells to antigens and germinal centre formation.

Acknowledgements
Patrick Matthias
Friedrich Miescher Institute, Basel, Switzerland
Peter Nielsen
Max Planck Institut für Immunbiologie, Freiburg, Germany
Robert Roeder
Laboratory of Biochemistry and Molecular Biology, The Rockefeller University, New York, NY, USA

References

[1] Kim, U. et al. (1996) Nature 383, 542–547.
[2] Schubart, D.B. et al. (1996) Nature 383, 538–542.
[3] Nielsen, P.J. et al. (1996) Eur. J. Immunol. 26, 3214–3218.
[4] Gstaiger, M. et al. (1995) Nature 373, 360–362.
[5] Luo, Y. et al. (1992) Cell 71, 231–241.
[6] Luo, Y. and Roeder, R.G. (1995) Mol. Cell. Biol. 15, 4115–4124.
[7] Schubart, D.B. et al. (1996) Nucleic Acids Res. 24, 1913–1920.
[8] Strubin, M. et al. (1995) Cell 80, 497–506.

Oct-2

Other names
OBF-2

Gene symbol
Pou2f2

Accession number
MGI: 101897

Area of impact
Transcription factor

General description

Oct-2 is a B lymphocyte-restricted member of the POU homeodomain family of transcription factors. The bipartite DNA-binding domain of Oct-2 interacts with a critical octamer motif found in the promoters of all immunoglobulin (Ig) genes, in the Ig heavy and κ light chain enhancers, and in the promoters of other B cell-specific genes such as B29 (Igβ), CD20, CD21 and c-lyn. Oct-2 is considered to be a key regulator of the terminal phase of B cell differentiation and Ig expression in mice. Oct-2 is expressed at all stages of B cell differentiation.

KO strain construction

The targeting vector was designed such that PGK*neo* and a stop codon were introduced upstream of DNA-binding (POU) domain. An HSV-*tk* cassette was inserted at the 5′ end of the construct. Any Oct-2 derivative expressed from the mutated allele would lack the C-terminal half of the protein, would be unable to bind DNA, and thus would be inactive. The targeting vector was originally introduced into D3 ES cells (from the 129 strain) and homozygous mutants were derived by standard techniques. The mutation has since been backcrossed for more than 10 generations to the C57BL strain. The null mutation was confirmed by Western blotting: no Oct-2 protein was detected in mutant cells.

Phenotype

Although mutant pups were born at the expected Mendelian frequency, homozygous mutation of Oct-2 resulted in neonatal lethality with 100% penetrance. Oct-2$^{-/-}$ pups became lethargic, dehydrated and cyanotic, and died within first few hours of life. The cause of death is still not known, but no anatomic or histologic anomalies were apparent upon gross examination. The lethality and homozygosity for Oct-2$^{-/-}$ have cosegregated for more than six generations of breeding of independent animals.

While the T cell compartment was spared, and Ig genes were rearranged and transcribed normally, a deficit was found in the B cell compartment which affected mature B cells. Mutants contained normal numbers of B cell progenitors but were slightly lacking in IgM$^+$ B cells. Mature, antigen-experienced, recirculating B cells were decreased in number. Plasma cell numbers were reduced, as were serum Ig levels in naive animals (all classes but IgA). Poor responses to Ti and Td antigens were observed following immunization. No B-1 B cells were found in the peritoneal cavities of mutant mice. *In vitro*, mutant cells showed poor proliferation when LPS or anti-R antibodies were used as mitogens. The CD40L response was normal. The block in proliferation mapped to mid-G_1; that is, Oct-$2^{-/-}$ B cells were unable to make the commitment to pass the "restriction point" and progress to late G_1 and S phases.

Comments

Ig genes, CD20, B29 and CD21 do not require Oct-2 for expression. However, Oct-2 is essential for the maturation of Ig-bearing B cells into Ig-secreting B cells. Recently, the expression of CRISP-3 and CD36 on mouse B cells was shown to be dependent on Oct-2, suggesting that these might be candidate genes through which Oct-2 influences B cell differentiation. Oct-2 has a late developmental role other than in the immune system that is critical for the viability of the perinatal mouse.

Acknowledgements
Lynn M. Corcoran
The Walter & Eliza Hall Institute of Medical Research, Royal Melbourne Hospital, Victoria, Australia

References

1 Corcoran, L.M. et al. (1993) Genes Dev. 7, 570–582.
2 Pfisterer, P. et al. (1994) EMBO J. 13, 1654–1663.
3 Corcoran, L.M. et al. (1994) Immunity 1, 635–645.
4 Scheerlinck, J.P.Y. et al. (1995) Immunol. Lett. 45, 215–217.
5 Konig, H. et al. (1995) Genes Dev. 9, 1598–1607.

OMP

Other names
Olfactory marker protein

Gene symbol
Omp

Accession number
MGI: 97436

Area of impact
Signal transduction, neurobiology

General description

The *Omp* gene is expressed in mature olfactory sensory neurons and in a small neuronal subpopulation in the CNS. OMP is a phylogenetically conserved protein of unknown function. Its promoter consists of at least three types of regulatory motifs: Olf-1 (EBF)-binding elements, NF-1-like and SP-1-like elements. Based on the knockout mouse phenotype, OMP is postulated to participate in odor detection/signal transduction in olfactory sensory neurons.

KO strain construction

The mouse OMP targeting construct contained 9 kb of 5′ and 1.5 kb of 3′ homology, and the PGK*neo* and HSV-*tk* cassettes in pBluescript KS(+). The entire coding region of the intronless *Omp* gene was deleted. W9.5 ES cells were targeted. Strains with the deleted *Omp* gene were of mixed 129Sv × C57BL6/J and inbred 129Sv backgrounds.

Phenotype

$OMP^{-/-}$ mice manifested altered physiological activity of olfactory sensory neurons. Electro-olfactograms (EOG) of adult OMP-null mice demonstrated a 20–40% decrease in the response magnitude to various odorants. The onset and recovery kinetics following isoamyl acetate stimulation were prolonged in the null mice. The slopes of the initial and decay phases of the EOG response were 53% and 43% smaller in the mutants. The ability of the mutants to respond to the second odor pulse of a pair was also impaired. Similar changes of physiological activity of the olfactory sensory neurons were observed in mice with the mutation on the mixed 129Sv × C57BL6/J and on the 129Sv genetic backgrounds. Altered neuronal activity of the olfactory neuroepithelium is the most likely cause of the mutant bulbar phenotype. However, the bulbar phenotype was observed only in mice with OMP deletion on the mixed genetic background. The olfactory bulbs of these mutant mice were 15% smaller and showed a lower level of tyrosine hydroxylase (TH), the rate-limiting enzyme of catecholamine biosynthesis, and neuropeptide cholecystokinin (CCK). TH and CCK content per bulb were reduced by 65% and 40–50%, respectively, in the mutant mice. A similar bulbar phenotype was observed in wild-type mice with chronic peripheral olfactory deafferentation or naris blockage.

Light and electron microscopic histological observations indicated that the olfactory neuroepithelium of OMP-null mice was morphologically normal, as was the ratio of immature to mature sensory neurons. The mutant mice showed behavioral evidence of anosmia. Despite the dependence of neonatal mice on olfactory function for nipple location, attachment and suckling, the mutants appeared behaviorally and anatomically normal at birth. Although olfactory cues modulate the development and expression of sexual and social behavior in mice, adult OMP-null mice bred, delivered, and raised pups that were themselves fertile. Their open field exploratory activity was normal.

Comments

Significant functional impairment of the olfactory neuroepithelium in the OMP-null mice implies that OMP is a novel modulator of the odor detection/ signal transduction cascade in olfactory sensory neurons.

Acknowledgements
Frank L. Margolis
University of Maryland at Baltimore, Baltimore, MD, USA

References

1 Buiakova, O.I. et al. (1996) Proc. Natl Acad. Sci. USA 93, 9858–9863.
2 Buiakova, O.I. et al. (1994) Genomics 20, 452–462.
3 Kudrycki, K. et al. (1993) Mol. Cell. Biol. 13, 3002–3014.

μ Opiate receptor

Gene symbol
Oprm

Accession number
MGI: 97441

Area of impact
Signal transduction, neurology

General description

The μ, δ and κ opiate receptors are 7-transmembrane domain G protein-linked receptors that are targets of morphine binding. Each of the genes encoding these receptors is expressed in neurons in several nociceptive circuits. The μ opiate receptor is postulated to mediate much of morphine-induced analgesia and physical dependence.

KO strain 1 construction[1]

The targeting vector was constructed using a 3.2 kb *Eco*RV/*Bgl*II 5′ fragment and a 5.1 kb *Eco*RI 3′ fragment (both derived from strain 129) which were subcloned into pPGK*neo*. An HSV-tk cassette was included 3′ of the *Eco*RI 3′ fragment. PGK*neo* replaced the first exon of the μ opiate receptor gene. The targeting vector was electroporated into AB1 ES cells (129SvEv) and positive clones were injected into C57BL/6J blastocysts to generate chimeras.

KO strain 2 construction[2]

The *Oprm* gene was inactivated in P1 ES cells (129Sv) by insertion of a neo^r cassette into exon 2. A positive clone was injected into C57BL/6 blastocysts and chimeras were mated to C57BL/6 mice.

Phenotype

Mutant mice were born at the expected frequency, were grossly normal and showed no histologic abnormalities in brain or spinal cord. κ and δ opiate receptors were expressed at near-normal levels. Mutant mice were indistinguishable under drug-free conditions from controls in several tests of learning, emotionality and locomotor skills. However, untreated knockout mice displayed shorter latencies on tail flick and hot plate tests for spinal and supraspinal nociceptive responses compared to wild-type mice. Treatment of mutant mice with morphine failed to significantly reduce nociceptive responses in these tests. Heterozygous mice exhibited shifts downward and to the right in dose–effect relationships for morphine analgesia.

Comments

The μ opiate receptor has an important role in mediating morphine-induced analgesia. Interactions of endogenous peptides with μ opiate receptors are likely to be involved in nociceptive responses in drug-free animals.

Acknowledgements
George R. Uhl
National Institute on Drug Abuse, Baltimore, MD, USA

Reference

[1] Sora, I. et al. (1997) Proc. Natl Acad. Sci. USA 94, 1544–1549.
[2] Matthes, H.W.D. et al. (1996) Nature 383, 819–823.

Otx1

Other names
Orthodenticle homolog 1

Gene symbol
Otx1

Accession number
MGI: 97450

Area of impact
Neurology

General description

Otx1, a homeobox gene, was isolated (with *Otx2*) as a mouse cognate of the *Drosophila* head gap gene *orthodenticle*. *Otx1* is expressed in the rostral brain with a caudal limit at the mes/metencephalic junction at the pharyngula stage, suggesting its role in the regional patterning of forebrain and midbrain. *Otx1* is also expressed uniquely during later neurogenesis in the rostral brain and optic system.

KO strain construction

The neomycin-resistance gene was inserted into the second exon that encodes the homeobox. It was flanked by 11.6 kb and 0.6 kb homologous sequences on the 5′ and 3′ sides, respectively. The genomic DNA was isolated from TT2 cells, and diphtheria toxin A fragment (DT-A) gene was used for negative selection. The ES cells used were TT2-derived from an F1 embryo derived from C57BL/6 and CBA. Chimeras were crossed with either C57BL/6 or CBA females; the strain is maintained in these backgrounds with heterozygotes.

Phenotype

Otx1 homozygous mutants were live-born in Mendelian ratio but all died shortly after birth for unknown reasons. Defects were subtle in the newborn skull and brain. The neocortex was smaller laterally, the third ventricle was narrowed, and the hippocampus was sometimes smaller. Cerebral cortical layers were poorly differentiated; the cortical plate and white matter were thin, and the subcortical plate was hardly visible. Defects were also found in the eyes. These defects were most likely to correspond to areas of Otx1-unique expression in later neurogenesis, and no defects were apparent in early patterning of rostral brain.

In the regions where both Otx1 and Otx2 are expressed, their functions may overlap. Indeed, *Otx1* and *Otx2* double heterozygous mutant mice exhibited marked defects throughout the fore- and midbrains, where defects were never found with a single mutation alone. Concomitantly, the isthmic region was expanded. *Wnt1*- and *En*-expression in midbrain were lost, though that in metencephalon was expanded.

Comments

The territory for midbrain and forebrain may be determined by *Otx* genes. Otx2 may suppress *Fgf8* expression, such that Fgf8 from the isthmus makes antero-posterior patterning in the midbrain.

Acknowledgements
S. Aizawa
University of Tokyo, Bunkyo-ku, Japan

References

[1] Suda, Y. et al. (1996) Genes to Cells 1, 1031–1044.
[2] Matsuo, I. et al. (1995) Genes Dev. 9, 2646–2658.
[3] Kuratani, S. et al. (1997) Dev. Dyn. 209, 139–155.

Otx2

Gene symbol
Otx2

Accession number
MGI: 97451

Area of impact
Development

General description

Otx2, a homeobox gene, was isolated along with *Otx1* as a mouse cognate of the *Drosophila* head gap gene *orthodenticle*. *Otx2* is expressed in visceral endoderm and epiblast before gastrulation, in anterior node derivatives during gastrulation and then in rostral brain with a caudal limit at mes/metencephalic junction. *Otx2* is also expressed uniquely during later neurogenesis in rostral brain and sensory organs.

KO strain 1 construction[1]

The neomycin-resistance gene was inserted into the *Kpn*I site of the second exon of the *Otx2* gene, interrupting the sequence between the first and second α helix of the homeobox. ES cells used were TT2 derived from an F1 embryo between C57BL/6 and CBA. Chimeras were crossed with either C57BL/6 or CBA females, and the strain is maintained in these backgrounds by mating between heterozygotes.

KO strain 2 construction[2]

A large part of the coding sequence of the *Otx2* gene was deleted and replaced with *lacZ* upstream of the start of translation, and including a neo^r cassette. *lacZ* is thus under the control of the *Otx2* promoter. The vector was introduced into HM-1 ES cells, which were injected into C57BL/6 blastocysts. The phenotype was examined on a mixed 129 × B6/D2 background.

KO strain 3 construction[3]

A deletion encompassing the entire homeodomain region was replaced by a PGK*neo* cassette. This insertion will truncate the Otx2 protein after the first exon, which contains 32 amino acids. The vector was introduced into R1 129 ES cells, which were aggregated with CD1 morulae. The phenotype was examined on a mixed CD1 × 129 background.

Phenotype

Homozygous *Otx2* mutants do not develop structures anterior to rhombomere 3; trunk structures such as posterior neural plate and somites are normally present. The embryos die around day 9 of gestation. The phenotype is quite similar to the knockout phenotype of Lim1, another component required for organization of the head. It is unclear whether the phenotype of Otx2$^{-/-}$ mice is due to defects in visceral endoderm, anterior node derivatives or neuro-ectoderm.

The *Otx2* mutation also displays a haplo-insufficient phenotype in the C57BL/6 genetic background. This includes craniofacial malformations designated as otocephaly: a series of graded defects are found in the cranial structures such as jaw, eye and rostral brain, but never in the post-cranial part of the body. Skull defects were restricted to pre-mandibular or trabecular components, and the mandible is specifically affected among the first arch components. In the trigeminal nerve, the ophthalmic branch is also specifically affected along with mesencephalic trigeminal neurons. Thus the defects are characteristic to components of mesencephalic neural crest origin; no defects are found in structures of rhombencephalic neural crest or cephalic mesoderm origin. Abnormalities are also found in anterior head structures such as eyes, olfactory system and hypophysis. All the affected structures correspond to the most anterior and most posterior domains of *Otx2* expression where the related *Otx1* gene is not or only weakly expressed.

Acknowledgements

S. Aizawa

University of Tokyo, Bunkyo-ku, Japan

References

1 Matsuo, I. et al. (1995) Genes Dev. 9, 2646–2658.
2 Acampora, D. et al. (1995) Development 121, 3279–3290.
3 Ang, S.L. et al. (1996) Development 122, 243–252.
4 Suda, Y. et al. (1996) Genes to Cells 1, 1031–1044.
5 Kuratani, S. et al. (1997) Dev. Dyn. 209, 139–155.

p18 NF-E2

Other names
AP-1-like transcription factor, p18 subunit of NF-E2 (nuclear factor-erythroid 2), Mafk

Gene symbol
Nfe2u

Accession number
MGI: 99951

Area of impact
Transcription factors

General description

p18 NF-E2 is a widely expressed basic leucine zipper polypeptide which forms half of the heterodimeric transcription factor NF-E2. (The other subunit is the larger hematopoietic-specific subunit, p45 NF-E2.) p18 NF-E2 shares extensive homology with members of the small *Maf* gene family. All small Maf polypeptides can form heterodimers with p45 NF-E2, but binding with p18 NF-E2 increases the affinity of the transcription factor for the NF-E2 DNA-binding site. The p18 NF-E2 subunit distinguishes the binding activity of NF-E2 from that of AP-1, since p18 NF-E2 specifically recognizes nucleotides not contained in the AP-1 core sequence.

KO strain construction

The targeting construct was based on the pPNT vector and contained both the PGK*neo*-resistance cassette and an HSV-*tk* negative selection cassette. The construct was designed to delete the entire third exon, including the bZIP domain. The construct was introduced into J1 (129) and D3 (129) ES cell lines and positive clones were introduced into C57BL/6 blastocysts. Chimeras were obtained from J1 clones and heterozygotes were intercrossed to obtain homozygous mutant mice. The mutation was both maintained in the 129Sv background and bred into a C57BL/6 background: no difference in phenotype was observed. The null mutation was confirmed by RT-PCR analysis of fetal liver RNA.

Phenotype

Mutant mice were indistinguishable from wild-type mice at all stages of development and in all parameters examined. Adult mutants remain healthy. No differences in the levels of either β-major or ζ-globin transcripts were observed. NF-E2-like DNA-binding activity was present in mutant fetal liver extracts, in spite of the absence of p18 NF-E2, suggesting that another Maf protein is able to substitute for the p18 NF-E2 subunit.

Comments

p18 NF-E2 is dispensable *in vivo*. The absence of phenotype in these mutant mice and the presence of an NF-E2 gel-shift band suggests that other Maf polypeptides are compensatory and redundant. These findings demonstrate the complex interactions between proteins acting on NF-E2-binding sites in hematopoietic cells.

Acknowledgements
Stuart Orkin
Divisions of Hematology/Oncology and Pediatric Oncology, Children's Hospital and Dana-Farber Cancer Institute, Harvard Medical School, Boston, MA, USA

Reference

[1] Kotkow, K.J. and Orkin, S.H. (1996) Proc. Natl Acad. Sci. USA 93, 3514–3518.

Other names
p38, CSBP (CSAID™ binding protein), RK, Mpk2, Hog1 (yeast)

Area of impact
Immunity and inflammation

General description

The MAP kinase p38 is a key component of a signal transduction pathway leading to the production of TNFα and IL-1β. Endotoxin-activated kinase p38 has also been implicated in the TNFα signaling pathway leading to MAPKAP kinase-2 activation, and the IL-1β signaling pathway leading to the production of IL-8. The production and biological activity of these master inflammatory cytokines are blocked by p38 inhibitors, indicating the potential role of p38 in chronic inflammatory diseases such as rheumatoid arthritis and inflammatory bowel disease.

KO strain construction

The targeting vector p38RV1 was designed to delete part of exon 3 and intron 4, inserting an anti-sense *neo* marker and creating a splice mutant. The targeting vector was introduced into AB2.1 and W9.5 ES cells (129SvEv), and was investigated on the C57BL/6 background.

Phenotype

The p38 knockout appeared to be an embryonic lethal in the 129SvEv × C57BL/6 background. Most embryos appeared to lose viability after day 10.5 pc, and very few embryos survived to day 12.5. Homozygous p38$^{-/-}$ ES cells had no detectable p38 mRNA or protein.

Comments

Viability of mice with the p38$^{-/-}$ mutation is being defined in alternative mouse backgrounds, and the mechanism of lethality is under investigation.

Acknowledgements
John S. Mudgett
Merck Research Laboratories, Rahway, NJ, USA

Reference

[1] The p38$^{-/-}$ mouse is not yet published.

p45 NF-E2

Other names
p45 subunit of NF-E2 (nuclear factor-erythroid 2), NF-E2

Gene symbol
Nfe2

Accession number
MGI: 97308

Area of impact
Hematopoiesis, transcription factors

General description

p45 NF-E2 is the hematopoietic-specific subunit of the heterodimeric basic leucine zipper (bZIP) transcription factor NF-E2. (The other subunit is the widely expressed p18 NF-E2.) NF-E2 was originally identified as the enhancer-binding protein in the β-globin locus control region (LCR). NF-E2 binds to an extended AP-1-related DNA sequence motif that has been identified within regulatory *cis*-elements of several erythroid-expressed genes. The expression of p45 NF-E2 is limited to cell lineages of the hematopoietic system (erythroid cells, megakaryocytes and mast cells). The murine p45 NF-E2 gene is composed of three exons, with the bZIP region required for protein dimerization and DNA binding located in the third exon.

KO strain construction

A PGK*neo*^{r} cassette was introduced in antisense orientation just upstream of the region encoding the bZIP domains within the third exon of the p45 NF-E2 gene. Although no portion of the gene was deleted as a result of the targeting, any product translated from the mutant allele would be unable to form heterodimers or bind DNA, rendering it inactive.

The targeting construct was electroporated into J1 ES cells derived from the 129Sv strain. Chimeras were generated on a C57BL/6 background. The KO strain has subsequently been maintained on an inbred 129SvEvTac background. No differences in phenotype have been observed.

The null mutation was confirmed by a lack of intact p45 NF-E2 mRNA and protein in mutant fetal liver as determined by Northern blot and immunoblot. A read-through transcript contained both *neo* sequences and multiple in-frame stop codons.

Phenotype

Mice lacking p45 NF-E2 underwent normal embryonic development and were produced in the expected Mendelian ratio. Heterozygotes were normal. However, within the first week of life, about 90% of mutant homozygotes died of massive hemorrhage related to an apparently complete lack of circulating blood platelets. Neonatal mutants showed severe anemia and dysmorphic erythrocyte changes. Mutant mice that survived the neonatal period continued to exhibit profound thrombocytopenia and chronic non-lethal bleeding. They also showed mild dyserythropoiesis (microcytic and hypochromic red blood cells) and massive splenomegaly but otherwise appeared normal. A small decrease in hemoglobin content per cell was noted but globin chain synthesis was balanced and the switching from fetal to adult globins proceeded normally. Surviving adult mutant mice were fertile.

Investigation of the thrombocytopenia in these mice showed that it resulted from a late arrest in megakaryocyte cytoplasmic maturation. Numerous polyploid megakaryocytes were found in the hematopoietic tissues but these cells had a large cytoplasm with a paucity of platelet-specific granules, and a failure to demarcate platelet fields. Small, platelet-like particles were found in association with splenic macrophages but it was unclear whether these reflected platelets generated in the normal manner or simply megakaryocyte debris. Thrombopoietin signaling, as measured by megakaryocyte proliferation, was not perturbed in mutant cells. No new DNA-binding complexes containing substitutes for p45 NF-E2 could be identified using gel shift assays with nuclear extracts from p45 NF-E2-null cells.

Comments

p45 NF-E2 is essential for platelet production but dispensable for erythrocyte development. These findings represent the first *in vivo* demonstration of a role for a lineage-restricted transcription factor in late megakaryocyte maturation and platelet differentiation. However, the relevant transcriptional target of this protein remains unknown. p45 NF-E2-null mice may be useful in blastocyst complementation studies to identify other genes required for platelet development.

Acknowledgements
Ramesh Shivdasani
Department of Medical Oncology, Dana-Farber Cancer Institute, Harvard Medical School, Boston, MA, USA
Stuart Orkin
Divisions of Hematology/Oncology and Pediatric Oncology, Children's Hospital and Dana-Farber Cancer Institute, Harvard Medical School, Boston, MA, USA

References

1. Andrews, N.C. et al. (1993) Nature 363, 722–728.
2. Shivdasani, R.A. et al. (1995) Cell 81, 695–704.
3. Shivdasani, R.A. and Orkin, S.H. (1995) Proc. Natl Acad. Sci. USA 92, 8690–8694.

Other names
NCF-1

Gene symbol
Ncf1

Accession number
MGI: 97283

Area of impact
Immunity and inflammation

General description

p47^{phox} is a cytosolic subunit required for the assembly of the NADPH oxidase in phagocytes. The NADPH oxidase is required for the generation of superoxide and its metabolites by phagocytes. Upon phagocyte stimulation, p47^{phox} is multiply phosphorylated and binds to membrane-bound components and other cytosolic components to form the functional NADPH oxidase enzyme complex. Chronic granulomatous disease (CGD) in humans results from a defect in NADPH oxidase activity in phagocytes which leads to severe bacterial and fungal infections and premature death.

KO strain construction

The targeting vector was constructed using pBluescript and a 1.5 kb fragment such that the pMC*neo* cassette was introduced into exon 7 of the mouse p47^{phox} gene (originally derived from a B6/CBA genomic DNA library). Exon 7 (amino acid 221) represents a region that has been shown to be necessary for human p47^{phox} function. No *tk* selection was used. The targeting vector was electroporated into D3 ES cells. Positive clones were injected into C57BL/6 blastocysts. Chimeras were bred to C57BL/6J females. The mutant strain has been further backcrossed into Tacomic C57 Black mice. The null mutation was confirmed by biochemical assays for superoxide production.

Phenotype

Mutant mice were born at the expected Mendelian frequency. Before the onset of severe infections, p47$^{phox-/-}$ mice had normal weight, fertility, and peripheral blood and differential leukocyte counts. However, p47$^{phox-/-}$ mice went on to spontaneously develop severe infections and to show a phenotype consistent with the absence of phagocyte superoxide production. Specifically, the mutant mice exhibited a high rate of spontaneous infection with *Staphylococcus xylosus* and saprophytic molds, and an abnormally exuberant peritoneal inflammatory response to intraperitoneal thioglycollate. No superoxide production in phagocytes was observed *in vitro*. Infections with *Burkholderia cepacia*, a common pathogen in human CGD, were lethal to p47$^{phox-/-}$ mice. However, there was no increase in morbidity or mortality following infection with *Staphylococcus aureus*. Successful transient gene therapy of mutant mice using retrovirally-transduced CD34^{+} stem cells was demonstrated.

Comments

This mouse is a good model for human p47^{phox} deficiency (CGD) and will allow for rigorous determination of the role of phagocytic superoxide production in a series of conditions such as atherosclerosis, reperfusion injury and carcinogenesis.

Acknowledgements
Steven M. Holland
National Institute of Allergy and Infectious Diseases, Bethesda, MD, USA

References

[1] Jackson, S.H. et al. (1995) J. Exp. Med. 182, 751–758.
[2] Mardiney III, M.M. et al. (1997) Blood 89, 2268–2275.

Other names
ISGF3γ (IFN-stimulated gene factor 3 gamma)

Gene symbol
Isgf3g

Accession number
MGI: 107587

Area of impact
Immunity and inflammation, transcription factors

General description

p48, an IRF family DNA-binding protein, forms the heterotrimeric complex ISGF3 with Stat 1α/β and Stat 2 upon type I and type II IFN stimulation. ISGF3 activates transcription of many IFN-inducible genes through DNA elements termed ISREs (IFN-stimulated response elements). ISGF3 also binds to IRF-E (IRF-element) in type I IFN gene promoters. p48, possibly as a component of ISGF3, activates type I IFN gene transcription by a positive feedback mechanism.

KO strain construction

The targeting construct contained 7 kb of the endogenous mouse *Isgf3g* gene derived from a C57BL/6 genomic DNA library. Exon 2, containing the initiation codon and a part of DNA-binding domain, was replaced with the neomycin-resistance cassette. The diphtheria toxin gene was used for negative selection. The targeting vector was introduced into the TT2 ES cell line derived from (129Sv × CBA) F1 mice. Positive clones were injected into eight-cell stage embryos of ICR mice, or into C57BL/6 blastocysts. Heterozygotes were obtained by breeding chimeric mice with ICR or A/J mice. The loss of p48 activity was confirmed by the absence of ISGF activities in electrophoretic mobility shift assays using the cell extracts from IFN-treated mutant embryonic fibroblasts (EFs).

Phenotype

p48$^{+/-}$ and p48$^{-/-}$ mice showed no apparent abnormalities in reproductive ability or behavior. The induction of many IFN-inducible genes by both types of IFNs was affected in EF cells from p48$^{-/-}$ mice. The induction of the 2′-5′ oligoadenylate synthetase gene and the double-stranded RNA-activated kinase gene by both types of IFNs was severely or completely abrogated in p48$^{-/-}$ EFs, whereas the induction of these genes was not affected in IRF-1$^{-/-}$ EFs. In contrast, induction of the guanylate-binding protein (GBP) gene by type II IFN was not affected in p48$^{-/-}$ EFs but was severely affected in IRF-1$^{-/-}$ EFs, demonstrating the non-redundant role of these factors. Induction of the GBP gene by type I IFN was substantially affected in p48$^{-/-}$ EFs. This induction was also weakly affected in IRF-1$^{-/-}$ EFs but completely abrogated in EFs lacking both p48 and IRF-1, indicating the cooperative interaction of p48 and IRF-1.

Anti-viral states against EMCV (encephalomyocarditis virus), VSV (vesicular stomatitis virus) and HSV (herpes simplex virus) induced with both types of IFNs were not well established in p48$^{-/-}$ EFs. Induction of type I IFN genes by NDV (Newcastle disease virus) infection was severely affected in EFs and macrophages from p48$^{-/-}$ mice *in vitro*. IFN levels in sera and organs after infection were also reduced in infected p48$^{-/-}$ mice *in vivo*. p48$^{-/-}$ mice were more sensitive to infection by viruses such as NDV and EMCV.

Comments

From a comparison of EFs lacking either p48 or IRF-1 or both genes, it is clear that p48 and IRF-1 regulate IFN-inducible genes in non-redundant manner in many cases, and that p48 plays a more critical role than IRF-1 in the establishment of IFN-induced anti-viral states. The study of p48$^{-/-}$ mice has also clarified the involvement of p48 in type I IFN gene induction.

Acknowledgements
Tohru Kimura
Tokyo Medical and Dental University, Tokyo, Japan

References

1 Kimura, T. et al. (1996) Genes to Cells 1, 115–124.
2 Harada, H. et al. (1996) Genes to Cells 1, 995–1005.

Other names
Transformation-related protein 53, tumor suppressor

Gene symbol
Trp53

Accession number
MGI: 98834

Area of impact
Apoptosis, tumorigenesis

General description

p53 is a transcriptional activator that may regulate the cell cycle, genome stability, and possibly apoptosis. p53 is also a tumor suppressor gene that is thought to play a role in the cellular response to DNA damage. Mutations in p53 are associated with a wide variety of tumors.

KO strain 1 construction[1]

The targeting strategy used a modified positive/negative selection vector in which a *neo* cassette interrupted exon 5 contained in a 3.7 kb genomic p53 gene fragment. A 106 nucleotide fragment of exon 5 (including the conserved p53 II domain) and 350 nucleotides of intron 4 were also deleted. AB1 ES cells were targeted and positive clones were injected into C57BL/6 blastocysts. Chimeras were bred to C57BL/6 females.

KO strain 2 construction[2]

Neo^r was used to replace exons 2–6 of the p53 gene. The vector was electroporated into E14 (129Ola) cells. Targeted ES cells were injected into C57BL/6 × CBA blastocysts and chimeras were mated to 129Ola. The +/+, +/− and −/− offspring were analyzed.

KO strain 3 construction[3]

The report cites studies done on cells from mice carrying a germ line disruption of the p53 gene which was the unpublished work of T. Jacks and R. Weinberg.

Phenotype

Studies of KO strain 1 p53$^{-/-}$ mice demonstrated that p53 was not essential for normal embryonic development, but that absence of the gene predisposed the mice to neoplasias, particularly lymphomas and sarcomas. Homozygotes developed tumors at 3–6 months of age and heterozygotes developed tumors at 10 months of age. KO strain 1 p53$^{-/-}$ mice also modeled some of the features of human Li–Fraumeni syndrome, a form of familial breast cancer with mutations in *TRP53*. If a tumor was introduced into these animals (for instance, by gamma irradiation), the response of the host to the tumor was abnormal, in that the transformed cells did not undergo growth arrest.

Thymocytes from KO strain 2 $p53^{-/-}$ mice were resistant to the induction of apoptosis by irradiation or etoposide but retained normal sensitivity to glucocorticoids and calcium. Time-dependent apoptosis in untreated cultures was unaffected. Heterozygous cells were partially resistant to irradiation and etoptoside. Similarly, immature thymocytes from KO strain 3 $p53^{-/-}$ mice died normally when exposed to compounds mimicking TCR engagement and to glucocorticoids but were resistant to the lethal effects of ionizing radiation. An increase in the expression of *Bcl-2* and a decrease in *Bax* expression has been described in $p53^{-/-}$ mice[4].

Comments

p53 is necessary for cell death induced by agents that cause DNA strand breakage but not for apoptosis induced by other means. p53 also acts to suppress tumor growth.

References

1 Donehower, L.A. et al. (1992) Nature 356, 215–221.
2 Clarke, A.R. et al. (1993) Nature 362, 849–852.
3 Lowe, S.W. et al. (1993) Nature 362, 847–849.
4 Miyashita, T. et al. (1994) Oncogene 9, 1799–1805.

Other names
Nerve growth factor receptor, low-affinity NGF receptor, LNGFR

Gene symbol
Ngfr

Accession number
MGI: 97323

Area of impact
Neurology

General description

Nerve growth factor (NGF) was the first neurotropin isolated. *In vitro* it supports both neuronal survival and process outgrowth. The injection of antibodies to NGF *in vivo* blocked the development of sensory and sympathetic neurons. NGF binds to the TrkA receptor with high affinity, and to the p75NGFR receptor with low affinity. The low-affinity receptor also binds other neurotropins, including BDNF, NT-3 and NT-4. These observations, together with the widespread expression of p75NGFR, preclude strong conclusions about the role of p75NGFR *in vivo.*

KO strain construction

Neo^r was inserted into exon 3 of the gene encoding p75NGFR, which is part of the ligand-binding site. *tk* was ligated to the 5′ flanking end. The vector was electroporated into J1 ES cells (129Svter). Clones showing homologous recombination were injected into BALB/c blastocysts. Chimeras were mated with 129 to obtain an inbred line.

Phenotype

p75NGFR$^{-/-}$ mice were viable and fertile. Immunohistochemical analyses of the footpad skin of mutant mice revealed markedly decreased sensory innervation by calcitonin gene-related peptide- and substance P-immunoreactive fibers. The defective innervation was correlated with loss of heat sensitivity, and associated with the development of ulcers in the distal extremities. Complicated by secondary bacterial infection, the ulcers progressed to toenail and hair loss. Crossing a human transgene encoding p75NGFR into the mutant animals rescued the absent heat sensitivity and the occurrence of skin ulcers, and increased the density of neuropeptide-immunoreactive sensory innervation of footpad skin. The mutation in the gene encloding p75NGFR did not decrease the size of sympathetic ganglia or the density of sympathetic innervation of the iris or salivary gland.

Comments

p75NGFR has an important role in the development and function of sensory neurons.

Acknowledgements
Kuo-Fen Lee
Salk Institute, La Jolla, CA, USA

References

[1] Lee, K.-F. et al. (1992) Cell 69, 737–749.
[2] Lee, K.-F. et al. (1994) Science 263, 1447–1449.
[3] Davies, A.M. et al. (1993) Neuron 11, 565–574.
[4] Lee, K.-F. et al. (1994) Development 120, 1027–1033.

Gene symbol
Pax2

Accession number
MGI: 97486

Area of impact
Development

General description

Pax2 is a member of the *Pax* gene family encoding transcription factors with a DNA-binding domain, known as the *paired* box, that is found in the *Drosophila paired* gene and its relatives. *Pax2* is expressed in the optic stalk, in the midbrain/hindbrain boundary and in the spinal cord from day 8.5 of development. In the spinal cord it is confined to longitudinal cells on both sides of the sulcus limitans. It is expressed in the developing kidney, earliest expression being detected in the pronephros and Wolffian duct. Later it is observed in the ductal and mesenchymal components of the metanephros and thus in both inducing and induced structures. In the metanephros it is also detected in the cortex.

KO strain construction

A 4 kb deletion from the *Not* 1 to the *Pvu*II site removed part of exon 1 including the ATG and a part of the second exon and replaced it with PGK*neo*. The targeting vector was introduced into R1 129 ES cells which were aggregated with outbred NMRI embryos. The phenotype was studied on a 129 × NMRI background.

Phenotype

Heterozygous mice have a reduced kidney size but are otherwise normal. Homozygous mice lack kidneys, ureters and genital tracts. There is a dysgenesis of both ductal and mesenchymal components of the developing urogenital system. The Wolffian and Müllerian ducts develop only partially and degenerate during embryogenesis. The ureters, the inducers of the metanephros, are absent and therefore kidney development does not take place. Mesenchyme of the nephrogenic cord fails to undergo epithelial transformation and is not able to form tubules in the mesonephros. Pax 2 is also required for the establishment of axonal pathways along the optic stalks and ventral diencephalon. In mutant brains, the optic tracts remain totally ipsilateral due to agenesis of the optic chiasma. Furthermore Pax $2^{-/-}$ mutants show extension of the pigmented retina into the optic stalks and failure of the optic fissure to close, resulting in coloboma. In the inner ear, Pax $2^{-/-}$ mice show agenesis of the cochlea and the spiral ganglion.

Acknowledgements
Peter Gruss
Max-Planck Institute for Biophysical Chemistry, Gottingen, Germany

References

[1] Torres, M. et al. (1995) Development 121, 4057–4065.
[2] Torres, M. et al. (1996) Development 122, 3381–3391.

Pax 4

Gene symbol
Pax4

Accession number
MGI: 97488

Area of impact
Development

General description

Pax 4 is a member of the *Pax* gene family encoding transcription factors with a DNA-binding domain, known as the *paired* box, that is found in the *Drosophila paired* gene and its relatives. The *Pax4* gene is expressed in the ventral neural tube on both sides of the floor plate in a subset of not yet characterized cells. Expression is also detected in the developing pancreas.

KO strain construction

After deletion of a 1.5 kb *Nhe*I fragment containing almost the entire paired domain, a *lacZ*PGK*neo* cassette was inserted in-frame with the *Pax4* ATG codon. The targeting vector was introduced into R1 129 ES cells, which were aggregated with outbred NMRI embryos. The phenotype was studied on a 129 × NMRI background.

Phenotype

Heterozygous animals do not exhibit any obvious abnormalities, survive to adulthood and are fertile. At birth, homozygous animals appeared normal and were indistinguishable from wild-type littermates. However after 48 hours, they exhibited growth retardation and dehydration despite being able to drink milk and died within 3 days of birth. The pancreas of a homozygous newborn lacks insulin-producing β cells and somatostatin-producing δ cells. Glucagon-producing α cells were still present and their numbers were higher than in wild-type animals.

Comments

The *Pax4* gene is essential for the differentiation of insulin-producing β cells in the pancreas.

Acknowledgements
Peter Gruss
Max-Planck Institute for Biophysical Chemistry, Gottingen, Germany

Reference
1 Sosa-Pineda, B. et al. (1997) Nature 386, 399–402.

Pax 7

Gene symbol
Pax7

Accession number
MGI: 97491

Area of impact
Development

General description

Pax7 is a member of the *Pax* gene family encoding transcription factors with a DNA-binding domain, known as the *paired* box, that is found in the *Drosophila paired* gene and its relatives. The *Pax7* gene is expressed early at E8.5 in all brain vesicles and later retracts to the mesencephalon (anterior limit at the posterior commissure). In the spinal cord, *Pax7* is expressed in the alar region excluding the roof plate. It is detected in cephalic neural crest and in rhombomeres 1, 3 and 5. *Pax7* is expressed in the dermomyotome and myotome. In adult brain *Pax7* is expressed in the superior colliculus. It is also expressed in the nasal region.

KO strain construction

Two mutations have been generated. In the first construct PGK*neo* is inserted in the first exon of the *paired* box (the paired box is composed of three exons). In the second construct the β-galactosidase gene is inserted in-frame into the *paired* box second exon. PGK*neo* was used as a selection cassette. The vector was introduced into both D3 and R1 ES cells.

Phenotype

Both KO constructs give the same phenotype. Heterozygous mice are normal and fertile. Homozygous mice are born and most of them die within the first 3 weeks after birth. They are growth retarded, sickly and only 2% survive longer than 3 weeks. The mice exhibit phenotypes in cephalic neural crest derivatives. In the nose the number of serous glands is reduced. The lateral inferior bone of the nasal capsule is missing and the lacrimal bone is affected. The maxilla is normal but reduced in size in the anterior to posterior direction. The animals also exhibit dilations in the distal part of the small intestine and may die from obstructions. The phenotype is quite mild as compared to the related gene *Pax3*, which is mutated in the mouse mutation *Splotch*.

Comments

There may be some overlap in function between Pax 3 and Pax 7.

Acknowledgements
Peter Gruss
Max-Planck Institute for Biophysical Chemistry, Gottingen, Germany

Reference

[1] Mansouri, A. et al. (1996) Development 122, 831–838.

Other names
Receptor for platelet-derived growth factor B, CD140b, PDGF beta chain

Gene symbol
Pdgfrb

Accession number
MGI: 97531

Area of impact
Neurology

General description

Platelet-derived growth factor (PDGF) is a mitogen for many cell types including fibroblasts, smooth muscle cells, macrophages, glia and endothelial cells. It exists as a dimer between two A or B chains, or an AB heterodimer. The receptor for PDGF is a protein tyrosine kinase. The α subunit binds both PDGF-A and B, whereas the β subunit binds PDGF-B. PDGF and its receptor are expressed in developing kidney and brain.

KO strain construction

A β-gal/neo^r fusion was inserted into the exon coding the second Ig domain of the PDGFβR. *tk* was fused to the 3′ end. The vector was electroporated into AB1 ES cells (129). Selected clones showing homologous recombination were injected into blastocysts. Chimeric mice were crossed to 129Sv or C57BL/6.

Phenotype

Mice homozygous for a targeted deletion of the PDGFβR exhibited hemorrhagic and thrombocytopenic effects, were severely anemic, and exhibited defects in kidney glomeruli due to lack of mesangial cells. Mice died at or shortly before birth. However, many cell types and tissues which express the receptor, including major blood vessels and the heart, appeared normal in the absence of the receptor. These results indicate that whereas the β receptor is essential in certain cell types during embryonic development, its broader role may be masked because of compensation by the α-subunit.

Comments

The glomerular and hemolytic defects seen in PDGFβR$^{-/-}$ mice are similar to those in the knockout of the PDGF-B gene.

Reference

[1] Soriano, P. et al. (1994) Genes Dev. 8, 1888–1896.

PDGF-A

Other names
Platelet-derived growth factor A

Gene symbol
Pdgfa

Accession number
MGI: 97527

Area of impact
Hematopoiesis, development

General description

PDGF-A is one of two polypeptide chains constituting members of the dimeric PDGF family of proteins (PDGF-AA, AB and BB). These factors act by binding to two related receptor tyrosine kinases, PDGFαR and PDGFβR. The PDGFαR is thought to bind both the A and B chains, and the B polypeptide is thought to bind to both the α and β receptors. PDGF-A and PDGF-B are encoded by separate genes. PDGF-A is expressed in the mammalian pre-implantation embryo, and following implantation in developing epithelium, muscle and nerve tissue. PDGFα receptors occur broadly on mesenchymal cells during development. The physiological function of PDGF-A is assumed to include connective tissue development and repair. PDGF is also implicated in pathological processes involving abnormal connective tissue cell proliferation.

KO strain construction

The replacement-type targeting vector was designed such that a 2.5 kb fragment encompassing exon 4 (encoding the N-terminal half of PDGF-A) was replaced by a PGK*neo* cassette and flanked by a total of 10 kb of homologous *Pdgfa* gene sequence obtained from a 129Sv mouse genomic library (Stratagene). No negative selection was used. The targeting vector was electroporated into E14.1 ES cells and positive clones were injected into C57BL blastocysts. Male chimeras were mated to C57BL females. Intercrossing resulted in homozygous PDGF-A$^{-/-}$ offspring of mixed 129Sv/C57BL genetic background. The null mutation was confirmed by Southern analysis.

Phenotype

PDGF-A$^{-/-}$ mice died at various time points during pre- and post-natal development. As 129Sv/C57BL hybrids, about 50% of PDGF-A$^{-/-}$ embryos died before E10. Embryos surviving this restriction point developed outwardly normally until birth, at which time they were slightly growth retarded (20% weight reduction). About half of these died within the first 1–2 post-natal days, whereas the remaining mutants lived for an additional 2–6 weeks. The growth deficiency became more severe post-natally, and at 3 weeks of age, mutants were generally half to a third of the weight of control littermates. These animals developed severe pulmonary emphysema due to failure of the development of alveolar smooth muscle cells. Analysis of the ontogeny of such cells in PDGF-A$^{-/-}$ mice revealed that their precursors fail to spread distally along the lung epithelium branches at the canalicular stage of lung development (approximately E17). At this time, the normal lung epithelium expresses PDGF-A and specific clusters of lung mesenchymal cells express PDGFαR. PDGF-A$^{-/-}$ mice also exhibited oligodendrocyte deficiency, intestinal villus dysmorphogenesis, and a lack of white adipose tissue. The pathogeneses of these defects are currently under investigation.

Comments

PDGF-A is essential for alveolar smooth muscle cell ontogeny. The phenotype of PDGF-A$^{-/-}$ mice is consistent with a role for PDGF-A in the spreading of PDGFαR-positive mesenchymal cells along PDGF-A-producing epithelial surfaces. The PDGF-A$^{-/-}$ mouse develops severe and general pulmonary emphysema with a complete lack of alveolar septum formation. The pathogenesis of this phenotype is distinct from that of human pulmonary emphysema which results from alveolar septum destruction. Analogies between PDGF-A- and PDGF-B-null phenotypes are apparent, in that both involve the aberrant distal spreading of smooth muscle, or myofibroblast, progenitors associated with PDGF ligand-producing epithelial (PDGF-A) or endothelial (PDGF-B) structures.

Acknowledgements
Christer Betsholtz
Department of Medical Biochemistry, University of Göteborg, Göteborg, Sweden

References

[1] Boström, H. et al. (1996) Cell 85, 863–873.
[2] Lindahl, P. et al. (1997) Development 124, 3943–3953.

PDGF-B

Other names
Platelet-derived growth factor B, c-sis proto-oncogene

Gene symbol
Pdgfb

Accession number
MGI: 97528

Area of impact
Hematopoiesis, development

General description

PDGF-B is one of two polypeptide chains constituting members of the dimeric PDGF family of proteins (PDGF-AA, AB and BB). The PDGF-A and PDGF-B polypeptides are derived from separate genes. The PDGF-B gene consists of seven exons, of which exons 4 and 5 encode the majority of the mature PDGF-B chain. PDGF-B is expressed in the developing embryo in certain vascular endothelia and in megakaryocytes, and in placenta trophoblasts. In the adult, PDGF-B is expressed in the platelets (megakaryocytes), in activated monocytes, in macrophages, and in certain populations of neurons. Biologically active PDGF-B occurs as homodimers (PDGF-BB), or as heterodimers with PDGF-A (PDGF-AB), and interacts with receptor tyrosine kinases PDGFαR and PDGFβR. PDGF receptors occur on various connective tissue cells. The physiological function of PDGF is assumed to include connective tissue development and repair. PDGF is also implicated in pathological processes involving abnormal connective tissue cell proliferation.

KO strain construction

The replacement-type targeting vector was designed to delete part of exon 3 and all of exon 4 of the *Pdgfb* gene. A 1.5 kb fragment encompassing exon 4 (encoding the N-terminal half of PDGF-B) was replaced by a PGK*neo* cassette and flanked by a total of 11.5 kb of homologous *Pdgfb* gene sequence obtained from a 129Sv mouse genomic library (Stratagene). No negative selection was used. The targeting vector was electroporated into E14.1 ES cells and positive clones were injected into C57BL blastocysts. Male chimeras were mated to C57BL females. Intercrossing resulted in homozygous PDGF-$B^{-/-}$ offspring of mixed 129Sv/C57BL genetic background. The null mutation was confirmed by Southern analysis.

Phenotype

PDFG-$B^{-/-}$ mice died perinatally following the abrupt onset of bleeding in multiple tissues. Bleeding was preceded by capillary microaneurysm formation. There was a complete loss of microvascular pericytes in tissues where bleeding occurred, such as in the brain and skin, but also in other tissues, such as the heart. There was also the loss of kidney glomerulus mesangial cells, leading to replacement of the glomerular capillary tuft by a microaneurysm-like structure. Loss of mesangial cells has also been reported in PDGFβR null mice. Additional phenotypes included subcutaneous edema, lack of urine collection in the urinary bladder, heart dilation and hypotrophy of the myocardium, and dilation of large arteries.

Comments

PDGF-B is crucial for the establishment of certain renal and circulatory functions. Since PDGF-B is expressed by vascular endothelial cells, and the PDGFβR by vascular wall precursors, all, or part, of the PDGF-B$^{-/-}$ phenotype is likely to reflect the loss of paracrine signaling in the developing vascular wall. The loss of pericytes is consistent with the failure of sprouting endothelium to attract PDGFβR-positive pericyte progenitors in the course of new capillary formation by angiogenic mechanisms.

Loss of pericytes, and microaneurysm formation and rupture, occur in the retina as late complications to type 1 diabetes mellitus. PDGF-B$^{-/-}$ embryos may therefore provide a model in which some aspects of diabetic microangiopathy could be studied.

Acknowledgements

Christer Betsholtz
Department of Medical Biochemistry, University of Göteborg, Göteborg, Sweden

References

1 Levéen, P. et al. (1994) Genes Dev. 8, 1875–1887.
2 Lindahl, P. et al. (1997) Science 277, 242–245.

Other names
Pancreatic and duodenal homeobox gene 1, insulin promoter factor 1 (Ipf1) in the mouse, somatostatin transcription factor (Stf-1) in the rat, homeobox gene 8 (XlHbox8) in *Xenopus*, islet duodenum homeobox gene 1 (Idx-1) in the rat

Gene symbol
Pdx1

Accession number
MGI: 102851

Area of impact
Hormone, development

General description

Pdx1 is the mammalian homolog of a homeobox gene, *XlHbox8*, first cloned in *Xenopus*. PDX-1 expression is restricted to endodermal derivatives of the posterior foregut (adult pancreatic β cells, a subset of cells of the duodenum and distal stomach). During embryogenesis, PDX-1 expression is initiated at the 10–12 somite stage in the dorsal and ventral walls of the primitive foregut, at the positions from which the pancreas later forms. Although PDX-1 is expressed in all pancreatic cell types during early development, its expression becomes restricted to adult insulin-secreting β cells of the pancreatic islets. PDX-1 expression is also maintained in the duodenal epithelium in adults. PDX-1 has been shown to bind and transactivate the insulin and somatostatin promoters.

KO 1 strain construction[1]

Ipf1/PDX-1 contains two exons. Two different constructs were generated, both of which deleted the second exon containing the DNA-binding homeodomain. (1) XBko contains 7 kb of 5′ homology (*Xba*I/*Bam*HI) and 1.5 kb of 3′ homology (*Pst* I/*Xba*I). The second exon was replaced with an MC1 neomycin-resistance cassette. (2) XSlacZko contains 9 kb of 5′ homology from the 5′ *Xba*I site to the *Sma*I site within the homeobox, and the same 1.5 kb of 3′ homology as XBko. A nuclear targeted β-galactosidase cassette was fused in-frame with *Pdx1*. Both constructs used 5′ PGKII-*tk* and 3′ MC1-*tk* cassettes. ES cells were 129-derived R1 cells. ES cells were injected into C57BL/6 blastocysts. Male chimeras were bred to Black Swiss females. Heterozygous (Black Swiss × 129) offspring were interbred to produce homozygous animals. Lines of mice corresponding to XBko and XSlacZko were generated.

KO 2 strain construction[2]

A mouse *Pdx1* gene targeting construct was made by removing a 3.1 kb fragment which included exon 2. The second exon was replaced with the gene for neomycin resistance, leaving 0.7 and 3.4 kb homology 5′ and 3′ to the

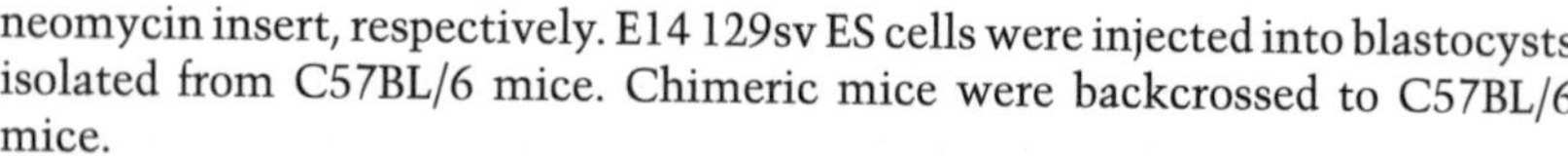

neomycin insert, respectively. E14 129sv ES cells were injected into blastocysts isolated from C57BL/6 mice. Chimeric mice were backcrossed to C57BL/6 mice.

Phenotype

The KO strain 1 mutations of *Pdx1* were neonatal lethal when homozygous and these mice selectively lacked a pancreas. There was no phenotype associated with animals heterozygous for either KO strain 1 targeting construct. The same phenotype was observed for both homozygous XBko and XSlacZko$^{-/-}$ mice. The mutant pups survived fetal development but died within a few days after birth. By one day post-partum, mutant animals showed growth retardation and dehydration. This is likely due to malnutrition resulting from the lack of digestion in the absence of a pancreas and the presence of malformations of the stomach–duodenal junction. The gastrointestinal tract and all other internal organs were macroscopically normal in appearance. No pancreatic tissue and no ectopic expression of insulin or pancreatic amylase could be detected in mutant embryos and neonates.

In KO strain 2 PDX-1$^{-/-}$ mice, the early inductive events leading to the formation of the pancreatic buds and the appearance of the early insulin and glucagon cells still occurred. However, the subsequent morphogenesis of the pancreatic epithelium and the progression of differentiation of the endocrine cells were arrested in KO strain 2 PDX-1$^{-/-}$ embryos. In contrast, the pancreatic mesenchyme grew and developed normally, both morphologically and functionally, and independently of the epithelium. The pancreatic epithelium was unable to respond to the mesenchymal-derived signal(s) which normally promote pancreatic morphogenesis. In addition to the defects in pancreas development, the rostral duodenum showed a local absence of the normal columnar epithelium, villi and Brunner's glands, which were replaced by a GLUT2-positive cuboidal epithelium resembling the bile duct lining. Distal to this abnormal epithelium, the numbers of enteroendocrine cells – specifically serotonin, secretin and CCK – in the duodenal villi were greatly reduced.

Comments

The phenotype observed in PDX-1-null mice shows that PDX-1 is not required for the initial generation and outgrowth of pancreatic progenitors but is necessary for continued outgrowth, differentiation and morphogenesis of the pancreas. The lack of a pancreas in PDX-1$^{-/-}$ mice is due to a defect in the pancreatic epithelium. PDX-1 seems to function both in the regionalization of the primitive gut endoderm and in the maturation of the pancreatic β-cell. Moreover, PDX-1 has an essential role in the morphogenesis and differentiation of the rostral duodenum.

Acknowledgements
Christopher V.E. Wright
Howard Hughes Medical Institute, Vanderbilt University School of Medicine, Nashville, TN, USA

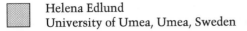

Helena Edlund
University of Umea, Umea, Sweden

References

[1] Offield, M.F. et al. (1996) Development 122, 983–995.
[2] Jonsson, J. et al. (1994) Nature 371, 606–609.
[3] Wright, C.V.E. et al. (1988) Development 105, 787–794.
[4] Guz, Y. et al. (1995) Development 121, 11–18.
[5] Peshavaria, M. et al. (1994) Mol. Endocrinol. 8, 806–816.
[6] Ahlgren, U. et al. (1996) Development 122, 1409–1416.

Other names
P1, cytolysin, pore-forming protein

Gene symbol
Pfp

Accession number
MGI: 97551

Area of impact
Immunity and inflammation, T cells

General description

Perforin is a cytolytic protein expressed by cytotoxic T cells and NK cells. In these cells, it is mainly located in cytoplasmic granules, which are directionally exocytosed during conjugation with a target cell. Perforin shares homology with the late component of the complement system C9, and, like C9, polymerizes in the target cell membrane to form pores. These pores, in combination with other granule components, induce the rapid death of target cells in a way that resembles apoptosis. The perforin gene spans a region of approx. 1.6 kb and contains three exons, only the last two of which code for the protein.

KO strain 1 construction[1]

The targeting construct was purely insertional and was obtained by inserting the neomycin-resistance gene into the *Bst*EII site of a 3.3 kb genomic *Eco*RV fragment containing most of exon 3 and a part of the preceding intron. The linearized construct was used to electroporate C57BL/6-derived BL6/III ES cells. Targeted ES cell clones were injected into BALB/c blastocysts and transferred to CD1 foster mothers. Male chimeras were crossed with C57BL/6 females. Because of the use of C57BL/6-derived ES cells, these perforin-deficient mice had a homogeneous C57BL/6 genetic background. Immunocytological staining of activated $CD8^+$ T cells confirmed the absence of perforin in homozygous mutant mice.

KO strain 2 construction[2]

The targeting construct was made by subcloning a 3.4 kb *Pst*I fragment derived from a genomic cosmid clone containing the complete perforin gene. The genomic fragment contained part of exon 2 and the complete exon 3. Exon 3 was interrupted by insertion of a neomycin cassette which disrupted the putative transmembrane domain of perforin, thought to consist of two amphipathic α helices spanning amino acid residues 167–221. An HSV-*tk* cassette was placed 3′ of the perforin sequences.

The targeting DNA was introduced into ES cells (D3, S129) by electroporation. Positive ES clones were injected into C57BL/6J blastocysts. Chimeras were crossed to obtain homozygous perforin-deficient offspring. The null mutation was confirmed by PCR analysis and Northern blotting of mRNA from activated spleen cells. Reduced amounts of perforin mRNA were present in perforin$^{+/-}$ heterozygotes. No perforin transcripts were detected in perforin$^{-/-}$ mice.

KO strain 3 construction[3]

The targeting vector was designed to eliminate exon 2 (containing the initiation codon and the cytolytic domain) of the perforin gene by replacing it with a neomycin-resistance cassette. The diphtheria toxin active subunit gene (DT-A) was appended 3′ of the replacement sequences. The original mouse genomic library was derived from ICR mouse liver. The targeting construct was electroporated into E14 ES cells. A positive clone was injected into blastocysts of RAG-2$^{-/-}$ mice to obtain chimeras. Analyses were carried out on CD8^{+} CTL lines established from spleen cells of perforin$^{-/-}$RAG-2$^{-/-}$ chimeric mice. The null mutation was confirmed by Southern blotting and by immunoblot.

KO strain 4 construction[4]

The targeting construct was created by inserting the neomycin-resistance gene fragment from pMC1*neo*polyA into the *Sma*I site of exon 2 of the mouse genomic perforin gene. The vector also contained the HSV-*tk* gene flanking the 3′ end of the perforin sequences. AB-1 ES cells were electroporated with the construct and positive clones were injected into C57BL/6 blastocysts followed by transfer to (C57BL/6 × CBA) F1 foster females. Male chimeras were bred to C57BL/6 females to obtain germ line transmission of the mutated perforin allele. The null mutation was confirmed by immunoblot of activated splenocytes.

Phenotype

The analyses of perforin-deficient mice generated by all four groups[1–4] were very compatible. Perforin$^{-/-}$ mice were viable, fertile, normal in appearance and had lymphoid organs of normal size. They displayed normal numbers of CD8^{+} T cells and NK-marker-positive cells. However, T cell-mediated cytotoxic activity against Fas-negative target cells was completely absent. Cytotoxic activity against other target cells was mediated by the interaction of Fas ligand with Fas. NK cell-mediated cytotoxicity was completely absent, even against Fas-expressing target cells. Perforin-deficient mice were not able to clear lymphocytic choriomeningitis virus (LCMV) infection; however, vaccinia virus, vesicular stomatitis virus and Semliki Forest virus were controlled normally. Upon LCMV infection, perforin-deficient mice developed weight loss and eventually died. Resistance against several, but not all, injected MHC class I-expressing, syngeneic tumor cell lines was reduced. The capacity of NK cells to eliminate injected MHC class I-negative RMA-S cells was also diminished. Furthermore, CD8^{+} T cell-mediated control of *Listeria monocytogenes* was almost completely abolished in the absence of perforin. The role of perforin-dependent cytotoxicity in rejection of allogeneic transplants was surprisingly limited: fully allogeneic heart grafts were rejected by perforin-deficient mice as efficiently as by normal control mice. Only grafts with a

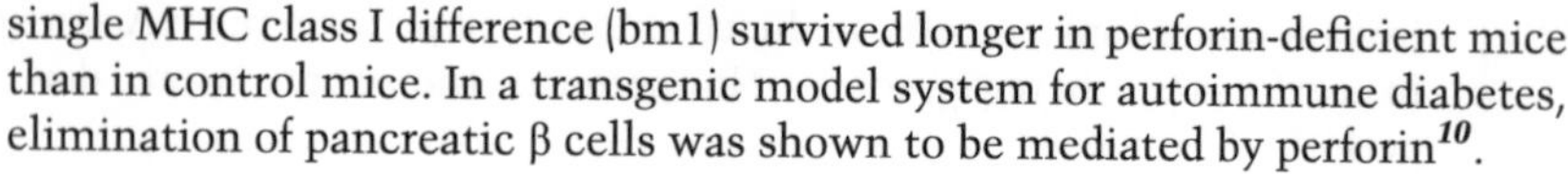

single MHC class I difference (bm1) survived longer in perforin-deficient mice than in control mice. In a transgenic model system for autoimmune diabetes, elimination of pancreatic β cells was shown to be mediated by perforin[10].

Comments

Perforin-deficient mice were crucial for the resolution of the controversial role of perforin in cell-mediated cytotoxicity. It is now well-established that T cells kill via both a perforin-dependent and a Fas-dependent pathway. The latter pathway can be activated only on Fas-expressing target cells. NK cells mediate cytotoxicity mainly via the perforin-dependent pathway. In addition, since T cell ontogeny and activation occur normally in perforin-deficient mice, these mutants were and will continue to be useful in defining the role of contact-dependent cytotoxicity vs. soluble factors in immune responses of interest.

Acknowledgements
David Kägi
Ontario Cancer Institute and Amgen Institute, Toronto, ON, Canada
Jurg Tschopp
Institute of Biochemistry, University of Lausanne, Epalinges, Switzerland

References

1 Kägi, D. et al. (1994) Nature 369, 31–37.
2 Lowin, B. et al. (1994) Proc. Natl Acad. Sci. USA 91, 11571–11575.
3 Kojima, H. et al. (1994). Immunity 1, 357–364.
4 Walsh, C.M. et al. (1994) Proc. Natl Acad. Sci. USA 91, 10854–10858.
5 Braun, M.Y. et al. (1996) J. Exp. Med. 183, 657–661.
6 Lowin, B. et al. (1994) Nature 370, 650–652.
7 Kägi, D. et al. (1995) Eur. J. Immunol. 25, 3256–3262.
8 van den Broek, M.F. et al. (1995) Eur. J. Immunol. 25, 3514–3516.
9 Schulz, M. et al. (1995) Eur. J. Immunol. 25, 474–480.
10 Kägi, D. et al. (1996) J. Exp. Med. 183, 2143–2152.

Other names
Paired-like homeobox 2a

Gene symbol
Pmx2a

Accession number
MGI: 97713

Area of impact
Neurology

General description

Phox2a is a neuron-specific homeobox gene of the paired superclass. It is expressed transiently or permanently in all (nor)adrenergic cells of the PNS and CNS and has been shown to regulate *in vitro* and *in vivo* the expression of dopamine β-hydroxylase (DBH), the key enzyme in norepinephrine synthesis. Phox2a expression extends beyond noradrenergic cells and delineates synaptic pathways of the autonomic nervous system.

KO strain construction

A 3.5 kb fragment encompassing the homeobox and all downstream sequences of the Phox2a gene were replaced by the *neo* gene driven by an HSV-*tk* promoter. 1.2 kb and 9 kb of homology were used 5′ and 3′ of *neo*, respectively. The HSV-*tk* gene was inserted at the 3′ end for double selection. The mutation was introduced into 129Ola-derived E14 ES cells. Chimeras were generated by injection into C57BL6 blastocysts and F2 progeny were analyzed on a heterogeneous 129/C57 background.

Phenotype

Phox2a$^{+/-}$ mice were viable and fertile. Phox2a$^{-/-}$ mice were born in a Mendelian ratio, but died within 24 hours after birth. In the CNS, the most rostral areas of Phox2a expression were absent in the mutants: the oculomotor and trochlear motor nuclei, and the locus coeruleus, the main noradrenergic center of the CNS. These regions were never detected with an independent marker, suggesting either that these cells changed their fate or that they died very early in the absence of Phox2a. The other areas of Phox2a expression in the CNS were apparently spared in the mutants. In the PNS, the enteric nervous system did not show any defect. The sympathetic nervous system was mostly normal, except for a morphological alteration in the superior cervical ganglion. By contrast, all parasympathetic ganglia of the head were missing, while paracardiac ganglia were not affected. The three cranial sensory ganglia (VIIth, IXth and Xth) that express Phox2a were severely atrophic at birth, while a normal complement of cells was found in these ganglia at E10.5. These ganglionic cells failed to show the normal transient expression of the DBH gene. They also failed to express the tyrosine kinase Ret, a subunit of the receptor for the neurotropic factor GDNF on which these cells are known to be dependent. The control of Ret expression by Phox2a in cranial ganglionic cells could explain their death by apoptosis at around E11.5. Thus, Phox2a appears to regulate both differentiation traits (i.e. the noradrenergic phenotype) and the survival of these cells.

Comments

The Phox2a$^{-/-}$ phenotype affects only some areas of Phox2a expression. A second homeoprotein with an identical homeodomain, called Phox2b, has a widely overlapping expression domain[2], suggesting a partial redundancy between Phox2a and Phox2b.

Acknowledgements
Jean-Francois Brunet
JFB IBDM, Marseille, France

References
1 Morin, X. et al. (1997) Neuron 18, 411–423.
2 Pattyn, A. et al. (1997) Development 124, 4065–4075.
3 Valarchè, I. et al. (1993) Development 119, 881–896.
4 Tiveron, M.-C. et al. (1996) J. Neurosci. 16, 7649–7660.

Other names
Double-stranded RNA-dependent protein kinase

Gene symbol
Prkr

Accession number
MGI: 97761

Area of impact
Immunity and inflammation, signal transduction

General description

PKR is an interferon-induced dsRNA-activated protein kinase of 65 kDa. This kinase is expressed at basal levels in most cells and is induced by interferon through the Jak/Stat pathway. Following binding to dsRNA, PKR is autophosphorylated and its kinase function activated. PKR phosphorylates the α subunit of the translation initiation factor eIF2, thereby inhibiting protein synthesis. PKR can also phosphorylate the inhibitor of NFκB, IκB, resulting in NFκB activation. Inactive mutants of PKR can act as dominant oncogenes in NIH 3T3 cell transformation assays.

KO strain construction

A 2.0 kb *Bal*I/*Pst* fragment containing parts of exons 2 and 3 (including the coding sequence from the first ATG up to codon 66 and the intron in between) was replaced by the 2.0 kb pPGK*neo*UMS cassette. A HSV-*tk* gene was inserted into the vector *Not*I site downstream of this segment to allow for counter-selection. The targeting vector was linearized at the *Sac*II site proximal to the long arm of the PKR genomic sequence. The targeting vector was electroporated into GS-1 ES cells derived from 129Sv(ev) blastocysts. Cells carrying the PKR mutation were injected into 3.5-day-old blastocysts of C57BL/6J mice and implanted into ICR foster mice. The resulting male chimeras gave germ line transmission of the mutated PKR allele when mated with C57BL/6J females. Heterozygous mice were intercrossed to generate homozygous PKR$^{-/-}$ offspring with a mixed 129Sv(ev) × C57BL/6J genetic background. The null mutation was confirmed by Southern blot, RT-PCR analysis, Northern blot, and assay of PKR autophosphorylation.

Phenotype

PKR$^{-/-}$ mice were physically normal, showed no alterations in behavior, were fertile and produced normal litters. No gross anatomical or histological changes to internal organs were observed. The induction of type 1 IFN genes by polyIC and virus was unimpaired but the anti-viral response to IFN-γ and polyIC was diminished. In mutant embryo fibroblasts, the induction of type I IFN by polyIC and the activation of NF-κB was impaired but could be restored by priming with IFN. IRF-1 and Gbp promoter constructs were unresponsive to

IFN-γ or polyIC, but the response could be restored by cotransfection with PKR. The activation of IRF-1 and NF-κB was diminished. The expression of several genes, including class I MHC, iNOS, the JC and RANTES chemokines, E-selectin and Fas, was found to require PKR for induction by IFN-γ or polyIC. Fibroblasts from PKR$^{-/-}$ mice were resistant to apoptotic cell death induced by polyIC, TNFα or LPS, but not to DNA-damaging agents. Although the peripheral lymphoid compartments in these mice appeared normal, they exhibited an exaggerated and prolonged contact hypersensitivity response to hapten, resulting from a defect in Th1 and Th2 cytokine production and response.

Comments

PKR$^{-/-}$ mice have proved useful in uncovering a selective role for PKR in signal transduction in response to specific inducers. PKR acts as a signal transducer for IFN-stimulated genes dependent on the transcription factors IRF-1 and NF-κB. However, PKR is not essential for responses to polyIC, and a polyIC-responsive mechanism independent of PKR is induced by IFN.

Because of their subtle immune phenotype, PKR$^{-/-}$ mice will be useful for investigating mechanisms of Th1 and Th2 switching. Although transdominant mutants of PKR are transforming oncogenes in NIH 3T3 assays, PKR KO mice on 129Sv and C57BL/6J backgrounds showed no increase in tumor incidence, indicating that PKR is not a tumor suppressor gene.

Acknowledgements
Bryan R.G. Williams
Department of Cancer Biology, The Cleveland Clinic Foundation, Cleveland, OH, USA

References

1 Yang, Y.-L. et al. (1995) EMBO J. 14, 6095–6106.
2 Kumar, A. et al. (1997) EMBO J. 16, 406–416.
3 Der, S. et al. (1997) Proc. Natl Acad. Sci. USA 94, 3279–3283.

Plakoglobin

Other names
γ-Catenin

Gene symbol
Jup

Accession number
MGI: 96650

Area of impact
Cardiology

General description

Plakoglobin (γ-catenin) is the first member discovered of the *armadillo* family of proteins which are characterized by domains composed of variable numbers of *arm* repeats. Plakoglobin consists of 13 *arm* repeats flanked by unique N- and C-terminal sequences, and is a constitutive component of plaques associated with diverse adhering junctions. These include (a) desmosomes, which anchor intermediate-sized filaments (IF), and (b) various microfilament anchoring junctions, such as the *zonulae adhaerentes* of epithelia and the belt plaques of endothelial adhering junctions. Plakoglobin is specifically bound to a defined domain in the C-terminal part of certain cadherins such as desmogleins desmocollins and classical cadherins.

KO strain 1 construction[1]

A plakoglobin genomic clone was isolated from a 129Sv mouse genomic library, and a targeting construct was assembled from a 11 kb genomic fragment. In this vector, a large part of exon 3, the following intronic sequences and 5′ region of exon 4 (encoding amino acids 70–160) were replaced by a neomycin gene cassette inserted in the same transcriptional orientation. An HSV-*tk* cassette was inserted at the 3′ end of the construct. The linearized targeting vector was electroporated into E14.1 ES cells, which were selected using G418 and gancyclovir. ES cell clones harboring the desired integration event were injected into C57BL/6 recipient blastocysts and these were transferred into pseudopregnant NMRI females to generate chimeric mice. Male offspring exhibiting extensive coat color chimerism were mated to C57BL/6 females, and genotypes were identified by PCR and Southern blot analysis.

KO strain 2 construction[2]

A 4 kb fragment including exons 3 and 4 and parts of exon 5 were replaced by pPGKbgeobpA. An HSV-*tk* gene was inserted at the 3′ end. R1 ES cells were electroporated and injected into C57BL/6 × DBA/2 blastocysts. Chimeric mice were backcrossed into 129Sv and C57BL/6 strains.

Phenotype

Homozygous plakoglobin$^{-/-}$ mutant animals die between days 12 and 16 of embryogenesis due to defects in heart function. Often, heart ventricles burst and blood flooded the pericard. This tissue instability correlated with the absence of desmosomes in heart, but not in epithelial organs. Instead, extended adherens junctions were formed in the heart, which contained desmosomal proteins. The general morphology of cardiac structures was unaltered in plakoglobin$^{-/-}$ embryos: atrial and ventricular chambers were present, and the endocardial cushions and ventricular trabecules were well-developed. However, the morphology of the intercalated discs was grossly altered as they did not show differentiation of desmosomes and adherens junctions. Typical desmosomes were no longer detectable, and instead adherens junctions with particularly prominent plaques were seen. Immunoelectron microscopy of hearts of plakoglobin$^{-/-}$ mice showed that desmoplakin occurred in all plaque-bearing structures, i.e. junctions resembling adherens junctions with myofibrillar bundles attached. This new form of junction varied in size, and included some extremely large ones with axes of up to 4.5 μm.

Surprisingly, desmosomes in epithelial cells, for instance in skin and gut, did form appropriately in mutant mice. The data demonstrate that functional heart desmosomes and proper sorting of adherens junction proteins are required during the second half of gestation, when the embryo becomes dependent on its own blood circulation and therefore on a functional cardiovascular system[1]. KO strain 2 plakoglobin$^{-/-}$ mice[2] developed skin blistering and subcorneal acantholysis reminiscent of human blistering disease. Moreover, in a C57BL/6 background some KO strain 2 plakoglobin$^{-/-}$ mice were able to progress further in development and died around birth.

Comments

More subtle alteration of the plakoglobin gene than the null mutation described here, or other types of interferences with plakoglobin, may impair heart function and play a role in human heart disease.

Acknowledgement
Walter Birchmeier and Patricia Ruiz
Max-Delbrück Center for Molecular Medicine, Berlin, Germany
Rolf Kemler
Max Planck Institut für Immunbiologie Stübeweg, Freiburg, Germany

References

[1] Ruiz, P. et al. (1996) J. Cell Biol. 135, 215–225.
[2] Bierkamp, C. (1996) Dev. Biol. 180, 780–785.
[3] Cowin, P. et al. (1986) Cell 46, 1063–1073.
[4] Garrod, D.R. (1993) Curr. Opin. Cell Biol. 5, 33–40.
[5] Hulsken, J. et al. (1994) J. Cell Biol. 127, 2061–2069.
[6] Witcher, L.L. et al. (1996) J. Biol. Chem. 271, 10904–10909.

Other names
Phospholamban

Gene symbol
Pln

Accession number
MGI: 97622

Area of impact
Cardiovascular

General description

PLB regulates the Ca^{2+}-ATPase in the cardiac muscle sarcoplasmic reticulum. It may also have a role in inotropic responses of cardiomyocytes to β-adrenergic stimulation. β-Adrenergic stimulation leads to increased cAMP levels and subsequent phosphorylation of regulatory substrates including PLB. Dephosphorylated PLB inhibits Ca^{2+}-ATPase activity. Phosphorylation of PLB releases this inhibitory effect and leads to an increased affinity of the sarcoplasmic Ca^{2+} pump for Ca^{2+}. PKC, cAMP-dependent protein kinase, and Ca^{2+}/calmodulin-dependent kinase can phosphorylate PLB at distinct serine and threonine sites.

KO strain construction

Parts of exon 2 containing the PLB-coding region were replaced by a pMC1*neo* cassette. HSV-*tk* was placed at the 3′ end. D3 ES cells derived from 129 mice were injected into C57BL/6J blastocysts.

Phenotype

PLB-null mice appeared healthy and had no gross abnormalities. However, these mice displayed enhanced myocardial contractility and intraventricular pressures without changes in heart rate. The affinity of the sarcoplasmic Ca^{2+}-ATPase to Ca^{2+} was enhanced and β-agonist stimulation (isoproterenol) was lost. PLB also regulated the sarcoplasmic Ca^{2+} pump in smooth muscle cells of the mouse aorta and could regulate KCl and receptor-mediated contractility in vascular smooth muscle[2].

Comments

PLB seems to function as a repressor of myocardial and vascular smooth muscle contractility, and is involved in the contractile response to β-adrenergic stimulation.

References
[1] Luo, W. et al. (1994) Circ. Res. 75, 401–409.
[2] Lalli, J. et al. (1997) Circ. Res. 80, 506–513.

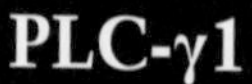

PLC-γ1

Other names
Phospholipase C gamma 1

Gene symbol
Plcg1

Accession number
MGI: 97615

Area of impact
Signal transduction

General description

In mammalian cells, there are 10 known phospholipase C (PLC) gene products, which are classified as PLC-β, PLC-γ and PLC-δ isoforms. These enzymes mediate the hydrolysis of phosphatidylinositol 4,5-bisphosphate to inositol 1,4,5-trisphosphate and diacylglycerol. The PLC-γ isoform family consists of two isozymes, PLC-γ1 and PLC-γ2, both of which have unique SH2 and SH3 domains and high sequence homology, but which exhibit different patterns of expression and distinct chromosome locations. They are activated by growth factor receptor tyrosine kinase and non-receptor tyrosine kinase.

KO strain construction

Two constructs, the targeting vectors (TV)-I and II, were generated by using the pPNT vector. In both vectors, genomic sequences encoding the X domain and both SH2 domains of PLC-γ1 were replaced with PGK*neo*. PGK*tk* was included for negative selection. TV-II also contained a *lacZ* reporter fused in-frame with the PLC-γ1 N-terminus. TV-I and TV-II were used to transfect the R1 (129) and TL1 ES cell lines, respectively. Chimeras were inbred onto the 129SvJ strain or onto the hybrid backgrounds 129SvJ × C57BL/6, 129SvJ × Black Swiss and 129SvJ × CD1.

Phenotype

The embryos with homologous disruption of *Plcg1* died at approximately embryonic day 9.0. Histological analysis indicates that PLC-γ1$^{-/-}$ embryos appeared normal at E8.5 but failed to continue normal development and growth beyond E8.5–9.0. This phenotype was consistent in the four different genetic backgrounds in which *Plcg1* was disrupted by the TV-I or TV-II targeting vectors. The mechanism of the lethal phenotype of disrupted *Plcg1* remains unclear. Heterozygotes were normal.

Comments

PLC-γ1 is essential for normal embryonic development. The absence of PLC-γ1 does not alter other signal transducing pathways. Other PLC genes do not compensate for the loss of PLC-γ1.

Acknowledgements
Graham Carpenter
Department of Biochemistry, Vanderbilt University School of Medicine, Nashville, TN, USA

Reference

[1] Ji, Q.S. et al. (1997) Proc. Natl Acad. Sci. USA 94(7), 2999–3003.

Other names
Plasminogen, plasmin, angiostatin

Gene symbol
Plg

Accession number
MGI: 97620

Area of impact
Hormone

General description

The plasminogen activation system is an intricate system of serine proteases, protease inhibitors and protease receptors, whose ultimate function is to govern the conversion of the abundant plasma protease zymogen plasminogen (Plg) to the active protease plasmin. Plasmin has a broad substrate specificity, and has been implicated in the activation of latent growth factors and procollagenases, degradation of extracellular matrix, and fibrin clearance in the context of physiologic and pathologic tissue remodeling, thrombus dissolution and cell migration.

KO strain 1 construction[1]

A PGK*hprt* minigene cassette flanked by a 4 kb fragment from the 5′ flanking region of the plasminogen gene, a 1 kb fragment from intron 2 sequences, and an HSV-*tk* cassette were used in the targeting vector. This vector deletes 9 kb of promoter and exon 1 and 2 sequences. Hprt-deficient E14TG2 ES cells were targeted and injected into C57BL/6 blastocysts. Mice were inbred into 129 and C57BL/6 backgrounds.

KO strain 2 construction[2]

Exons 15–17 containing the catalytic site of Plg (amino acids His605 and Asp648) were replaced by a PGK*neo* cassette in D3 ES cells. An HSV-*tk* gene was placed at the 3′ end of the construct. Targeted ES cells were injected into C57BL/6 blastocysts.

Phenotype

Plg$^{-/-}$ mice had no detectable Plg mRNA, Plg protein or Plg enzymatic activity. Plg$^{-/-}$ mice were born, survived to adulthood, and reproduced. Embryonic development of Plg$^{-/-}$ mice continued to term in the absence of sibling-derived or maternal Plg, demonstrating that loss of the plasminogen-activation system is compatible with development and reproduction. However, Plg$^{-/-}$ mice develop a plethora of progressive and ultimately fatal pathologies. These included severe wasting, rectal prolapse, and spontaneous fibrin-rich, ulcerated and inflammatory lesions of epithelial surfaces of the gastrointestinal, reproductive and respiratory tracts, the cornea, and the conjunctiva (ligneous conjunctivitis). Fibrin-rich lesions also developed in the liver, lung, adrenals, pancreas, thymus, ovaries and other organs, often with associated necrosis. $Plg^{-/-}$ mice presented delayed and aberrant skin and corneal wound healing after experimental injury, as a consequence of impaired dermal and corneal keratinocyte migration[3]. Mammary gland involution was impaired in $Plg^{-/-}$ mice, reducing the reproductive success of multiparous females. Moreover, $Plg^{-/-}$ mice had a defect in spontaneous clot dissolution. The phenotype of $Plg^{-/-}$ mice was comparable to ligneous conjunctivitis, an inherited, often fatal, autosomal recessive disease in humans, that has recently been associated with Plg deficiency. In experimental pathological settings, $Plg^{-/-}$ mice displayed accelerated atherosclerosis in a hypercholesterolemic mouse background, but demonstrated delayed dissemination of metastatic lung carcinoma, melanoma and mammary adenocarcinoma, and are resistant to excitotoxic neuronal degeneration. Loss of fibrinogen rescues $Plg^{-/-}$ mice from all the spontaneous pathologies documented so far, restores normal life-expectancy, and corrects skin and corneal wound healing times, indicating that fibrinolysis is the essential, non-redundant physiological function of the plasminogen activation system[4].

Comments

Plg has a central role in multiple post-developmental physiological processes that involve tissue remodeling and cell migration. The non-redundant function of Plg in these processes may be restricted to the timely degradation of intravascular and extravascular fibrin(ogen). Plg has a broad substrate range in pathologic tissue remodeling. $Plg^{-/-}$ mice resemble t-$PA^{-/-}$u-$PA^{-/-}$ mice.

Acknowledgements
Thomas H. Bugge and Jay L. Degen
Childrens Hospital Research Foundation, Cincinnati, OH, USA

References

1 Bugge, T.H. et al. (1995) Genes Dev. 9, 794–807.
2 Ploplis, V.A. et al. (1995) Circulation 92, 2585–2593.
3 Romer, J. et al. (1996) Nature Med. 2, 287–292.
4 Bugge, T.H. et al. (1996) Cell 87, 709–719.

Other names
Proteolipid protein, DM20, DXNds2, DXMit9

Gene symbol
Plp

Accession number
MGI: 97623

Area of impact
Neurology

General description

The isoproteins PLP and DM20 are synthesized by oligodendrocytes as the two major integral proteins of myelin membranes of the CNS. The *Plp* gene has a length of 17.4 kb and is located on chromosome Xq22.3. It contains seven exons encoding 276 amino acids. DM20 is an alternative splice product of the PLP primary transcript in which 105 bp at the 3′ end of exon 3 are deleted due to the activation of a cryptic splice site.

KO strain construction[1]

The targeting construct contained 6.7 kb of murine genomic *Plp* sequence with a deletion of 105 bases at the 3′ end of exon 3. The *neo* gene (pMCI*neo*polyA, Stratagene) was inserted in reverse orientation into intron 3. The gene (pIC19rMC1tk) was added to the 5′ end of the vector. E14 ES cells (129) were injected into C57BL/6J blastocysts. Chimeric offspring were intercrossed with C57BL/6J mice.

Double KO mutant strain construction[2]

PLP-deficient mice were crossed with heterozygous shiverer mice ($MBP^{+/-}$ [3]). The resulting F1 generation was intercrossed to obtain mice homozygous for the PLP/MBP deficiency ($PLP^{-/-}MBP^{-/-}$).

Phenotype

The ultrastructure of the multilayer myelin sheath of all axons in the CNS of hemizygous male or homozygous female $PLP^{-/-}$ mice was highly disordered. The apposition of the extra-cytoplasmic surfaces, and thereby the intraperiod dense line, was lacking. The disrupted assembly of the myelin sheath led to a profound reduction of conductance velocities of CNS axons, impairments in neuromotor coordination, and behavioral changes.

Homozygous double $PLP^{-/-}MBP^{-/-}$ mutant mice had ameliorated shiverer symptoms and rare seizures. They showed normal development and no premature death of oligodendrocytes, had a normal reproduction rate and a lifespan of at least 18 months. Light and electron microscopy revealed that oligodendrocytes of PLP/MBP-deficient mice were able to ensheath many CNS axons with an atypical myelin containing only few lamellae which, in contrast to the myelin of shiverer mice, spirally wrapped axons with a larger diameter. The lamellae formed a loose and uncompacted myelin. They lacked the major dense line but adhered tightly on their extra-cytosolic surfaces and formed distinct electron-dense intermediate lines. The paranodal complexes of the double mutant mice resembled in their morphology the "pseudonodes" described in the CNS of the shiverer mutant. The PNS myelin of the double mutant exhibited a normal periodic structure like the nearly normal PNS of MBP-deficient mice.

Acknowledgements
Wilhelm Stoffel
University of Cologne, Cologne, Germany

References

[1] Boison, D. and Stoffel, W. (1994) Proc. Natl Acad. Sci. USA 91, 11709–11713
[2] Stoffel, W. et al. (1997) Cell Tissue Res. 289, 195–206.
[3] Jackson Laboratory, Bar Harbor.
[4] Boison, D. et al. (1995) J. Neurosci. 15, 5502–5513.
[5] Rosenbluth, J. et al. (1996) J. Comp. Neurol. 371, 336–344.

Gene symbol
Pms2

Accession number
MGI: 104288

Area of impact
DNA repair

General description

The mammalian MutL (*E. coli*) homolog PMS2 functions in the repair of DNA replication errors in all tissues examined and in processes involved in meiotic recombination. PMS2 is a homolog of the yeast PMS1 protein and shows 35% identity and 56% similarity at the protein level.

KO strain construction

A neomycin gene was used to replace exon 2 in D3 ES cells. This exon encodes a highly conserved portion of all MutL-like proteins. The deletion of exon 2 produced a null allele as demonstrated by immunoblot analysis. C57BL/6 female mice were used in breeding with the chimeric male.

Phenotype

PMS2-deficient animals were prone to the spontaneous development of lymphoma and uterine sarcoma. Microsatellite instability was observed in spermatozoa, tail DNA and tumor DNA from these animals. Increased base substitution was also evident in cultured embryonic fibroblasts. While female mice were fertile, PMS2-deficient male mice were sterile and produced only abnormal spermatozoa. Analysis of spermatocytes from PMS2-deficient animals indicates abnormalities in chromosome synapsis in prophase of meiosis I.

Comments

Analysis of PMS2-deficient mice suggests links between DNA mismatch repair, genetic recombination and chromosome synapsis in meiosis.

Acknowledgements
R. Michael Liskay
Oregon Health Sciences University, Portland, OR, USA

Reference

[1] Baker, S.M. et al. (1995) Cell 82, 309–319.

PPARα

Other names
Peroxisome proliferator-activated receptor α isoform, mPPAR

Gene symbol
Ppara

Accession number
MGI: 104740

Area of impact
Metabolism

General description

PPAR belongs to the nuclear receptor superfamily. At least three forms of PPAR are known to exist in mice. These include PPARα, PPARγ and PPARδ. PPARα is the predominant form in liver, kidney and heart. Leukotriene B_4 was found to be a ligand for PPARα. PPARs require the dimerization partner retinoic acid X-receptor in order to *trans*-activate target genes. The physiological roles of PPAR appear to be related to the modulation of lipid metabolism.

KO strain construction

The targeting vector contained 6.4 kb of genomic DNA spanning introns 6–8. A *neo* cassette was inserted and replaced 83 bp of the coding region of exon 8. An HSV-*tk* gene was inserted at the 3′ end of the construct. Targeted J1 129Sv ES cells were injected into C57BL/6 blastocysts, mutant mice were either of a mixed background (Sv129ter × C57BL/6N) or inbred (Sv129ter).

Phenotype

Mice homozygous for the mutation lacked expression of PPARα protein and yet were viable and fertile and exhibited no detectable gross phenotypic defects. Furthermore, these animals did not respond to the prototypical peroxisome proliferators clofibrate and Wy-14,643, and lacked detectable hepatomegaly, proliferation of peroxisomes or induction of mRNAs encoding the peroxisomal and microsomal lipid-metabolizing enzymes such as CYP4A3, CYP4A1, thiolase, BIEN and L-FABP.

Acknowledgements
Susanna S.T. Lee
Department of Biochemistry, The Chinese University of Hong Kong, Shantin, New Territories, Hong Kong

References
[1] Lee, S. et al. (1995) Mol. Cell Biol. 15, 3012–3022.
[2] Lee, S. et al. (1996) Acad. Sci. 804, 524–529.

Other names
Protective protein/cathepsin A

Gene symbol
Ppgb

Accession number
MGI: 97748

Area of impact
Metabolism, signal transduction

General description

PPCA is a serine carboxypeptidase. It can associate with two glycosidases, β-galactosidase and neuraminidase, and modulates their intralysosomal stability and activity. PPCA also functions as a lysosomal cathepsin at acidic pH and as a deaminase/esterase at neutral pH. PPCA has a role in the inactivation of certain neuropeptides such as substance P, oxytocin and endothelin I. A defect in human PPCA causes the autosomal recessive disease galactosialidosis, which is characterized by a defect in the activities of both β-galactosidase and neuraminidase. Galactosialidosis patients exhibit coarse facies, ocular cherry red spots, vertebral changes, foamy bone marrow cells, vacuolated lymphocytes, bone deformity, hepatosplenomegaly, severe defects in the heart and kidneys, and psychomotoric defects.

KO strain construction

A hygromycin-resistance gene was inserted into exon 2 of the mouse *Ppgb* gene. The position of the hygromycin-gene insertion prevents translation. HSV-tk was placed at the 3′ end of the construct. Targeted 14 ES cells were injected into C57BL/6 blastocysts.

Phenotype

PPCA-null mice were viable. Shortly after birth, these mice developed a disease that resembles the human lysosomal storage disorder galactosialidosis. Cells were vacuolated and there was excessive excretion of sialyloligosaccharides in the urine. PPCA-null mice progressively deteriorated and developed organ failures. They had swollen limbs and eyelids, disheveled coats and broad faces. Transplantation of bone marrow cells that expressed the human PPCA gene under an erythrocyte-specific promoter corrected the phenotype.

Comments

These experiments demonstrate the feasibility of using transgenic bone marrow cells and bone marrow transplantation to correct lysosomal storage diseases.

Reference

[1] Zhou, X.Y. et al. (1995) Genes Dev. 9, 2623–2634.

Other names
Progesterone receptor

Gene symbol
Pgr

Accession number
MGI: 97567

Area of impact
Hormone, development

General description

In most target tissues, the PR gene is induced by estrogen via its cognate nuclear receptor, termed the estrogen receptor. The PR gene encodes two isoforms of the PR (A and B) for which an *in vivo* functional role has yet to be assigned. Although the PR gene has been cloned from many species, the downstream target genes for this nuclear receptor have yet to be identified.

KO strain construction

The PR gene targeting vector was designed to insert the neomycin-resistance gene (PGK*neo*bpolyA) into the first exon of the mouse *Pgr* gene, downstream from the initiating codons ATG(A) and ATG(B) that encode the A and B forms of the PR. This insertion site was chosen to effectively disrupt the transcription of both forms of PR. The construct contains 7 kb of PR homologous sequence of which the short arm of homology is 1.5 kb and the corresponding long arm of homology is 5.5 kb. For positive/negative selection, the HSV-*tk* gene was located in the 5′ region of the targeting construct. The construct was introduced into AB-1 ES cells. The PR-null mutation was carried in the 129SvEv × C57BL/6 hybrid and 129SvEv inbred genetic backgrounds.

Phenotype

Both male and female mice homozygous for the PR-null mutation developed to adulthood at the predicted Mendelian and sex ratios. Although PR$^{-/-}$ male mice did not exhibit an overt phenotype, removal of PR function in the female resulted in extensive abnormalities in a number of reproductive systems. These included an inability to exhibit a sexual behavorial response, a dysfunctional uterus as evidenced by a defect in the elaboration of the decidual response as well as a hypersensitivity to estrogen and progesterone administration, and an impairment in mammary gland ductal proliferation and differentiation in response to pregnancy levels of hormones.

Comments

This animal model represents a tool to explore the involvement of PR, as distinct from ER, in a number of important reproductive systems in the female.

Utilizing this mouse model, investigators will be able to evaluate PR's role in reproduction and mammary gland development, including tumorigenesis, at the morphological, cellular, molecular and genetic levels.

Acknowledgements
Bert W. O'Malley
Baylor College of Medicine, Houston, TX, USA

References

[1] Lydon, J.P. et al. (1995) Genes Dev. 9, 2266–2278.
[2] Lydon, J.P. et al. (1996) J. Steroid Biochem. Mol. Biol. 56, 67–77.
[3] Mani, S.K. et al. (1996) Mol. Endocrinol. 10, 1728–1737.

Other names
Prolactin receptor, PRL receptor, RPRL receptor

Gene symbol
Prlr

Accession number
MGI: 97763

Area of impact
Hormone, signal transduction

General description

Prolactin acts by binding and inducing the dimerization of a membrane-bound receptor, which in turn activates an associated tyrosine kinase, Jak2, and the transcription activator Stat5. Prolactin also activates several other signal transduction pathways. Prolactin is best known for its action on mammary development and lactation, but has numerous other actions, including effects on reproduction, behavior, immune responses, water and electrolyte balance, and growth and development.

KO strain construction

A targeting construct was prepared with 7.5 kb of overall homology, in which a 1.5 kb fragment containing exon 5 was replaced with the similarly-sized *tk*-neomycin-resistance gene cassette, resulting in an in-frame stop codon. Any resulting mRNA would encode a protein of only 44 amino acids. E14.1 mouse ES cells were transfected with the targeting vector prepared from a genomic 129Sv clone and injected into blastocysts from C57BL/6 animals mated to C57BL/6 males. Chimeric animals were mated to C57BL/6 or 129Sv partners.

Phenotype

Heterozygous females in their first lactation showed almost complete failure to lactate, which was due to greatly reduced mammary development. In subsequent pregnancies, or when the first pregnancy was delayed to 20 weeks of age, partial ability to lactate was attained, and almost all pups survived. These results demonstrate that two functional alleles of the prolactin receptor gene are required for correct lactation, and that this phenotype in heterozygotes is attributable primarily to a deficit in the degree of mammary gland development. Some but not all heterozygous mothers were observed to show unusual maternal behavior, often scattering the pups throughout the cage or even burying them.

Homozygous females were sterile, owing to a number of reproductive deficiencies. First, all PRLR$^{-/-}$ females were sterile despite regular mating. Secondly, mating did not produce a pseudopregnancy in PRLR$^{-/-}$ females. Thirdly, irregular mating patterns of half of the females indicated an alteration of estrous cyclicity. Pre-implantation development of embryos was examined and multiple abnormalities were observed in the $PRLR^{-/-}$ females: fewer eggs were fertilized, oocytes at the germinal vesicle stage were released from the ovary, and fragmented embryos were found. Reduced numbers of eggs were seen and fertilized eggs developed poorly to the blastocyst stage in $PRLR^{-/-}$ animals. Transplantation experiments confirmed that embryos were not able to develop in the uterus of $PRLR^{-/-}$ females, suggesting that the uterus of these animals is refractory to implantation. Homozygous males were able to reproduce, although some animals showed a delay in acquiring reproductive capacity.

Comments

This work establishes the prolactin receptor as a key regulator of mammalian reproduction, and provides the first total ablation model to study the role of the prolactin receptor and its ligands in the numerous other functions attributed to this hormone.

Acknowledgements
Paul A. Kelly
INSERM Unite 244, Paris, France

References

[1] Ormandy, C.J. et al. (1997) Genes Dev. 11, 167–178.
[2] Boutin, J.M. et al. (1988) Cell 53, 69–77.

Other names
Prion protein, Prn-p

Gene symbol
Prn-p

Accession number
MGI: 97769

Area of impact
Neurology

General description

The prion diseases are neurodegenerative conditions which are transmissible by inoculation. They are sometimes inherited in an autosomal dominant manner. The prion consists principally of a modified form of the host-encoded glycoprotein PrP. PrP is a membrane protein which is expressed at high levels in brain. Its normal function is unknown. Post-translationally altered forms of the PrP protein (designated PrP^{sc}) accumulate in prion diseases such as CJD, BSE and scrapie.

KO strain 1 construction[1]

Codons 4–187 of the 254 codon open reading frame of the *Prn-p* gene were replaced by the *neo* gene under the control of the HSV-*tk* promoter. The targeting vector was introduced into AB1 ES cells (129SvEv) and positive clones were injected into blastocysts to generate chimeras. Chimeras were mated to C57BL/6J females.

KO strain 2 construction[2]

A replacement vector permitting positive–negative selection in *hprt*-deficient HM-1 cells (derived from 129Ola) was used. A 2.1 kb fragment containing the entire coding region (exon 3) was replaced by PGK*hprt*. HSV-*tk* was included for negative selection. The 5′ arm of homology was 1.8 kb; the 3′ arm of homology was 3.7 kb. The mutation was studied on both an inbred background (129Ola) and an outbred segregating background (129Ola × BALB/c).

KO strain 3 construction[3]

The entire coding region of the *Prn-p* gene (isolated from the 129Sv strain) was replaced with PGK*neo*. PGK*tk* was included for negative selection. The targeting vector was electroporated into J1 ES cells and positive clones were injected into C57BL/6J blastocysts. Resulting chimeras were mated with C57BL/6J mice.

Phenotype

In general, $PrP^{-/-}$ mice developed normally and did not exhibit either behavioral or neurological abnormalities in early to midlife. $PrP^{-/-}$ mice did not succumb to prion infection and did not support the clandestine replication of prions. Neurophysiological defects in synaptic inhibition and long-term potentiation were observed in KO strain 1 $PrP^{-/-}$ mice which could be rescued by breeding to transgenic mice expressing human PrP. KO strain 2 $PrP^{-/-}$ mice were viable but appeared to develop a late onset neurological phenotype. At 70 weeks of age, KO strain 3 $PrP^{-/-}$ mice showed progressive symptoms of ataxia with impaired motor coordination. There was an extensive loss of Purkinje cells in the vast majority of cerebellar folia.

Comments

PrP is not essential for either normal development or CNS function, but is required for host susceptibility to prion infection and propagation. PrP may contribute to the long-term survival of Purkinje neurons.

Acknowledgements
David Melton
Institute of Cell and Molecular Biology, Edinburgh University, Edinburgh, UK

References

[1] Büeler, H. et al. (1992) Nature 356, 577–582.
[2] Moore, R.C. et al. (1995) Bio/Technology 13, 999–1004.
[3] Sakaguchi, S. et al. (1995) J. Virol. 69, 7586–7592.
[4] Sakaguchi, S. et al. (1996) Nature 380, 528–531.

Other names
Presenilin 1, S182

Gene symbol
Ad3h

Accession number
MGI: 107441

Area of impact
Development

General description

Mutations in the human PS1 gene segregate with the majority of early-onset cases of familial Alzheimer's disease. PS1, endoproteolytically processed *in vivo*, is a multipass transmembrane protein homologous to the *Caenorhabditis elegans* sel-12 protein that facilitates signaling mediated by the Notch/lin12 family of receptors. PS1 is ubiquitously expressed, including in the neuro-epithelium, the pre-somitic mesoderm and somites during mouse embryonic development.

KO strain construction

In the PS1 targeting construct, a 1.9 kb *Xho*I/*Eco*RI fragment containing exon 4 (amino acids 30–113) and flanking intronic segments of the PS1 gene was replaced by the neomycin gene (PGK*neo*). Targeted AB2.129 ES cells were injected into C57BL/6J recipient blastocysts. Chimeras were bred with C57BL/6J females, and the phenotype examined on a 129 × B6 hybrid background.

Phenotype

Homozygous mutant mice do not survive beyond the first day after birth, but PS1-null embryos were present at expected Mendelian frequencies at various stages of gestation ranging from E8.5 to 18.5. The most striking phenotype observed in PS1-null mice was a severe perturbation in the development of the axial skeleton. Compared to control, PS1-null embryos are smaller and possess a stubby tail. Histochemical analysis of the skeleton revealed defects in the vertebral column and the ribs; the limbs appeared normal. Morphological analysis of E9.5 PS1-null embryos showed irregularly shaped somites and misalignment of somites across the midline. These abnormal somite patterns are highly reminiscent of somite segmentation defects described in mice with functionally inactivated *Notch1* or *Dll*1 alleles. Moreover, expression of mRNA encoding *Notch1* and *Dll*1 is markedly reduced in the presomitic mesoderm of PS1-null embryos. Somite polarity is not maintained in PS1-null embryos; sclerotome condensation failed to occur. In addition, the spinal ganglia were fused over multiple segments along the cranio-caudal axis of the vertebral column. Although the cellularity and cytoarchitecture of the developing brain of PS1-null embryos appeared normal, all PS1-null embryos after

E11.5 exhibited hemorrhages limited to the brain and/or spinal cord. The hemorrhages are present beneath the primordial dura and leptomeninges, within the ventricles and in neural parenchyma, rarely with focal necrosis.

Comments

Because *Notch1* and *Dll*1 mRNAs are downregulated in the paraxial mesoderm of PS1-null embryos, it is suggested that PS1 serves to regulate the spatiotemporal expression of *Notch1* and *Dll*1.

Acknowledgements
Philip Wong
The Johns Hopkins University School of Medicine, Department of Pathology, Division of Neuropathology, Baltimore, MD, USA

References

[1] Sherrington, R. et al. (1995) Nature 375, 754–760.
[2] Levitan, D. et al. (1995) Nature 377, 351–354.
[3] Thinkaran, G. et al. (1996) Neuron 17, 181–190.
[4] Conlon, R.A. et al. (1995) Development 121, 1533–1545.
[5] Hrabe de Angelis, M. et al. (1997) Nature 386, 717–721.

PTH-PTHrP

Other names
Parathyroid hormone/parathyroid hormone-related peptide receptor

Gene symbol
Pthr

Accession number
MGI: 97801

Area of impact
Metabolism, hormone, development

General description

The PTH-PTHrP receptor binds and responds equally to the N-terminal portions of two ligands, parathyroid hormone (PTH) and parathyroid hormone-related peptide (PTHrP). PTH regulates Ca^{2+} homeostasis in kidneys and bones. PTPrP is produced in multiple tissues during embryogenesis and adult life and functions as a paracrine factor. Various actions have been ascribed to PTHrP, including similar effects as PTH on calcium homeostasis, inhibition of bone resorption, or growth and differentiation of many cells types during development. One of the main functions of PTHrP is to delay chondrocyte differentiation to allow enhanced chondrocyte proliferation. Expression of the PTH-PTHrP receptor correlates with PTH target organs (bone, kidney) and PTHrP expression (e.g. growth plates of bones). Other receptors of PTH and PTHrP exist.

KO strain construction

Exons E2 to T encoding most of the receptor were deleted by introduction of a neomycin-resistance cassette. Targeted ES cells were injected into C57BL/6 blastocysts.

Phenotype

Most PTH-PTHrP receptor-null animals died in midgestation beginning from E9.5. The surviving mice were proportionally smaller in size and displayed accelerated chondrocyte differentiation, hypertrophic chondrocytes, reduced growth plates in long bones, abnormal mineralization of bones, and abnormalities throughout the whole skeleton (dulled skull, foreshortened mandibulae, overall reduced size of the skeleton). *In vitro*, the bones were resistant to Sonic hedgehog and PTHrP-induced expansion of less-differentiated chondrocytes and suppression of differentiation of chondrocytes into hypertrophic, type X collagen producing chondrocytes. Crossing into a Black Swiss background partially suppressed early lethality but all newborn mice died within minutes after birth.

Comments

A constitutively active PTH-PTHrP receptor mutation is the cause of the rare short-limbed dwarfism syndrome, Jansen-type metaphyseal chondrodysplasia, which is defined by severe ligand-independent hypercalcemia, hypophosphatemia, and abnormal endochondral bone formation[2].

References

1 Lanske, B. et al. (1996) Science 273, 663–666.
2 Schipani, E. et al. (1995) Science 268, 98–100.

PTHrP

Other names
Parathyroid hormone-related protein or peptide

Gene symbol
Pthlh

Accession number
MGI: 97800

Area of impact
Metabolism

General description

Expression of PTHrP in the developing rodent is seen as early as the compacted morula stage and widespread expression continues into adult life. Expression is prominent in the epithelia of many tissues, including the developing lung, tooth, skin, kidney and colon. Other cells types such as chondrocytes and osteoblasts in developing bone and neurons of the cerebral cortex, hippocampus and cerebellar cortex are also positive for gene expression. PTHrP appears to function as a polyhormone and various actions have been ascribed to different portions of the protein. The first 34 amino acids have effects similar to parathyroid hormone (PTH) on the skeleton and kidney and overall calcium homeostasis. In contrast, a mid-molecule portion from amino acid 107 to 139 has been shown to inhibit bone resorption. The C-terminal portion of the molecule is PTH-like. Other postulated physiological roles for this peptide hormone are in the growth and differentiation of many cell types during fetal development and in the adult, as well as in lactation and smooth muscle function.

KO strain construction

The mouse PTHrP gene consists of five exons, the major coding exon being exon 4. In order to inactivate this gene in ES cells, a replacement construct was produced. A PGK*neo* gene was ligated into a 9.5 kb fragment of PTHrP genomic DNA from BALB/c mice, replacing exon 4. To produce chimeras, D3 ES cells were injected into F2 blastocysts generated from a (C57BL6/J × C57BL/10ScSn) F1 × F1 cross. Chimeras were crossed with C57BL6/J mice to give a mixed background. In addition, chimeras were crossed with 129Sv mice to create a pure 129Sv background.

Phenotype

PTHrP-null mutants animals died post-natally; some died within hours, while others survived for up to 24 hours. All null mutant animals exhibited abnormal features, including shortened limbs and mandible, protruding tongue and domed skull. These features are typical of an osteochondrodysplasia. All mutant animals exhibited gross skeletal abnormalities due to abnormal endochondral bone development and ossification of the costal cartilages of the ribs, a site that does not normally convert bone. The abnormal conversion of costal cartilages to bone is variable and is most likely due to genetic background variation and does not reflect survival; that is, those animals with the least affected ribs were not those that survived longest. Histological examination of day E18.5 null mutants revealed that all tissues appeared normal with the exception of the long bones and perhaps the lungs. The diameter of the epiphyseal plates of the long bones was greatly reduced, due to a diminution in the size of the proliferating zone. The lungs of the null mutant animals appeared to be underdeveloped and less well-expanded compared to those from normal heterozygous littermates. PTHrP has been shown to play a role in lung development and surfactant production. Both PTHrP and the PTHrP-PTH receptor are expressed in type II pneumocytes, the cells that produce surfactant. Cocultures of mutant lung fibroblasts and pneumocytes exposed to PTHrP and PTH showed increased phosphatidylcholine synthesis, and cultured type II pneumocytes exposed to PTHrP showed an increase in the number of lamellar bodies. Moreover, homozygous animals had a significantly lower level of circulating ionized calcium than their heterozygous and wild-type littermates, but had a higher total body calcium content. The lower ionized calcium may in part be explained by the greater calcium transport across the placenta[5]. Lungs from homozygous animals expressed surfactant proteins A, B and C at normal levels. Type II pneumocytes contained lamellar bodies and alveolar spaces contained secreted surfactant.

Comments

The results obtained from null mutant mice that died shortly after birth imply that, during fetal development, the main function of PTHrP is the control of chondrocyte differentiation and proper bone formation even though PTHrP is expressed in many other developing tissues including the lung. These animals provide models to study the role of PTHrP in lung function and transplacental calcium transport.

Acknowledgements
Vicki Hammond
University of Melbourne, Victoria, Australia

References

[1] Karaplis, A.C. et al. (1994) Genes Dev. 8, 277–289.
[2] Hammond, V.E. et al. (1996) J. Bone Mineral Res. 11, S437.
[3] Mangin, M. et al. (1990) Gene 95, 195–202.
[4] Philbrick, W.M. et al. (1996) Physiol. Rev. 76, 127–173.
[5] Tucci, J. et al. (1996) J. Mol. Endocrinol. 17, 159–164.

PU.1

Other names
Spi-1, Sfpi-1

Gene symbol
Sfpi1

Accession number
MGI: 98282

Area of impact
Hematopoiesis, transcription factors

General description

PU.1, a member of the ets transcription factor family, is specifically expressed in the hematopoietic system. The PU.1 protein contains the ets DNA-binding domain, a sequence of about 85 amino acids that is highly conserved among all ets family members. The ets domain recognizes a purine-rich DNA sequence containing the core sequence 5′ GGAA 3′. PU.1 is the product of the Spi-1 proto-oncogene. The Spi-1 locus is the site of integration of the spleen focus-forming virus in 95% of murine erythroleukemias induced by Friend virus complexes. It has been shown to act as a transcriptional activator, and is expressed at particularly high levels in the monocytic, granulocytic and B-lymphoid lineages. Numerous presumptive target genes have been identified in each of these lineages.

KO strain 1 construction

The targeting vector was designed to delete a portion of exon 5 that encodes the ets DNA-binding domain. This segment was replaced by a PGK*neo* fusion gene. The *neo* gene was flanked by 2 and 12 kb of PU.1 homologous sequence on its 5′ and 3′ ends, respectively. To enrich homologous recombination events, the PGK-*tk* cassette was used as a negative selectable marker. The targeting vector was electroporated into the CCE.1 ES cell line derived from 129Sv mice. PU.1 heterozygous ES clones were injected into C57BL/6 blastocysts and the mutation was transmitted onto the BL/6 genetic background. The null mutation was confirmed by Southern blotting.

KO strain 2 construction

A plasmid (pBluescript KS+) containing a 3 kb genomic fragment (exons 4 and 5) from the PU.1 gene was used for the targeting construct. The plasmid was digested with *Bss*H2 and the HSV-*tk-neo* gene was blunt end ligated into this site within exon 5 in the antisense orientation, disrupting the DNA-binding domain. The HSV-*tk* gene was inserted downstream. The targeting vector was electroporated into D3jm p47 ES cells. Positive clones were injected into C57BL/6 blastocysts. Chimeras were bred to C57BL/6 mice and heterozygotes were crossed to obtain homozygous mutants. The null mutation was confirmed by Southern blotting.

Phenotype KO strain 1

This mutation of the PU.1 gene resulted in late embryonic lethality at E17 to E18 of gestation. The mutation caused a complete block in the development of lymphoid and myeloid lineages in the fetal liver. Importantly, the mutation did not arrest erythrocytic or megakaryocytic development. By generating chimeric animals using either PU.1 homozygous mutant ES cells or PU.1 hematopoietic progenitors, PU.1 was demonstrated to function in an exclusively cell-autonomous manner to regulate the development of the lymphoid-myeloid system. Multipotential lymphoid-myeloid progenitors ($AA4.1^+$, Lin^-) were significantly reduced in $PU.1^{-/-}$ embryos and failed to differentiate into B lymphoid and myeloid cells *in vitro*. These results suggested that the lymphoid and myeloid lineages develop in the fetal liver from a common hematopoietic progenitor not shared with erythrocytes and megakaryocytes. Finally, the *Ikaros* gene was expressed in PU.1 mutant embryos, suggesting the PU.1 and Ikaros factors are independently required for specification of embryonic lymphoid cell fates.

Phenotype KO strain 2

This disruption of the PU.1 gene on both chromosomes in ES cells appeared to have no effect on these cells in the undifferentiated state. When these cells were induced to differentiate along the macrophage lineage, there was a complete block in the production of mature macrophages compared to normal and PU.1 single knockout ES cells. The double knockout ES cells stained positively with the MOMA-2 antibody (which recognizes a marker on early phagocytic cells and blood monocytes), suggesting the development of an early monocytic precursor. The development of erythroid cells appeared normal when the double knockout ES cells were induced to differentiate along the erythroid lineage.

Mice homozygous for this mutation of the PU.1 DNA-binding domain were born alive at the expected Mendelian frequency but died within 48 hours of severe septicemia. The analysis of these neonates revealed a lack of mature macrophages, neutrophils, osteoclasts, dendritic cells, B cells and T cells, although erythrocytes and megakaryocytes were present. The absence of lymphoid commitment and development in null mice was not absolute, since mice maintained on antibiotics began to develop normal-appearing T cells 3–5 days after birth. Both CD4 and CD8 double positive and single positive cells were detected. Mature B cells were undetectable in the PU.1 mice as was any evidence of Ig rearrangement, although cells expressing B220, BP1 and HSA were seen. This suggested that cells bearing early B cell markers were present but that the development of mature B cells was blocked. Within the myeloid lineage, despite a lack of macrophages in the older antibiotic-treated animals, a few chloroacetate esterase-positive cells with the characteristics of neutrophils began to appear by day three after birth. In addition a few $Gr\text{-}1^+$/$Mac\text{-}1^+$ cells were seen in these older PU.1-null mice, consistent with the development of neutrophils (although their numbers were far less than those seen in normal littermates). No evidence of osteoclasts was observed and these mice exhibited the classical hallmarks of osteopetrosis.

Comments

The mutations of the PU.1 gene described above are distinct and generate mouse phenotypes with many similarities but also two significant differences. Importantly, both mutations in the PU.1 gene result in a block to lymphopoiesis and myelopoiesis in the fetal liver and thymus. However, the mutation in KO strain 2 mice[2] results in neonatal lethality whereas the mutation in KO strain 1 mice[1] results in late embryonic lethality. KO strain 2 animals can be kept alive for up to 17 days with antibiotic treatment; these animals showed evidence of delayed T cell development.

Evidence from the study of KO strain 1 mice suggests that the PU.1 gene represents a unique and critical regulator of all lymphoid and myeloid lineages in the hematopoietic system. However, evidence from the study of KO strain 2 mice suggests that while the PU.1 protein appears not to be essential for myeloid and lymphoid lineage commitment, it is absolutely required for the normal differentiation of B cells, macrophages and osteoclasts.

Acknowledgements
Harinder Singh
University of Chicago, Chicago, IL, USA
Richard Maki
The Burnham Institute,
La Jolla, CA, USA

References

1 Scott, E. et al. (1994) Science 265, 1573–1577.
2 McKercher, S. et al. (1996) EMBO J. 15, 5647–5658.
3 Olson, M.C. et al. (1995) Immunity 3, 703–714.
4 Scott, E. et al. (1997) Immunity 6, 437–447.
5 Henkel, G.W. et al. (1996) Blood 88, 2917–2926.
6 Tondravi, M.M. et al. (1997) Nature 386, 81–84.
7 Klemsz, M. et al. (1990) Cell 61, 113–124.

Other names
Recombination activating gene 1

Gene symbol
Rag1

Accession number
MGI: 97848

Area of impact
Immunity and inflammation, B and T cells

General description

RAG-1 is one of the two subunits of the V(D)J recombinase. Along with RAG-2, it is expressed in B and T cells and is required for recombination of the immunoglobulin (BCR) and T cell receptor (TCR) genes.

KO strain 1 construction[1]

A targeting vector was constructed which had the neomycin-resistance gene inserted into the *Rag1* gene, resulting in the deletion of 1356 bp of the coding sequence at the 5′ end (deletion from position 482 to 1837). The vector was electroporated into AB1 ES cells (129Sv strain), and recombinant ES cell clones were injected into C57BL/6J blastocysts. Chimeric mice were backcrossed with CD1 mice. The homozygous mutation of the *Rag1* gene was confirmed by Southern blotting.

KO strain 2 construction[2]

The neomycin-resistance gene was inserted into an *Eco*RV site in the *Rag1* gene at position 330 of the amino acid sequence. This insertion generates a frameshift mutation. The vector was electroporated into J1 ES cells (129Sv strain). ES cell clones were injected into C57BL/6 blastocysts. Homozygous mutants were identified by Southern blotting.

Phenotype

RAG-$1^{-/-}$ mutant mice were healthy and fertile. The mice exhibited greatly reduced cellularity in all primary and secondary lymphoid organs. Analysis of the spleen, bone marrow and thymus revealed a complete absence of mature B and T lymphocytes, and serum IgM was not detectable. Southern blot analysis of Abelson-transformed cell lines demonstrated that V(D)J recombination was absent. Despite the fact that RAG-1 has been demonstrated to be expressed in the CNS, RAG-1-deficient mice did not demonstrate any neuroanatomical or behavioral abnormalities.

Flow cytometric analysis of thymocytes and bone marrow cells revealed that both B and T lymphocyte development was arrested at a very early stage, after lineage commitment but before rearrangement. B cell precursors were arrested at the B220$^+$CD43$^+$ stage (fraction C). T cell precursors were arrested at the CD4$^-$CD8$^-$ stage. Developmentally arrested pro-T cells did express small amounts of CD3 on their surface, and could be induced to develop to the pre-T cell stage by anti-CD3 antibodies, or by insertion of a TCR transgene. The RAG-1 knockout phenotype is similar to the phenotype of the RAG-2 knockout mouse.

Comments

The RAG-1 knockout phenotype demonstrates the crucial role that RAG-1 plays in the initiation and/or the catalysis of V(D)J recombination in B and T lymphocytes. The RAG-1 mouse phenotype is similar to the SCID mouse, but the RAG-1-deficient mice do not demonstrate leakiness in their phenotype in the same way that SCID mice do. RAG-deficient mice represent a useful tool for the study of lymphocyte development and the role of specific immunity in a variety of pathogenic states.

Acknowledgements
Eugenia Spanopoulou
Mount Sinai School of Medicine, New York, NY, USA
Lynn Corcoran
Walter & Eliza Hall Institute of Medical Research, Royal Melbourne Hospital, Victoria, Australia

References

[1] Mombaerts, P. et al. (1992) Cell 68, 869–877.
[2] Spanopoulou, E. et al. (1994) Genes Dev. 8, 1030–1042.
[3] Strasser, A. et al. (1994) Nature 368, 457–460.

Other names
Recombination activating gene 2

Gene symbol
Rag2

Accession number
MGI: 97849

Area of impact
Immunity and inflammation, B and T cell

General description

The variable regions of both BCRs and TCRs are assembled during the early stages of B and T lymphocyte differentiation, respectively. Germ line variable (V), diversity (D) and joining (J) gene segments are brought together during V(D)J recombination using conserved flanking recombination sequences (RS). It is generally accepted that the same V(D)J recombination process occurs in both B and T cells to produce rearranged antigen receptors, and that a single recombinase enzymatic activity is responsible for this process in both types of lymphocytes. V(D)J recombinase activity has been found only in immature lymphoid cells. *Rag1* and *Rag2* are genes that have been shown to confer the ability to rearrange transfected V(D)J substrates *in vitro*. These highly conserved genes are closely linked and are highly expressed only in precursor T and B lymphocytes. It is thought that RAG-1 and RAG-2 encode a lymphoid-specific recombinase enzymatic activity.

KO strain construction

The mouse RAG-2 protein is encoded by a single exon. The targeting vector contained a 4 kb genomic *Rag2* fragment in which an 850 bp region coding for 286 amino acids of the RAG-2 protein was replaced by pMC1*neo*. The replacement also resulted in a translational frameshift of the remaining downstream *Rag2* sequences. The HSV-*tk* gene under the control of the PGK promoter was included 5′ of the genomic sequences. The targeting vector was transfected into CCE ES cells and positive clones were injected into C57BL/6 or MF1 blastocysts to generate chimeras from which homozygous RAG-$2^{-/-}$ mutant mice were obtained.

Phenotype

Mice heterozygous for the mutated *Rag2* allele had a normal phenotype. Homozygous RAG-$2^{-/-}$ mice were born at the expected ratio, were healthy and showed no gross anomalies. Both males and females were fertile. No changes were observed to a wide variety of non-immune tissues. However, if maintained in non-barrier facilities, mutant mice were smaller than wild type and prone to infection. The thymus in RAG-$2^{-/-}$ mice was absent or very small, with 10- to 100-fold fewer cells than in the wild type. The lack of cellularity became more pronounced with age. While spleens were comparable in size to wild type, they contained 5–10-fold fewer cells. $sIgM^+$ cells were absent from bone marrow, spleen and peritoneum, and no IgM was detectable in serum of mutant mice. However, $sIgM^-/B220^{dull}/CD43^+/30F1^{dull}/BP\text{-}1^-$ cells were present in significant numbers in RAG-$2^{-/-}$ bone marrow and spleen, leading to the conclusion that an absence of RAG-2 leads to a complete block in early B cell differentiation at the point at which Ig gene rearrangement commences. Neither double nor single positive T cells could be detected in RAG-$2^{-/-}$ thymi, nor $TCR\alpha\beta^+$, $TCR\gamma\delta^+$ or $CD3^+$ lymphocyte populations in spleen or thymi. However, a small population of $CD8^+$ single positive splenocytes expressing only the CD8α chain, and a small population of $CD4^+$ single positive splenocytes, were found in most RAG-$2^{-/-}$ mice; these cells did not express TCR or CD3 chains. The majority of cells in RAG-$2^{-/-}$ thymi co-expressed Thy-1 antigen and the α chain of the IL-2 receptor, indicating that T cell development was blocked at the stage immediately preceding TCR rearrangement. DJ_H rearrangment in the Ig J_H locus could not be detected in RAG-$2^{-/-}$ bone marrow. Similarly, no rearrangement of the TCRδ locus (the first TCR gene to rearrange) could be detected in RAG-$2^{-/-}$ thymocytes. The RAG-$2^{-/-}$ mutation did not affect the development of NK cells, macrophages or granulocytes. Experiments with Abelson murine leukemia virus (A-MuLV)-transformed RAG-$2^{-/-}$ B cell lines showed a complete lack of V(D)J recombination activity. V(D)J recombination could be restored in RAG-$2^{-/-}$ pre-B cells by transfection of a functional RAG-2 expression construct.

Comments

RAG-2 is required for the initiation of V(D)J recombination in both B and T lymphocytes. The blockages in T and B cell differentiation in RAG-$2^{-/-}$ mice occur at earlier stages than those in SCID mice. RAG-2 function and V(D)J recombinase activity are not required for development of cells other than lymphocytes.

Reference

[1] Shinkai, Y. et al. (1992) Cell 68, 855–867.

Other names
Receptor associated protein, mouse heparin-binding protein-44 (HBP-44), rat RAP: domain of the Heyman nephritis antigen (HNAC)

Gene symbol
Lrpap1

Accession number
MGI: 96829

Area of impact
Metabolism

General description

RAP is a soluble protein of about 40 kDa that is always detected in affinity-purified A2MR/LRP (α_2-macroglobulin receptor/LDL receptor-related protein) preparations. Similarly to A2MR/LRP, RAP is expressed by many different cell types in nearly every organ and tissue. RAP binds with high affinity to the LDL receptor and thereby blocks the binding of all known ligands such as chylomicron remnants or α_2-macroglobulin. RAP localizes to the ER and Golgi apparatus and might function as a chaperone in the biosynthesis of LRP. ER-retained RAP could function as a molecular chaperone for A3MR/LRP (and other receptors) by transiently interacting with newly synthesized A2M/LRP to maintain the receptor in an inactive ligand-binding state.

KO strain 1 construction[1]

An 8 kb *Xho*I fragment containing exons 3–8 of the RAP mouse gene was used as a replacement construct. A 0.7 kb fragment containing 68 bp of exon 6 and part of intron 6 was replaced by a 1.8 kb *Bgl*II fragment encoding the hygromycin B phosphotransferase gene driven by the phosphoglycerate kinase promoter (antisense orientation). The E14 ES cell line (129Ola) was injected into C57BL/6 blastocysts and these were implanted into F1 (C57BL × CBA/J) foster mice. Chimeric mice were mated to C57BL/6 mice to obtain heterozygous Agouti pups. Homozygous mice used in the experiments were a mixture of the 129Ola and C57BL/6J strains.

KO strain 2 construction[2]

A neomycin-resistance cassette was inserted in antisense orientation into the exon which contains the leader sequence of RAP. Two copies of the HSV-*tk* gene were placed at the 5′ end. JH-1 and AB1 ES cells were targeted and mutated ES cells injected into C57BL/6 blastocysts.

Phenotype

RAP-null mice are viable and fertile and appear normal. LRP levels are reduced in the brain and liver of the mutant mice leading to an impairment of

α_2-macroglobulin clearance in the liver. The plasma levels of total cholesterol and triglycerides and the amounts of the apoplipoproteins apoB100, apoB48, apoE and apoAI were comparable among RAP-null and wild-type control mice. Chylomicron remnant clearance was impaired in $RAP^{-/-}LDLR^{-/-}$ double knockout mice and LDL and chylomicron remnants were significantly increased. These double mutants had also high levels of cholesterol and triglycerides in the plasma on a high-fat diet. Moreover, double mutant mice had increased plasma levels of apoB100, apoB48 and apoE but not of apoAI. RAP also binds to the endocytic receptor gp330, the VLDL receptor, and weakly to the LDL receptor (LDLR).

Comments

RAP promotes expression of LRP *in vivo* and might stabilize LRP in the secretory pathway.

Acknowledgements
Fred Van Leuven
Centre for Human Genetics, University of Leuven, Leuven, Belgium

References
[1] Van Leuven, F. et al. unpublished results.
[2] Willnow, T.E. et al. (1995) Proc. Natl Acad. Sci. USA 92, 4537–4541.
[3] Van Leuven, F. et al. (1995) Genomics 25, 492–500.
[4] Strickland, D.K. et al. (1991) J. Biol. Chem. 266, 13364–13369.
[5] Bu, G. et al. (1995) EMBO J. 14, 2269–2280.

Other names
Retinoic acid receptor α (all isoforms)

Gene symbol
Rara

Accession number
MGI: 97856

Area of impact
Development

General description

The effects of retinoic acid (RA) are mediated through two families of receptors (RARs and RXRs) that belong to the superfamily of ligand-inducible transcriptional regulatory factors, which includes steroid/thyroid hormone and vitamin D3 receptors. For each RAR, isoforms are produced by a combination of differential promoter usage and alternative splicing. The high degree of conservation of a given receptor across vertebrates, and their specific patterns of expression during embryogenesis and in adult tissues, has suggested that each RAR performs a specific function. The RARα gene is unique among the RARs in being almost ubiquitously expressed in the embryo and adult with highest expression in skin and regions of the brain.

KO strain 1 construction[1]

An 11 kb *Eco*RI/*Spe*I genomic fragment was used. A unique *Not*I site was created after the CCA encoding the proline at amino acid 19 of the B region, into which a 1.7 kb *Not*I fragment containing the GTI-II enhancer-driven neomycin gene was cloned. This mutation should disrupt expression of both isoforms of the receptor. The mutation was introduced into 129Sv ES cells, which were injected into C57BL/6 blastocysts. The phenotypes were examined on 129 × B6 hybrid or 129 inbred backgrounds.

KO strain 2 construction[2]

A *neo* cassette was inserted into exon 1, replacing part of the endogenous exon. This mutation should disrupt only the α1 isoform of RARα. The mutation was introduced into 129Sv ES cells, which were injected into C57BL/6 blastocysts. The phenotypes were examined on 129 × B6 hybrid or 129 inbred backgrounds.

Phenotype

Mice homozygous for mutation of the α1 isoform are viable and fertile. However, targeted disruption of the entire RARα gene resulted in early postnatal lethality and testis degeneration.

Comments

The observed RARα-null phenotype suggests a high degree of functional redundancy among the RARs.

Acknowledgements
Pierre Chambon
IGBMC, Illkirch, France
En Li
Massachusetts General Hospital, Cambridge, MA, USA

References

[1] Lufkin, T. et al. (1993) Proc. Natl Acad. Sci. USA 90, 7225–7229.
[2] Li, E. et al. (1993) Proc. Natl Acad. Sci. USA 90, 1590–1594.

Other names
Retinoic acid receptor β

Gene symbol
Rarb

Accession number
MGI: 97857

Area of impact
Development

General description

The effects of retinoic acid (RA) are mediated through two families of receptors (RARs and RXRs) that belong to the superfamily of ligand-inducible transcriptional regulatory factors, which includes steroid/thyroid hormone and vitamin D3 receptor. For each RAR, isoforms are produced by a combination of differential promoter usage and alternative splicing. The high degree of conservation of a given receptor across vertebrates, and their specific patterns of expression during embryogenesis and in adult tissues, has suggested that each RAR performs a specific function. RARβ is expressed during development in the CNS, in the respiratory, urogenital and digestive tracts, and in the developing limbs.

KO strain 1 construction[1]

A 0.9 kb *Sst*I/*Xba*I fragment, which contains exon 6 sequences encoding the first zinc-finger motif of the DNA-binding domain, was deleted and replaced with a PGK*neo* cassette. The vector was introduced into R1 129 ES cells, which were injected into C57BL/6 blastocysts. The phenotype was analyzed on a mixed B6 × 129 background.

KO strain 2 construction[2]

A 7 kb *Hin*dIII genomic fragment was used, spanning exons E8 to E11 encoding the ligand binding domain of the receptor. The 1.3 kb *Bgl*II fragment containing exons E9 and most of exon E10 was replaced with a PGK neomycin-resistance gene cassette. The mutation was introduced into 129Sv ES cells, which were injected into C57BL/6 blastocysts. The phenotype was examined on 129 × B6 hybrid or 129 inbred background.

Phenotype

Homozygous RARβ mutants are growth-deficient, but are fertile and have a normal longevity. KO strain 1 RARβ$^{-/-}$ mice have a low frequency of cranial ganglia defects[1]. KO strain 2 RARβ$^{-/-}$ mice display a retrolenticular membrane with high frequency (90%), which arises from the persistence and hyperplasia of the primary vitreous body[3]. The teratogenic effects of RA still occur in mutant embryos.

Comments

RARβ is apparently functionally redundant with either RARα or RARγ. Compound mutants demonstrate a number of novel phenotypes.

Acknowledgements
Pierre Chambon
IGBMC, Illkirch, France

References

1 Luo, J. et al. (1996) Mech. Dev. 55, 33–44.
2 Grondona, J.M. et al. (1996) Development 122, 2173–2188.
3 Ghyselinck, N.B. et al. (1997) Int. J. Dev. Biol. 41, 425–447.

Other names
Retinoic acid receptor γ

Gene symbol
Rarg

Accession number
MGI: 97858

Area of impact
Development

General description

The effects of retinoic acid (RA) are mediated through two families of receptors (RARs and RXRs) that belong to the superfamily of ligand-inducible transcriptional regulatory factors, which includes steroid/thyroid hormone and vitamin D3 receptors. For each RAR, isoforms are produced by a combination of differential promoter usage and alternative splicing. The high degree of conservation of a given receptor across vertebrates, and their specific patterns of expression during embryogenesis and in adult tissues, has suggested that each RAR performs a specific function. RARγ transcripts are restricted to the presomitic caudal region of 8.0 day pc embryos, and to the frontonasal mesenchyme, pharyngeal arches, sclerotomes and limb bud mesenchyme at 8.5–11.5 days pc. At later stages, RARγ transcripts are found in pre-cartilaginous condensations (12.5 days pc) with subsequent restriction to cartilage and differentiating squamous keratinizing epithelia (13.5 days pc).

KO strain construction

A 6 kb *Eco*RI genomic fragment was used. A GTI-II *neo* cassette was cloned into the unique *Kpn*I site present in exon 8, which contains the B region common to all RARγ isoforms. The mutation was introduced into 129Sv ES cells, which were injected into C57BL/6 blastocysts. The phenotype was examined on 129 × B6 hybrid or 129 inbred background.

Phenotype

RARγ mutants were born alive but exhibited growth deficiency, early lethality and male sterility due to squamous metaplasia of the seminal vesicle and prostate. Congenital defects included Harderian gland agenesis, tracheal cartilage malformations, and homeotic transformations along the rostral axial skeleton. *In utero* treatment with retinoic acid failed to produce lumbosacral truncations in RARγ mutant mice, indicating that RARγ mediates this teratogenic effect of RA.

Comments

The relatively mild RARγ-null phenotype suggests a high degree of functional redundancy among the RARs, which is confirmed by more severe phenotypes observed in double mutants.

Acknowledgements
Pierre Chambon
IGBMC, Illkirch, France

References

[1] Lohnes, D. et al. (1993) Cell 73, 643–658.
[2] Lohnes, D. et al. (1994) Development 120, 2723–2748.
[3] Mendelsohn, C. et al. (1994) Development 120, 2749–2771.

Other names
Retinoblastoma 1

Gene symbol
Rb1

Accession number
MGI: 97874

Area of impact
Neurology, hematopoiesis, development

General description

The *Rb1* gene encodes a nuclear protein that suppresses tumor development. It is widely expressed in all tissues. The Rb protein is a substrate for the cell cycle enzyme Cdc kinase 2. Rb becomes phosphorylated and then dephosphorylated during the cell cycle and binds to several oncoproteins.

KO strain 1 construction[1]

neo^r was inserted into exon 20, a region frequently mutated in tumors. The vector was electroporated into AB1 ES cells (129SvEv). Following positive–negative selection, the ES clones were injected into C57BL/6 blastocysts. Chimeras were crossed with C57BL/6.

KO strain 2 construction[2]

Stop codons were inserted into exon 3 of the *Rb1* gene. *neo*^{r} was placed in the intron. The vector was used to electroporate D3 ES cells (129). Chimeric mice from C57BL/6 blastocysts were crossed with C57BL/6 mice to generate progeny for breeding and analysis.

Phenotype

Animals which contained a null mutation in the retinoblastoma gene showed widespread cell death in the erythropoietic peripheral and central nervous systems and died at 13–15 days of gestation. Mice which were heterozygous for the deletion were grossly normal, but showed some propensity to develop pituitary adenocarcinomas, rather than retinoblastomas. The results from these and other Rb animals suggested that the retinoblastoma protein mediates a withdrawal from the cell cycle, which is a prerequisite to normal terminal differentiation. In the absence of Rb, cells continued to synthesize DNA, which led to the induction of apoptosis in some cells. Between E11.5 and 13.5, lens fiber cells showed greatly elevated levels of apoptosis in Rb$^{-/-}$ animals; this apoptosis was mediated by p53 (as shown with double mutants p53$^{-/-}$Rb$^{-/-}$[3]).

Erythropoiesis in the early liver E (13) was impaired in Rb$^{-/-}$ mice. The number of CFU-Es were normal in both KO strains, which suggests a defect in the maturation of erythroid progenitors. Widespread neuronal death was seen in the developing nervous system of $Rb^{-/-}$ mice, as well as ectopic cell divisions. The mutant phenotype was rescued by crossing the $Rb^{-/-}$ mice with a wild-type Rb cDNA transgenic line.

Comments

Rb plays a critical role in early development, but is not required for cell division or survival in early embryos.

References

[1] Lee, E.Y. et al. (1992) Nature 359, 288–294.
[2] Jack, S.T. et al. (1992) Nature 359, 295–300.
[3] Morgenhesser, S.D. et al. (1994) Nature 371, 72–77.

Other names
RBP-2N

Gene symbol
Rbpsuh

Accession number
MGI: 96522

Area of impact
Development

General description

RBP-Jκ is a transcription factor with unique DNA-binding properties. It regulates both viral and cellular genes in mammals and is the homolog of the *Drosophila Suppressor of Hairless* [Su(H)] gene. Su(H) is a key downstream component of the *Notch* receptor signaling pathway in development in *Drosophila.*

KO strain construction

A PGK*neo* cassette was inserted into exon 7 in the opposite transcriptional orientation. The regions surrounding the integrase motif, which is important for DNA-binding activity, should be disrupted by this integration. The vector was introduced into D3 129Sv ES cells, which were injected into C57BL/6 blastocysts. The mutant phenotype was studied on a mixed 129 × B6 background.

Phenotype

Homozygous mutant embryos died by E10.5, and were growth retarded as early as E8.5. As well as generalized retardation, embryos showed defective chorio-allantoic fusion, which probably led to embryonic lethality. Somites were poorly formed and disorganized. The phenotype observed is similar to but more severe than the phenotype of *Notch1* mutant, consistent with RBP-Jκ being in the *Notch* signaling pathway. Further analysis of markers of neurogenesis in both RBP-Jκ and *Notch1* mutant embryos has revealed that there is a neurogenic phenotype. Homologs of many of the genes involved in neurogenesis in *Drosophila* are misregulated in a manner consistent with conservation of the *Notch* signaling pathway from flies to mice.

Comments

Full analysis of the role of RBP-Jκ in *Notch* signaling will require tissue-specific knockouts, in which later developmental events can be analyzed.

References

[1] Oka, C. et al. (1995) Development 121, 3291–3301.
[2] de la Pompa, J. et al. (1997) Development 124, 1139–1148.

Gene symbol
Rel

Accession number
MGI: 97897

Area of impact
Immunity and inflammation, signal transduction

General description

c-Rel is a member of a family of genes encoding the NF-κB/Rel family of transcription factors. c-Rel dimerizes with other members of this family to form active transcription factor complexes which bind to κB sites in the enhancers of genes responsible for activation of immunoregulatory cytokines, cell surface receptors and acute phase proteins. In adult vertebrates, c-Rel expression is largely restricted to hematopoietic organs, with the highest expression found in B and T lymphocytes. c-Rel regulates the expression of genes in B and T cells that are involved in cell division and immune function. c-Rel expression levels change in a stage-specific manner during B cell development.

KO strain construction

A targeting vector was constructed which replaced exons 4–9 (encoding amino acid residues 145–588 of c-Rel) with the neomycin-resistance gene. The targeting vector was electroporated into the ES cell line W9.5 (129SvJ), and recombinant ES cells were injected into C57BL/6 blastocysts. Southern and Western blot analyses were used to confirm the homozygous null mutation of the *Rel* gene.

Phenotype

The c-Rel$^{-/-}$ mice were born at the expected Mendelian frequency, and were grossly normal in appearance. Lymphocyte development and hematopoiesis in the mutant mice were normal. Splenic B cell stimulation by LPS, CD40 ligand and IgM-specific mitogenic antibodies failed to induce proliferation. The colony formation frequency of B cells was diminished. T cells also failed to proliferate following ConA stimulation or anti-CD3 activation, a defect which was not overcome by costimulation with CD28. Stimulated T cell cultures exhibited markedly reduced levels of IL-2; the addition of exogenous IL-2 restored the proliferative capacity of T cells. IL-2, 4, 5 and 6 all failed to promote B cell proliferation. The c-Rel$^{-/-}$ mice also displayed defects in antibody production. Immune responses to immunization by the T cell-independent antigen NP-LPS showed that IgG3 levels were about 30% of wild-type levels. IgG1 levels in response to NP-KLH, which is a T cell-dependent antigen, were 100-fold lower for c-Rel-deficient mice. Other Rel family members were expressed at normal levels.

Comments

Mice lacking the c-Rel proto-oncogene exhibit defects in lymphocyte proliferation, humoral immunity and IL-2 expression. c-Rel is required for synthesis of IgG1 and IgG2a, and appears to be important for T cell-dependent humoral responses.

Acknowledgements
Frank Köntgen
The Walter and Eliza Hall Institute of Medical Research, The Royal Melbourne Hospital, Victoria, Australia

Reference

1 Köntgen, F. et al. (1995) Genes Dev. 9, 1965–1977.

Other names
p65

Gene symbol
Rela

Accession number
MGI: 103290

Area of impact
Immunity and inflammation, transcription factors

General description

NF-κB transcription factors are key regulators of genes involved in immune responses, inflammation and stress. All members share the Rel homology domain which is important for dimerization, nuclear translocation and DNA binding. RelA is a member of this family and typically exists in many different cell types as a heterodimer with the p50 subunit of the NF-κB family. Studies have shown that RelA is the transcriptional activator within the p50/RelA heterodimer. RelA can also heterodimerize with other members of the NF-κB family such as p52 and c-Rel.

KO strain 1 construction[1]

The targeting vector contained a 2.1 kb upstream genomic fragment (from 129Sv) encompassing exons 2–5 and part of exon 6 (codons 2–155) of the RelA gene, and a downstream fragment of 12 kb which started at codon 287 in exon 8 and included exon 11. These fragments were located on either side of the PGK*neo* cassette. The targeting vector was electroporated into J1 ES cells and positive clones were injected into C57BL/6J blastocysts. Male chimeras were crossed with C57BL/6J females to obtain heterozygotes, which were interbred to obtain RelA$^{-/-}$ homozygous mice.

Homologous recombination of this construct was expected to lead to truncation of the RelA polypeptide at residue 155, deleting 300 residues of the Rel domain. No RelA protein was detected in Western blots of extracts of RelA$^{-/-}$ embryonic fibroblasts using a C-terminal antibody against RelA.

KO strain 2 construction[2]

The targeting vector contained 7 kb of the mouse RelA gene (derived from C57BL/6) encompassing exons 1–6. pMC1*neo* was inserted into exon 1 at an *Nco*I site 3 bp downstream of the initiation codon. The HSV-*tk* gene was included 3′ of the RelA sequences. The targeting vector was transfected into CCE ES cells and positive clones were injected into C57BL/6 blastocysts to generate chimeras. Chimeras were mated with C57BL/6 mice to obtain heterozygotes, which were intercrossed to obtain RelA$^{-/-}$ mice. The null mutation was confirmed by Southern blotting and Northern analysis of transcripts in liver.

Phenotype

RelA-deficient mice died *in utero* between 14.5 and 16 days of gestation, apparently as a result of large-scale hepatocyte death. RelA$^{-/-}$ embryonic fibroblasts were impaired in TNF-mediated (NF-κB-dependent) induction of IκBα and GM-CSF, although basal transcription of these genes was normal. Fibroblasts and macrophages from RelA-deficient mice were shown to undergo cell death in the presence of TNFα, unlike RelA$^{+/+}$ cells. This effect was found to be dependent on TNF receptor 1. Studies of mice deficient in both RelA and TNF receptor 1 showed that the absence of TNF receptor 1 rescued embryonic lethality in RelA-deficient animals, since the double mutant animals were born with no obvious developmental abnormalities[5]. Thus, the embryonic lethality in RelA$^{-/-}$ animals is most likely due to TNF cytotoxicity to hepatocytes.

Fetal liver cells of E13.5 RelA$^{-/-}$ embryos (from heterozygous RelA$^{+/-}$ matings) transplanted into SCID mice resulted in normal T and B cell development in the recipients. T cells were able to mature to the Thy-1^{+}/TCRαβ^{+}/CD3^{+}/CD4^{+} or CD8^{+} stage. While RelA$^{-/-}$ B cells could differentiate to the IgM^{+}B220^{+} stage, and were able to secrete some immunoglobulins, secretion of IgG1 and IgA was reduced. Signal transduction appeared to be impaired in RelA$^{-/-}$ lymphocytes, since both mutant T and B cells exhibited decreased proliferation in response to stimulation by anti-CD3, anti-CD3 plus anti-CD28, ConA, anti-IgM, PMA plus calcium ionophore, or LPS. IL-2 production by RelA$^{-/-}$ spleen cells in response to these stimuli was normal.

Comments

RelA is required for mouse embryonic development and is a critical regulator of TNF-inducible genes. RelA is involved in the secretion of certain Ig isotypes and in signal transduction required for lymphocyte proliferation.

Acknowledgements
Amer Beg
Columbia University, New York, NY, USA

References

1 Beg, A.A. et al. (1995) Nature 376, 167–170.
2 Doi, T.S. et al. (1997) J. Exp. Med. 185, 953–961.
3 Beg, A.A. and Baltimore, D. (1996) Science 274, 782–784.
4 Baeuerle, P.A. and Henkel, T. (1994) Annu. Rev. Immunol. 12, 141–179.
5 Alcamo, E., Beg, A.A. and Baltimore, D. (unpublished results).

Gene symbol
Relb

Accession number
MGI: 103289

Area of impact
Immunity and inflammation, transcription factors

General description

RelB is a member of the NF-κB/Rel family of transcription factors which play important roles in the expression of genes involved in acute-phase reactions and the immune response. All members of the NF-κB/Rel family contain the Rel homology domain, a conserved sequence of about 300 amino acids which is important for dimerization, DNA binding and nuclear localization. RelB associates with p50 or p52 to form heterodimers, but apparently not with c-Rel or RelA (p65). RelB heterodimers are thought to be involved in regulating certain NF-κB-dependent genes that are constitutively expressed in lymphoid tissues. RelB is induced in activated fibroblasts but is constitutively expressed in bone marrow-derived lymphoid dendritic cells and a subset of thymic medullary epithelial cells.

KO strain 1 construction[1]

These RelB$^{-/-}$ mice arose as the result of the chance integration of an MHC class I gene into the *Relb* locus. The *Relb* gene was inactivated by the insertion of an H-2K^{k} transgene just downstream of the exon encoding the Rel domain, and just upstream of the C-terminal transactivation sequences. Downstream *Relb* sequences were intact. The transgenic line was backcrossed to C57BL/6 mice.

The disrupted *Relb* gene was shown to be a null mutation by RT-PCR analysis of spleen and thymus RNA and by Northern blotting. No normal or aberrant transcripts or splice products were detected in mutant mice. No RelB protein was detected by immunoblot of extracts of thymic medulla, splenic white pulp or various other tissues.

KO strain 2 construction[2]

The targeting vector contained *Relb* genomic sequences from D3 ES cell DNA. The PGK*neo* selection cassette replaced a 500 bp deletion in exon IV at the start of the DNA-binding subdomain of the Rel domain. PGK-*tk* was present 3′ of the *Relb* sequences. The 7.2 kb long arm 5′ of the *neo* cassette contained exon III and part of exon IV, while the short arm was 3.6 kb and contained exons V, VI and part of exon VII. The targeting vector was electroporated into D3 ES cells and positive clones were injected into C57BL/6 blastocysts. Male chimeras were bred to C57BL/6 females and heterozygous progeny were intercrossed to obtain RelB$^{-/-}$ mice.

The null mutation was confirmed by Western blotting of thymic extracts. No RelB protein was detected in tissues of $RelB^{-/-}$ mice and constitutive κB-binding activity was greatly reduced. No increase in the expression of the remaining wild-type allele occurred in $RelB^{+/-}$ heterozygotes. The expression of other NF-κB/Rel family members and of IκBα was not affected in $RelB^{-/-}$ mice.

Phenotype

$RelB^{-/-}$ mice were born at the expected Mendelian ratio and appeared normal at birth. However, 2–6 weeks later, most mutant mice began to appear progressively more ill, their fertility was decreased, and some died prematurely. Histologic examination showed alterations of the spleen, bone marrow, lymph nodes, lung and liver of all $RelB^{-/-}$ mice of age 8 days and older. Myeloid hyperplasia in the bone marrow and extramedullary hematopoiesis resulting in splenomegaly were observed. T and B lymphocyte development were normal. Strikingly, there was an apparent loss of lymphoid dendritic cell function in $RelB^{-/-}$ mice, as suggested by poor generation of immune responses (especially CTL), a failure in thymic negative selection of autoreactive T cells, and an absence of secondary lymphoid tissues (attributed in part to a requirement for dendritic cells in organizing lymphoid tissues). Possibly related to this loss of dendritic cells was the absence of thymic medulla and the UEA-1-positive subset of medullary epithelium. A second major phenotype in these mice was the presence of multiorgan inflammation, which increased in severity with time. The inflammation appeared to be due to the overproduction of cytokines/chemokines in non-lymphoid tissues, especially fibroblasts.

Comments

RelB is crucial for the normal function of the hematopoietic system, possibly involved in determining lineage commitment (particularly with respect to dendritic cells). Other members of the NFκB/Rel family are unable to compensate for the absence of RelB. RelB may mediate functions relating to the transition from acute inflammation to adaptive immunity.

Acknowledgements
David Lo
Scripps Research Institute, La Jolla, CA, USA

References

[1] Burkly, L. et al. (1995) Nature 373, 531–536.
[2] Weih, F. et al. (1995) Cell 80, 331–340.
[3] Lo, D. et al. (1992) Am. J. Pathol. 141, 1237–1246.
[4] DeKoning, J. et al. (1997) J. Immunol. 158, 2558–2566.
[5] Xia, Y. et al. (1997) Am. J. Pathol. 151, 375–387.

c-Ros

Other names
c-Ros 1

Gene symbol
Ros1

Accession number
MGI: 97999

Area of impact
Development

General description

c-Ros has transforming potential when mutated. The gene was initially identified in a transfection tumorigenicity assay using NIH 3T3 cells, and independently, as oncogene of the acutely transforming chicken retrovirus UR2. The proto-oncogene encodes a tyrosine kinase receptor which is closely related to the sevenless receptor in *Drosophila*. c-Ros is expressed in the epithelium of the Wolffian duct and its derivatives in the fetus. In the adult, expression is observed in males and associated with the epithelium of the epididymis.

KO strain construction

Two targeting vectors were generated: (a) insertion of the neomycin-resistance gene into an exon encoding extracellular sequences; (b) deletion of exon sequences encoding the tyrosine kinase domain which were replaced by a neomycin-resistance gene.

The mutations were introduced into E14-1 ES cells derived from 129Ola: the phenotype was analyzed on a mixed 129Ola/C57BL/6 background.

Phenotype

The homozygous mutant animals are viable and healthy. Homozygous mutant males are sexually active but infertile; female reproduction is not affected. c-Ros is not required in a cell autonomous manner for male germ cell development or function. The gene therefore does not directly affect sperm generation. The primary defect in mutant animals was located in the epididymis, where c-Ros controls appropriate development of epithelia, particularly their regionalization and terminal differentiation. The epididymal defect does not interfere with production or storage of sperm but, rather, with sperm maturation and the ability of the sperm to function *in vivo*.

Acknowledgements
Carmen Birchmeier
Max-Planck-Centrum für Molekulare Medizin, Berlin, Germany

Reference

1 Riethmacher-Sonnenberg, E. et al. (1996) Genes Dev. 10, 1184–1193.

Other names
Retinoic X receptor α

Gene symbol
Rxra

Accession number
MGI: 98214

Area of impact
Development

General description

RxRα is a nuclear receptor for 9-*cis* retinoic acid. It is also a heterodimeric partner required *in vitro* for high-affinity DNA binding for many nuclear receptors such as RARs, thyroid hormone receptors, vitamin D receptor or PPARs.

KO strain 1 construction[1]

A 1 kb *Eco*RI/*Xba*I genomic fragment containing exon 4 (encoding most of the DNA-binding domain) was replaced with a PGK*neo*polyA+ cassette (in reverse orientation). No RXRα protein was detected in mutants, demonstrating that this is a null mutation. The mutation was introduced into 129Sv ES cells, which were injected into C57BL6 blastocysts. The phenotype was examined on a 129 × B6 hybrid or 129 inbred background.

KO strain 2 construction[2]

A PGK*neo* cassette was introduced in an antisense orientation between an *Eco*RV site in the third exon and an *Xba*I site in the next intron. This removes part of the B/C1 coding region and the splice donor region. The vector was introduced into J1 129 ES cells which were injected into C57BL/6 blastocysts. The mutant phenotype was examined on a 129 × B6 background.

Phenotype

Homozygous embryos die *in utero* between 11.5 days pc and 17.5 days pc from defects in cardiac function. The mutant hearts show ventricular wall hypoplasia. This phenotype is associated with a premature differentiation and a reduced proliferation of subepicardial ventricular myocytes, and is identical to that found in vitamin A deficient (VAD) embryos. There are also ocular defects: thickening of the corneal stroma, shortened ventral retina, closer position of the eyelids, persistent retrolenticular mesenchyme. A similar defect occurs in VAD embryos, and RARβ/RARγ double mutants. There is a high degree of synergism between the RXRα and RAR mutations, as double mutants bearing the RXRα-null mutation and null alleles in one of the RARs recapitulate a large number of the defects observed in RAR compound mutants.

Comments

The synergy between RXRα and RAR mutations strongly supports the idea that RXRα/RAR heterodimers are the functional units transducing the retinoid signal *in vivo.*

Acknowledgements
Pierre Chambon
IGBMC, Illkirch, France

References

1 Kastner, P. et al. (1994) Cell 78, 987–1003.
2 Sucov, H.M. et al. (1994) Genes Dev. 8, 1007–1018.
3 Kastner, P. et al. (1997) Development 124, 313–326.

Other names
Retinoid X receptor β

Gene symbol
Rxrb

Accession number
MGI: 98215

Area of impact
Development

General description

RXRβ is a nuclear receptor for 9-*cis* retinoic acid. It is also a heterodimerization partner required *in vitro* for high-affinity DNA binding of many other nuclear receptors, such as RARs, TRS, VDR, PPAR, FXR and LXR.

KO strain construction

1.5 kb of genomic DNA containing the sequences encoding the DNA-binding domain (part of exon 3 and exon 4) was replaced with a PGK*neo*polyA^+ cassette. The mutation was introduced into 129Sv ES cells, which were injected into C57BL/6 blastocysts. The phenotype was examined on 129 × B6 hybrid or 129 inbred background.

Phenotype

Most homozygotes are viable although a fraction exhibit perinatal lethality. Males are sterile, due to oligo-astheno-teratozoospermia. Failure of sperm release occurs within the germinal epithelium, and most of the few spermatozoa produced exhibit abnormal acrosomes and tails. RXRβ mutants exhibit a progressive accumulation of unsaturated triglycerides within their Sertoli cells. In old mutant males, a progressive degeneration of the germinal epithelium occurs, ending with the formation of acellular lipid-filled tubules. RXRβ is specifically expressed in Sertoli cells, suggesting that the primary defect responsible for the abnormal spermatogenesis occurs in these cells.

Comments

The spermatogenesis defects in $RXR\beta^{-/-}$ mutants differ from those occurring in vitamin A deficiency, and thus likely reflect a function for RXRβ in a signaling pathway different from that of retinoic acid.

Acknowledgements
Pierre Chambon
IGBMC, Illkirch, France

Reference

[1] Kastner, P. et al. (1996) Genes Dev. 10, 80–92.

Other names
Retinoid X receptor γ

Gene symbol
Rxrg

Accession number
MGI: 98216

Area of impact
Development

General description

Retinoid X receptor γ belongs to the RXR subfamily of the nuclear receptors for retinoids, which are activated by 9-*cis* retinoic acid. Expression of this gene starts during mouse embryonic development at stage 10.5 days pc and persists throughout adult life. There are two isoforms, γ1 and γ2, which exhibit different expression patterns. In adult mice, γ1 is present in brain and lungs, whereas γ2 is present in adrenals, kidney and liver. Both isoforms are highly expressed in heart and muscle. In the brain, the expression of the RXRγ gene is strikingly strong in caudate putamen, yet significant in pituitary and cerebellum. Other brain regions such as hippocampus and hypothalamus also express RXRγ, albeit at much lower levels.

KO strain 1 construction[1]

In the targeting vector, the neomycin cassette replaced a 2.5 kb RXRγ genomic fragment containing exons 3 and 4 (DNA-binding domain). The regions of homology consisted of 5.5 kb and 2.5 kb sequences at the 5′ and 3′ sides of *neo*, respectively. The GTI.II-*tk* cassette allowed negative selection. The genetic background of the mutant strain was 129Sv × C57BL/6.

KO strain 2 construction[2]

Part of intron/exon3 was deleted and replaced with a *lacZ* PGK*neo* cassette. The splice acceptor sequence from the intron/exon junction was placed in-frame with *lacZ*, resulting in a RXRγ–β-galactosidase fusion. The mutation was introduced into 129Sv ES cells, which were injected into C57BL/6 blastocysts. The phenotype was examined on a mixed 129 × B6 hybrid background.

Phenotype

Homozygous mutants are viable, normal in appearance and fertile.

Comments

The phenotype may be masked due to functional redundancies with RXRα and RXRβ, which have widespread expression patterns during mouse development and in the adult.

Acknowledgements
Pierre Chambon
IGBMC, Illkirch, France
Ming-Yi Chiang
Salk Institute, La Jolla, CA, USA

References

[1] Krezel, W. et al. (1996) Proc. Natl Acad. Sci. USA 93, 9010–9014.
[2] The KO strain 2 mouse is unpublished (M.Y. Chiang).
[3] Dolle, P. et al. (1994) Mech. Dev. 45, 91–104.
[4] Kastner, P. et al. (1997) Development 124, 313–326.
[5] Mangelsodorf, D.J. et al. (1992) Genes Dev. 6, 329–344.

RyR1

Other names
Ryanodine receptor type-1, SKRR

Gene symbol
Ryr1

Accession number
MGI: 99659

Area of impact
Metabolism, signal transduction

General description

Two Ca^{2+}-release channels, inositol 1,4,5-triphosphate (IP3) receptors and ryanodine receptors (RyR), regulate intracellular Ca^{2+} stores and cytosolic Ca^{2+} concentrations. RyRs are homotetrameric complexes with a structure that spans the sarcoplasmic reticulum and the transverse tubules in the skeletal muscle. Three RyR subtypes encoded by three different genes exist: RyR1 in the skeletal muscle, RyR2 in the cardiac muscle and RyR3 in the brain. RyR1 is an ~5000 amino acid Ca^{2+} channel involved in excitation–contraction coupling in which depolarization is sensed by the tubule voltage sensor which then activates the Ca^{2+}-release channel. RyR2 mediates cardiac-type excitation–contraction coupling and is also expressed in the brain. RyR3 is widely expressed (smooth muscle, germ cells, cerebrum, cerebellum) and can induce intracellular Ca^{2+} release in response to increased Ca^{2+} concentrations at low sensitivity.

KO strain construction

A neomycin-resistance gene was inserted into exon 2. HSV-*tk* was placed at the 3′ end of the construct. Targeted J1 ES cells were injected into C57BL/6 blastocysts.

Phenotype

Neonatal RyR1-null mice failed to breathe or move and died perinatally, probably due to respiratory failure. They exhibited severe abnormalities of skeletal muscle, muscle degeneration, abnormal rib cage, abnormal curvature of the spine, arched vertebral column, thin limbs and a thick neck. Contractile muscle responses to electrical, but not to caffeine, stimulation were abolished in $RyR1^{-/-}$ mice.

Comments

RyR1 is involved in muscle differentiation and regulates excitation-contraction coupling in the skeletal muscle. $RyR1^{-/-}$ mice somewhat resemble myogenin-null and muscular dysgenic mice.

References

[1] Takeshima, H. et al. (1994) Nature 369, 556–559.
[2] Takeshima, H. et al. (1995) EMBO J. 14, 2999–3006.
[3] Giannini, G. et al. (1995) J. Cell Biol. 128, 893–904.

RyR3

Other names
Ryanodine receptor type 3

Gene symbol
Ryr3

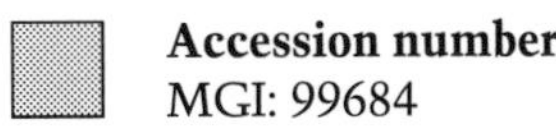

Accession number
MGI: 99684

Area of impact
Metabolism, signal transduction

General description

Two Ca^{2+}-release channels, inositol 1,4,5-triphosphate (IP3) receptors and ryanodine receptors (RyR), regulate intracellular Ca^{2+} stores and cytosolic Ca^{2+} concentrations. RyRs are homotetrameric complexes with a structure that spans the sarcoplasmic reticulum and the transverse tubules in the skeletal muscle. Three RyR subtypes encoded by three different genes exist: RyR1 in the skeletal muscle, RyR2 in the cardiac muscle and RyR3 in the brain. RyR1 is an ~5000 amino acid Ca^{2+} channel involved in excitation–contraction coupling in which depolarization is sensed by the tubule voltage sensor, which then activates the Ca^{2+}-release channel. RyR2 mediates cardiac-type excitation–contraction coupling and is also expressed in the brain. RyR3 is widely expressed (skeletal and smooth muscle, germ cells, cerebrum, cerebellum) and can induce intracellular Ca^{2+} release in response to increased Ca^{2+} concentrations at low sensitivity.

KO strain construction

A 0.9 kb fragment containing the putative promoter region and exon 1 were replaced with a neomycin-resistance gene. HSV-*tk* was placed at the 3′ end of the construct. Targeted J1 ES cells were injected into C57BL/6 blastocysts.

Phenotype

RyR3-null mice were viable, fertile, and grew normally. Excitation–contraction coupling was normal in muscle cells. Lymphocyte functions (ConA, IL-2 and LPS activation) and smooth muscles appeared normal. However, RyR3-null mice displayed 2-fold increased spontaneous locomotor activities and the Ca^{2+}-induced Ca^{2+} release from intracellular stores differed in sensitivity from the response in wild-type cells. In Ryr1$^{-/-}$Ryr3$^{-/-}$ double knockout mice[2], the Ca^{2+}-induced Ca^{2+} release from intracellular stores was completely abolished. Caffeine and choline chloride-induced Ca^{2+} releases were also blocked. Muscular defects were more severe in double knockout mice than in RyR1$^{-/-}$ mice. Ryr1$^{-/-}$Ryr3$^{-/-}$ double mutant mice also displayed severe muscular degeneration with swollen mitochondria and large vacuoles.

Comments

RyR3 appears to have a specific function in certain neurons of the CNS.

References

[1] Takeshima, H. et al. (1996) J. Biol. Chem. 271, 19649–19652.
[2] Ikemoto, T. et al. (1997) J. Physiol. 501.2, 305–312.

Sc1

Other names
Extracellular matrix protein, hevin

Gene symbol
Ecm2

Accession number
MGI: 108110

Area of impact
Neurology, immunity

General description

Sc1 is an extracellular matrix (ECM) protein enriched in brain, although it is present in many tissues. In the brain, it is found associated with astrocytes. In the human immune system, Sc1 is found associated with the high endothelial venule and has been called "hevin" to indicate this fact. It is closely related to SPARC, and has been shown to be anti-adhesive for endothelial cells and upregulated in astrocytes following focal mechanical trauma. The precise function of this protein is unknown.

KO strain construction

The construct was designed to replace exon 2 of the Sc1 gene with PGK*neo*. Exon 2 removal deleted the signal sequence for secretion, and created an out-of-frame mutation that introduced stop codons in all frames. Thus, this knockout generates a complete null mutation. The ES cells were of 129 origin and chimeras were bred to 129SvJ.

Phenotype

Mice homozygous for the targeted Sc1 gene did not produce any detectable protein in the nervous system. However, they appeared normal and were fertile. The phenotype of these mice is currently being examined.

Acknowledgements
Peter J. McKinnon
St. Jude's Hospital, Memphis, TN, USA

References

1. McKinnon, P.J. and Margolskee, R.F. (1996) Brain Res. 709, 27–36.
2. McKinnon, P. J. et al. (1996) Genome Res. 6, 1077–1083.
3. Johnston, I.G. et al. (1990) Neuron 2, 165–176.
4. Girard, J.-P. and Springer, T.A. (1996) J. Biol. Chem. 271, 4511–4517.

Other names
Sarcomeric mitochondrial creatine kinase

Gene symbol
Ckmt2 (proposed)

Area of impact
Metabolism

General description

CK isoenzymes catalyze the reversible conversion of phosphocreatine (PCr) plus ADP into creatine plus ATP. Cytosolic muscle CK exists as dimers composed of the M (muscle) and B (brain) subunits from which three isoenzymes, MM, BM and BB are derived. The cytosolic M and B subunits are encoded by the M-CK and B-CK genes. In addition, two mitochondrial CK isoforms (Mi-CK), called ubiquitous and sarcomeric Mi-CK (UbCKmit and ScCKmit), exist which can form octameric (317–377kDa) and dimeric (75–89 kDa) structures. Mi-CKs are localized at the inner mitochondrial membrane and are linked to oxidative phosphorylation. CK replenishes ATP levels during periods of energy consumption in the heart and skeletal muscle. In addition to the role of CK as an energy buffer during muscle contraction, CK functions in energy transport. PCr can be produced by two metabolic pathways catalyzed by the muscle CK homodimers and mitochondrial CKs. PCr diffuses to the myofibrillar M bands and serves to replenish ATP through action of the MM-CK. UbCKmit and the cytosolic B-CK isoenzymes are expressed in a variety of tissues with high and alternating energy requirements, including the brain, retina, smooth muscle, uterus, placenta, and spermatozoa. ScCKmit is co-expressed with M-CK in cardiac and skeletal muscle.

KO strain construction

Parts of intron 2 and exon 3 containing the ATG translation start site were replaced by a hygromycinB-polyA-resistance cassette. HSV-*tk* was placed 5′ to the *neo* cassette. Targeted E14 ES cells were injected into C57BL/6 blastocysts.

Phenotype

ScCKmit$^{-/-}$ mice were viable and fertile and exhibited no obvious abnormalities. Unlike M-CK$^{-/-}$ mice, muscles from ScCKmit$^{-/-}$ mice exhibited no changes in morphology, performance, force/relaxation or oxidative phosphorylation. The levels of high-energy phosphate metabolites and the expression of UbCKmit were also not changed in the null mutant mice.

Comments

The phenotype of the ScCKmit$^{-/-}$ mouse indicates that muscular function is not absolutely dependent on the presence of ScCKmit. These results challenge the concept that Ckmit function and aerobic respiration are necessarily coupled.

Reference

[1] Steeghs, K. et al. (1997) J. Neurosci. Meth. 71, 29–41.

SDF-1/PBSF

Other names
pre-B cell growth stimulating factor, stromal cell-derived factor

Gene symbol
Sdf1

Accession number
MGI: 103556

Area of impact
Hematopoiesis

General description

SDF-1/PBSF is a CXC chemokine which augments the proliferation of B cell progenitors and is chemotactic for lymphocytes, monocytes and $CD34^+$ hematopoietic progenitors. SDF-1/PBSF mRNA is constitutively expressed in bone marrow stromal cells, embryos and in many adult organs, including brain, thymus, heart, lung, liver, spleen and small intestine.

KO strain construction

The *Sdf1* gene (derived from strain 129) was inactivated in exon 2 which contains three cysteine residues and the presumptive receptor-binding region. Exon 2 was replaced by the *neo* gene and the HSV-*tk* gene was attached 5′ of the genomic sequences. The targeting vector was electroporated into E14-1 ES cells and positive clones were injected into blastocysts to generate chimeric mice. The background of the mutant mice was C57BL/6 × 129Sv.

Phenotype

About half of the $SDF\text{-}1/PBSF^{-/-}$ embryos died by E18.5, and the surviving mutant neonates died within one hour of birth. In the mutant fetal livers at E18.5, pro-B ($B220^+CD43^+$) and pre-B ($B220^+CD43^-$) cells and the numbers of CFU-IL-7 were severely reduced compared to the wild-type fetal liver. The numbers of granulocytes, monocytes, erythroid-myeloid mixed colonies (CFU-Mix), erythroid colonies (CFU-E) and myeloid colonies in the E18.5 mutant livers were almost normal. At E18.5, developing granulocytic cells were easily observed in the bone marrow cavities of control embryos but were virtually absent from mutant bone marrow. The numbers of CFU-Mix, CFU-GM and CFU-IL-7 were greatly reduced. However, the structure of cortical bones and trabeculae, the formation of bone marrow cavities, and the number of osteoblastic cells lining the cortical bone and trabeculae were normal. The hearts of $SDF\text{-}1/PBSF^{-/-}$ embryos had a defect of the membranous portion of the ventricular septum from E13.5 to the neonate stage.

Comments

SDF-1/PBSF is an essential cytokine for B lymphopoiesis during embryonic development. SDF-1/PBSF is also required for myelopoiesis in the bone marrow but not in the fetal liver, suggesting that SDF-1/PBSF may support colonization of the bone marrow by hematopoietic precursors. Finally, SDF-1/PBSF is required for heart ventricular septum formation.

Acknowledgements
Takashi Nagasawa
Research Institute, Osaka Medical Center for Maternal and Child Health, Osaka, Japan

References

1 Nagasawa, T. et al. (1996) Nature 382, 635–638.
2 Tashiro, K. et al. (1993) Science 261, 600–603.
3 Nagasawa, T. et al. (1994) Proc. Natl Acad. Sci. USA 91, 2305–2309.
4 Bleul, C. et al. (1996) J. Exp. Med. 184, 1101–1109.
5 Aiuti, A. et al. (1997) J. Exp. Med. 185, 111–120.

Other names
SAPK/ERK/kinase-1, c-Jun N-terminal kinase-1 (JNKK1), Map kinase kinase 4 (MKK4)

Gene symbol
Serk1

Accession number
MGI: 107199

Area of impact
Signal transduction

General description
Distinct and evolutionarily conserved signal transduction cascades mediate cell survival or death in response to developmental and environmental cues. Stress-activated protein kinases or Jun-N-terminal kinases (SAPKs/JNKs) are activated in response to a variety of cellular stresses, such as changes in osmolarity and metabolism, DNA damage, heat shock, ischemia or inflammatory cytokines. SEK1 (JNKK1/MKK4) is a member of the mitogen-activated protein kinase kinase group of dual specificity protein kinases. It functions as a direct activator of SAPKs/JNKs in response to environmental stresses or mitogenic factors.

KO strain 1 construction[1]
The transposon Tn5 neomycin-resistance gene (*neo*) driven by a eukaryotic promoter (PGK) was cloned into a *Bgl* II site in exon 2 of the murine *Serk1* gene. E14K ES cells (derived from strain 129J) were electroporated with the targeting construct. Mutations of both *Serk1* alleles were established from SEK1$^{+/-}$ ES cells by selection in a high dose of G418. SEK1 protein was absent in SEK1$^{-/-}$ ES cells. To generate SEK1$^{-/-}$ somatic chimeras, two independently established SEK1$^{-/-}$ ES cell clones were injected into RAG-2$^{-/-}$ blastocysts.

KO strain 2 construction[2]
An internal 3.5 kb *Xba*I/*Apa*I genomic fragment was replaced with the PGK*neo* gene cassette. The deleted region encompassed two exons encoding amino acids 211–268 of the murine SEK1 sequence. SEK1$^{+/-}$ ES cells of the W9.5 line (129SvJ) were injected into BALB/c blastocysts and resulted in chimeric mice that transmitted the disrupted SEK1 allele through the germ line.

Phenotype
Heterozygous SEK1$^{+/-}$ mice appeared to be healthy, since they were fertile and of normal size. However, SEK1$^{-/-}$ mice could not be obtained from intercrosses of heterozygous SEK1$^{+/-}$ mice. Homozygous SEK1$^{-/-}$ animals died between E10.5 and 12.5 due to a defect in liver formation. Hematopoiesis and vasculogenesis were normal in SEK1$^{-/-}$ embryos, indicating that SEK1 may mediate a signal for hepatocyte survival during liver morphogenesis[3].

Deletion of SEK1 in ES cells prevented SAPK/JNK activation in response to heat shock and the protein synthesis inhibitor anisomycin. In contrast, UV irradiation and sorbitol-induced osmotic shock continued to activate SAPKs/JNKs to normal levels in SEK1$^{-/-}$ ES cells. Activation of p38/Mpk2/CSBP in SEK1$^{-/-}$ ES cells was normal in response to anisomycin and sorbitol. Induction of MAPKs by phorbol myristyl acetate (PMA) and Ca^{2+} ionophore was also unaffected in SEK1$^{-/-}$ ES cells. These results show that different stresses utilize distinct SEK1-dependent and SEK1-independent intracellular signaling pathways for SAPK/JNK activation. A novel SAPK/JNK activator called SEK2/MKK7/JNKK2 has recently been cloned[4].

SEK1$^{-/-}$RAG-2$^{-/-}$ chimeric mice had normal numbers of mature T cells but fewer immature CD4$^+$CD8$^+$ thymocytes. The SEK1$^{-/-}$ mutation did not affect the induction of apoptosis in response to environmental stresses (γ-irradiation, UV irradiation, sorbitol, anisomycin or heat shock) in ES and T cells. However, SEK1$^{-/-}$ thymocytes were more susceptible to CD95 (Fas)- and CD3-mediated apoptosis than wild-type cells. SEK1$^{-/-}$ peripheral B cells displayed normal responses to IL-4, IgM and CD40 crosslinking. However, SEK1$^{-/-}$ peripheral T cells showed decreased proliferation and IL-2 production after CD28 costimulation and PMA/Ca2$^+$ ionophore activation[5]. Moreover, it was found that SEK1- and SEK2-regulated pathways for SAPK/JNK activation were developmentally regulated in T cells[5].

Comments

SEK1 signaling and SEK1-regulated SAPK/JNK activation are not essential for the induction of apoptosis in lymphocytes and ES cells. SEK1 may mediate survival signals during T cell development and hepatogenesis. SEK1 is an important effector molecule that relays CD28 signals resulting in IL-2 production and T cell proliferation.

During *Drosophila* embryogenesis, a dorsal cell sheet movement called dorsal closure allows establishment of the dorsal epidermis[6]. In the mutant hemipterous (hep), spreading of the epithelia is blocked. The HEP protein is homologous to the SEK2/MKK7 group of mitogen-activated protein kinase kinases (MAPKKs).

Acknowledgements
Hiroshi Nishina
Department of Physiological Chemistry, Graduate School of Pharmaceutical Sciences, University of Tokyo, Tokyo, Japan
Josef Penninger
Amgen Institute, Toronto, ON, Canada

References

[1] Nishina, H. et al. (1997) Nature 385, 350–353.
[2] Yang, D. et al. (1997) Proc. Natl Acad. Sci. USA 94, 3004–3009.
[3] Nishina, H. et al. (1998) unpublished results.
[4] Yao, Z. et al. (1997) J. Biol. Chem. 272, 32378–32394.
[5] Nishina, H. et al. (1997) J. Exp. Med. 186, 941–953.
[6] Glise, B. et al. (1995) Cell 83, 451–461.
[7] Sanchez, I. et al. (1994) Nature 372, 794–798.

E-selectin

Other names
ELAM-1 (endothelial-leukocyte adhesion molecule 1), LECAM-2 (leukocyte-endothelial cell adhesion molecule 2), CD62E

Gene symbol
Sele

Accession number
MGI: 98278

Area of impact
Immunity and inflammation

General description

Selectins (including E-selectin, L-selectin and P-selectin) are adhesion receptors mediating the interaction between leukocytes, platelets and the endothelium. E-selectin is expressed on the surface of endothelial cells after induction by inflammatory cytokines, such as IL-1, TNFα and LPS. Non-inflammatory expression has also been observed in dermal, placental and bone marrow endothelium. E-selectin counter-receptors include PSGL-1, ESL-1, CLA and other entities bearing sialylated Lewis X/A. E-selectin has been shown to mediate leukocyte rolling on activated endothelium at inflammatory sites, may also support tumor cell adhesion during hematogenous metastasis, and may play roles in angiogenesis and hematopoiesis.

KO strain 1 construction[1]

The 1.9 kb *Bam*HI/*Nsi* I genomic sequence of E-selectin, comprising the signal peptide, lectin domain (LEC) and part of the EGF domain, was deleted and replaced by a hygromycin B-PGK cassette. The replacement vector included 3.2 kb and 3.9 kb of genomic DNA flanking the hygromycin B-PGK cassette. E-selectin single KO mice arose from ES cells in which mutations in P- and E-selectins were on different chromosomes. Chimeric mice and heterozygous offspring were of C57BL/6 and 129Sv mixed background.

KO strain 2 construction[2]

The positive/negative selection targeting construct used isogeneic (129Svter) E-selectin genomic lambda phage clones, PGK*neo*^{r} and MC1-HSV-*tk*. In the mutant allele, sequences between *Bam*HI (exon 1) and *Sac*I (exon 3) (including the signal sequence through the middle of the lectin domain) were replaced by PGK*neo*^{r} (in reverse transcriptional orientation with respect to E-selectin). Mutant alleles from two independently targeted clones of J1 (129Svter) ES cells were backcrossed onto C57BL/6J, BALB/cJ and 129SvEv backgrounds. Lack of E-selectin expression by the mutant allele was documented at both the mRNA and protein levels.

KO strain 3 construction[3]

The targeting vector contained an MC1*neo* gene inserted after amino acid 162 within the lectin domain, 12 kb of 5′ and 3′ homology to the E-selectin locus, and the HSV-*tk* gene for negative selection. The construct disrupted the critical carbohydrate-binding domain. W9.5 ES cells were transformed with the construct. Positive ES clones were injected into C57BL/6J blastocysts to generate chimeras. Heterozygotes were intercrossed to generate homozygous E-selectin$^{-/-}$ mutants.

No E-selectin mRNA of wild-type size was identified by Northern blot on lung RNA from mutants treated with LPS to induce expression of endothelial selectins. Similarly, RT-PCR analysis confirmed the lack of mRNA containing intact EGF or LEC sequences in a variety of tissues examined 4 hours after IL-3 treatment. No E-selectin protein was synthesized in E-selectin$^{-/-}$ mice as demonstrated by immunoprecipitation of [^{35}S]methionine labeling of hearts, and by flow cytometric analysis.

Phenotype

E-selectin-null mice were born and survived at expected Mendelian ratios, appeared healthy and reproduced without difficulty. Basal and induced expression levels of other endothelial and leukocyte adhesion molecules, including P-selectin, VCAM-1, ICAM-1, L-selectin and CD11b were not altered in E-selectin$^{-/-}$ mice. White blood cell counts were comparable to those in wild-type mice, as was the response to thioglycollate-induced peritonitis. The recruitment of neutrophils to skin wound sites, as well as healing of skin wounds, were also normal. Microvascular endothelial cells isolated from E-selectin$^{-/-}$ mice and expanded in culture performed normally in several *in vitro* assays of capillary tube formation.

Although E-selectin$^{-/-}$ mice did not display overt inflammatory or immunological deficiencies, intravital microscopy revealed that endogenous circulating leukocytes displayed decreased firm adhesion to dermal microvessels activated by human TNFα. Function-blocking antibodies decreased leukocyte-mediated damage in a rabbit model of lung inflammation, but E-selectin-null mice did not show changes in leukocyte recruitment in other inflammatory settings. However, significant inhibition of neutrophil emigration in models of inflammation was observed in E-selectin$^{-/-}$ mice when the function of P-selectin was blocked by anti-murine P-selectin antibody.

Double E-selectin$^{-/-}$L-selectin$^{-/-}$ KO strain construction[4]

L-selectin is expressed on most leukocytes. A PGK-puromycin-resistance cassette was inserted in place of a 2.3 kb segment of the L-selectin gene, including all of the lectin domain and most of the EGF domain. Flanking segments of 1.8 kb and 4.3 kb were added 5′ and 3′. The targeting construct was transfected into 129Sv-derived ES cells (D3) which already contained P-selectin and E-selectin mutations in trans[4]. After puromycin selection, surviving clones were injected into C57BL/6 recipient blastocysts.

Double E-selectin$^{-/-}$L-selectin$^{-/-}$ KO mutant phenotype

Homozygotes were viable and fertile. Other aspects of the mutant mouse phenotype are under investigation.

Comments

E-selectin is not required for normal mouse development. Neutrophil migration requires at least one endothelial selectin, and E-selectin and P-selectin appear to overlap in function.

Acknowledgements
Richard Hynes
Center for Cancer Research, Massachusetts Institute of Technology, Cambridge, MA, USA
David Milstone and Michael Gimbrone
Department of Pathology, Brigham & Women's Hospital, Boston, MA, USA
Denisa Wagner
Center for Blood Research Inc., Boston, MA, USA

References

[1] Frenette, P.S. et al. (1996) Cell 84, 563–574.
[2] Bevilacqua, M.P. et al. (1989) Science 243, 1160–1165
[3] Labow, M.A. et al. (1994) Immunity 1, 709–720.
[4] Robinson, S.D. et al. (1997) Microcirculation 4, 131.
[5] Subramaniam, M. et al. (1997) Am. J. Pathol. 150, 1701–1709.
[6] Mayadas, T.N. et al. (1993) Cell 74, 541–554.
[7] Gerritsen, M.E. et al. (1996) Lab. Invest. 75, 175–184.

Other names
CD62L, LAM-1, MEL-14, Leu-8, TQ1, LEC.CAM-1, LECAM-1, Lyam-1

Gene symbol
Sell

Accession number
MGI: 98279

Area of impact
Immunity and inflammation, adhesion

General description

L-selectin is a member of the selectin family of adhesion molecules constitutively expressed by all classes of leukocytes. L-selectin binds carbohydrate groups on several counter-receptors expressed on endothelial cells, thereby mediating the initial interaction of leukocytes with the vascular endothelium. It acts as a "homing receptor" for lymphocytes and facilitates the binding of these cells to the specialized high endothelial venules (HEVs) of peripheral lymph nodes. L-selectin also recruits leukocytes and mediates their "rolling" at sites of tissue injury or inflammation.

KO strain 1 construction[1]

The targeting vector was derived from a genomic fragment containing L-selectin exons 2–5 obtained from a 129Sv library. The exon encoding the L-selectin lectin domain was replaced with the neomycin-resistance cassette. The targeting vector was electroporated into E14-1 ES cells and positive clones were injected into C57BL/6 blastocysts. Chimeric males were bred with C57BL/6 females to generate heterozygotes. 129 × C57BL/6 mice heterozygous for the targeted allele were intercrossed and L-selectin$^{+/-}$ and L-selectin$^{-/-}$ mice were obtained at the expected Mendelian frequency. The complete loss of L-selectin cell surface expression was confirmed by flow cytometric and immunofluorescence analyses of blood leukocytes, and by functional assays in which mutant lymphocytes were unable to attach to HEVs in *in vitro* assays.

KO strain 2 construction[2]

The targeting construct was designed such that a 5.5 kb fragment of the genomic L-selectin gene (derived from 129SvJ) encoding the L-selectin domain was replaced by a neomycin-resistance cassette in the reverse transcriptional orientation. The deletion included part of exon 3 and all of exons 4–6. The targeting vector was electroporated into D3 ES cells and positive clones were injected into C57BL/6 blastocysts. Chimeras were mated to C57BL/6 mice and offspring heterozygous for the mutated allele were intercrossed to obtain homozygous L-selectin$^{-/-}$ mice.

The null mutation was confirmed by flow cytometry of mutant PBL and immunostaining for expression of L-selectin on cells from mutant thymus, spleen and lymph nodes. No L-selectin was detected on the surface of these cells.

KO strain 3 construction[3]

A 6.6 kb fragment of L-selectin genomic DNA containing exons 2–6 was used to construct a replacement-type vector. Exon 3, which contains the lectin domain, was disrupted by the insertion (in the same transcriptional orientation) of a neomycin-resistance cassette. An HSV-*tk* cassette was placed 3′ of the genomic sequences. The targeting construct was introduced into J1 or AB1 ES cells and positive clones were injected into C57BL/6J blastocysts. Chimeras were mated to C57BL/6J mice and offspring heterozygous for the mutated allele were intercrossed to obtain homozygous L-selectin$^{-/-}$ mice. There was no difference in phenotype between mutant mice derived from J1 ES cells and those derived from AB1 ES cells.

The null mutation was confirmed by Northern analysis and a lack of expression of L-selectin on leukocytes of mutant mice.

KO strain 4 construction[4]

A PGK-puromycin-resistance cassette was inserted in place of a 2.3 kb segment of the L-selectin gene, including all of the lectin domain and most of the EGF domain. Flanking segments of 1.8 kb and 4.3 kb were added 5′ and 3′. The targeting construct was transfected into 129Sv-derived ES cells (D3) which already contained null mutations of the P- and E-selectin genes in *cis*[5]. Clones surviving puromycin selection were injected into C57BL/6 recipient blastocysts. The L-selectin$^{-/-}$ mice arose from ES cells in which the L-selectin mutation resided on the opposite chromosome to the P- and E-selectin mutations.

Phenotype

KO strain 1 L-selectin$^{-/-}$ mice were viable and fertile but demonstrated a complete lack of lymphocyte migration across HEVs of peripheral lymph nodes (PLNs). Migration across mesenteric lymph node and Peyer's patch HEVs was also impaired. This migratory defect resulted in a 70–90% decrease in the number of lymphocytes within PLN and a 30–55% increase in spleen cellularity. Elevated serum IgM and IgG1 levels and augmented humoral immune responses to T cell-independent and -dependent antigens following i.p. immunization were observed. In contrast, s.c. immunization with T cell-dependent antigens resulted in a reduced IgM response and a severely delayed IgG response. However, secondary responses were normal or increased in L-selectin$^{-/-}$ mice regardless of the immunization route.

KO strain 1 L-selectin-deficient mice also demonstrated greatly reduced leukocyte emigration at sites of inflammation. Leukocyte rolling was reduced ~70% 1 hour following exteriorization of the mesentery. As a result, KO strain 1 mice had markedly reduced leukocyte influx during peritonitis, impaired delayed type hypersensitivity (DTH) responses, delayed skin allograft rejection and increased resistance to LPS-induced toxic shock[8,9].

Studies of KO strain 2 L-selectin$^{-/-}$ mice showed that they were also impaired in cutaneous DTH responses when tested after 4 days, and primary T cell proliferative responses and IL-2, IL-4 and IFN-γ production were impaired when examined at 5 days post-immunization. However, at 9 days post-immunization, normal responses were observed, suggesting that mechanisms exist that, with time, can compensate for L-selectin deficiency. In KO strain 2 mice, primary humoral responses to a protein antigen and memory T cell responses were normal.

Studies of KO strain 3 L-selectin$^{-/-}$ mice confirmed the impaired contact hypersensitivity responses to reactive haptens. No induction of contact hypersensitivity responses was observed at later times after immunization. Epidermal Langerhans cells were normal in number and behavior, and T cells, neutrophils and monocytes were capable of entering inflamed skin sites. However, while antigen presentation and effector mechanisms were intact, essentially no antigen-specific T cells could be found in the draining peripheral lymph nodes after a contact challenge.

KO strain 4 L-selectin$^{-/-}$ mice were viable and fertile and similar in phenotype to KO strain 1 mice.

Comments

These studies confirm a critical role for L-selectin in the normal recirculation of lymphocytes, and in mediating leukocyte emigration into tissues at sites of inflammation. L-selectin is apparently essential for normal DTH responses. In the absence of L-selectin, antigen-specific T cells may be unable to home to the appropriate lymph nodes for activation. L-selectin appears to be required for the generation of primary T cell responses, but not necessarily for humoral or memory responses.

Since L-selectin can also serve as a signal-transducing molecule, some phenotypic aspects may result, in part, from a loss of receptor signaling. However, the phenotype of this mouse is less dramatic than that observed following mAb blockade of L-selectin ligand binding, which suggests that some of the mAb effects may result from signaling through L-selectin.

Acknowledgements
Douglas A. Steeber
Department of Immunology, Duke University Medical Center, Durham, NC, USA
Richard Hynes
Center for Cancer Research, Massachusetts Institute of Technology, Cambridge, MA, USA

References

1 Arbones, M.L. et al. (1994) Immunity 1, 247–260.
2 Xu, J. et al. (1996) J. Exp. Med. 589–598.
3 Catalina, M.D. et al. (1996) J. Exp. Med. 184, 2341–2351.
4 Robinson, S.D. et al. (1997) Microcirculation 4, 131.
5 Frenette, P.S. et al. (1996) Cell 84, 563–574.
6 Steeber, D.A. et al. (1996) J. Immunol. 157, 1096–1106.
7 Steeber, D.A. et al. (1996) J. Immunol. 157, 4899–4907.

[8] Tedder, T.F. et al. (1995) J. Exp. Med. 181, 2259–2264.
[9] Tang, M.L.K. et al. (1997) J. Immunol. 158, 5191–5199.
[10] Steeber, D.A. et al. (1997) J. Immunol. 159, 952–963.
[11] Tedder, T.F. et al. (1995) FASEB J. 9, 866–873.

Other names
PADGEM, GMP-140, CD62P

Gene symbol
Selp

Accession number
MGI: 98280

Area of impact
Immunity and inflammation

General description

P-selectin is a member of the selectin family of leukocyte adhesion receptors and interacts with neutrophils, monocytes, eosinophils, T cells and platelets. P-selectin is stored in the α granules of platelets and Weibel–Palade bodies of endothelial cells, and is rapidly presented on the cell surface after activation of these two cell types. Like the other selectins, the lectin domain of P-selectin is the binding site for carbohydrate-containing ligands. The expression of E-selectin, found only in endothelial cells, is induced by inflammatory cytokines. L-selectin is expressed on most leukocytes. All three selectins bind carbohydrates and participate in recruitment of leukocytes.

KO strain construction[1]

A 62 bp fragment of the intron sequence preceding exon 3, and 111 bp of the protein encoding exon 3 (which represents the last 10 of the 41 amino acid signal peptide and 27 amino acids of the lectin domain), were replaced with PGK*neo*. Flanking HSV-*tk* cassettes were placed at both ends of the targeting vector. The targeting vector was transfected into D3 ES cells derived from the 129Sv strain. Positive clones were injected into C57BL/6 blastocysts, so that the mutant mice were of mixed 129Sv/C57BL/6 background. Animals were also backcrossed to C57BL/6, BALB/c and 129Sv.

The null mutation of the P-selectin gene was confirmed by Northern blot analysis of lung and liver RNA from mice treated with LPS, by immunofluorescence analysis of lung tissue and activated platelets, and by immunoprecipitation of [^{35}S]-cysteine-labeled proteins from inflamed lung tissue. Expression of other selectins was not affected.

Phenotype

Homozygous P-selectin-deficient mice were born at the expected Mendelian frequency, appeared grossly normal, and were viable and fertile. P-selectin-deficient mice showed no obvious alterations in lymphocyte differentiation, distribution or adhesion to high endothelial venules in peripheral lymph nodes. Normal levels of platelets, reticulocytes and bone marrow precursors were found. However, P-selectin-deficient mice had moderate peripheral neutrophilia; that is, mutant mice had abundant circulating neutrophils that were unable to leave the blood effectively at early stages of inflammation.

Mutant mice showed no leukocyte rolling in the venules of the exteriorized mesentery, and severely compromised leukocyte rolling in the dorsal skin microvasculature. Leukocyte accumulations in several murine models of acute and chronic inflammation were decreased. Neutrophil and macrophage accumulation were reduced in the peritoneal cavity following an intraperitoneal injection of thioglycollate; a lag of 1–2 hours in neutrophil recruitment was observed. $CD4^+$ T lymphocytes, monocytes and neutrophils were decreased but no changes in vascular permeability or edema at sites of contact hypersensitivity were noted. Reductions in cerebrospinal fluid neutrophil accumulation and blood–brain barrier permeability following cytokine-induced meningitis were observed. Early neutrophil accumulation was reduced following a cutaneous full thickness skin wound.

P-selectin-deficient mice crossed to mice lacking the low-density lipoprotein receptor (LDLR) formed significantly smaller diet-induced fatty streaks in the cusp region of aortae, suggesting that P-selectin is important in the development of early atherosclerotic lesions. *In vitro*, P-selectin deficiency in platelets abrogated neutrophil interactions with thrombin-stimulated platelets.

Double P-selectin$^{-/-}$E-selectin$^{-/-}$ KO strain construction[2]

The 1.9 kb *Bam*H1/*Nsi*1 genomic sequence of E-selectin, comprising the signal peptide, lectin domain and part of the EGF domain, was deleted and replaced by a hygromycin B-PGK cassette. The replacement vector included 3.2 kb and 3.9 kb of genomic DNA flanking the hygromycin B-PGK cassette. Double KO mice arose from ES cells in which P- and E-selectin mutations were on the same chromosome. Chimeric mice and heterozygous offspring were of C57BL/6 and 129Sv mixed background.

Phenotype

The homozygous P-selectin$^{-/-}$E-selectin$^{-/-}$ double mutant mice were viable and fertile. The mice displayed severe leukocytosis (3.9-fold elevation) composed mainly of mature neutrophils (16-fold elevation), but the numbers of monocytes, eosinophils and lymphocytes were also increased. Granulocytopoiesis was increased both in the bone marrow and the spleen as evaluated by differential counts of bone marrow precursors, histologic examinations, and analysis of progenitors of both the bone marrow and spleen. Circulating hematopoietic cytokines (IL-3 and GM-CSF) were found to be elevated in double mutants. Severely impaired leukocyte rolling in inflamed venules (induced by TNFα) and compromised neutrophil influx, as seen in thioglycollate-induced peritonitis, made these animals susceptible to bacterial skin infections. These spontaneous skin infections were prevented by broad-spectrum antibiotics. Compared to wild-type or mice lacking a single endothelial selectin, double KO mutant mice experienced severe and prolonged (at least 3 days) defects in neutrophil recruitment into skin wounds, as well as delays in wound closures. In a meningitis model, cytokine-induced neutrophil recruitment was almost completely ablated.

Double P-selectin$^{-/-}$L-selectin$^{-/-}$ KO strain construction[3]

A PGK-puromycin-resistance cassette was inserted in place of a 2.3 kb segment of the L-selectin gene, including all of the lectin domain and most of the EGF domain. Flanking segments of 1.8 kb and 4.3 kb were added 5′ and 3′. The targeting construct was transfected into 129Sv-derived ES cells (D3) which already contained P-selectin and E-selectin mutations in *trans*[2]. After puromycin selection, surviving clones were injected into C57BL/6 recipient blastocysts.

Phenotype

Homozygotes were viable and fertile. Mice deficient in P-selectin and L-selectin showed moderate peripheral neutrophilia similar to that seen in P-selectin-deficient mice. Other aspects of the phenotype are under investigation.

Triple P-selectin$^{-/-}$L-selectin$^{-/-}$E-selectin$^{-/-}$ KO strain construction[3]

A PGK-puromycin-resistance cassette was inserted in place of a 2.3 kb segment of the L-selectin gene, including all of the lectin domain and most of the EGF domain. Flanking segments of 1.8 kb and 4.3 kb were added 5′ and 3′. The targeting construct was transfected into 129Sv-derived ES cells (D3) which already contained null mutations of the P-selectin and E-selectin genes in *cis*[2]. After puromycin selection, surviving clones were injected into C57BL/6 recipient blastocysts. The triple KO mice arose from ES cells in which the L-selectin mutation resided on the same chromosome as the P- and E-selectin mutations.

Phenotype

Homozygous triple mutant mice were viable and fertile. They displayed severe leukocytosis comparable to that seen in P- and E-selectin double knockout mice. Triple mutant mice also mimicked the P- and E-selectin double knockouts by being susceptible to bacterial skin infections. Other aspects of the phenotype are under investigation.

Comments

P-selectin is important in leukocyte interactions with the walls of blood vessels and essential for extravasation of neutrophils in early inflammation. P-selectin is not required for normal development, including vasculogenesis and angiogenesis, or for the production of platelets. Mutant mice were unexpectedly healthy under conditions of standard animal husbandry, implying overlapping function with other selectins.

Absence of more than one endothelial selectins severely affects several aspects of leukocyte dynamics and indicates that these selectins are as important for leukocyte functions as are leukocyte β2 integrins. It is unclear whether the abnormalities in hematopoiesis arise from subclinical low-grade infections, or whether there is a direct function of the selectins in hematopoiesis.

Acknowledgements
Denisa Wagner
Center for Blood Research Inc., Boston, MA, USA
Richard Hynes
Center for Cancer Research, Massachusetts Institute of Technology, Cambridge, MA, USA

References

[1] Mayadas, T.N. et al. (1993) Cell 74, 541–554.
[2] Frenette, P.S. et al. (1996) Cell 84, 563–574.
[3] Robinson, S.D. et al. (1997) Microcirculation 4, 131.
[4] Johnson, R.C. et al. (1995) Blood 86, 1106–1114.
[5] Subramaniam, M. et al. (1995) J. Exp. Med. 181, 2277–2282.
[6] Tang, T. et al. (1996) J. Clin. Invest. 97, 2485–2490.
[7] Johnson, R.C. et al. (1997) J. Clin. Invest. 99, 1037–1043.

Other names
Seizure-related gene 6

Gene symbol
Sez6

Accession number
MGI: 104745

Area of impact
Neurology

General description

Sez6 is a brain-specific gene which was originally isolated using differential hybridization. SEZ-6 is a transmembrane protein with five short consensus repeats (SCRs) (C3b/C4b binding site) and two repeated sequences which are partly similar to the CUB domain (C1r/s-like repeat). The *Sez6* gene is expressed mainly in the neurons of the olfactory bulb, olfactory tubercle, piriform cortex, striatum, cortex (layer II, V and the deepest part of layer VI) and the CA1 region of the hippocampus.

KO strain construction

The 5′ region of the *Sez6* gene was cloned from a 129SvJ mouse genomic phage library (Stratagene). Genomic fragments surrounding the second exon (104–764) were used for the construct. The vector contained 7.6 kb of the 5′ region, a neomycin-resistance gene driven by PGK promoter, 1.3 kb of the 3′ region, and a diphtheria toxin A fragment. CCE ES cells, derived from the mouse strain 129SvEv, were used for generating chimeric mice. Chimeric mice were crossed with C57BL/6 mice. Heterozygotes were mated to generate SEZ-6$^{-/-}$ mice.

Phenotype

Mutant mice were produced at the expected Mendelian frequency, suggesting that there was no developmental or survival disadvantage for homozygous mutant mice as compared with their wild-type or heterozygous littermates. SEZ-6$^{-/-}$ mice were indistinguishable from their wild-type littermates. Homozygous mutant males and females could be bred, and females had normal-sized litters which nursed normally. Histochemical examination showed that homozygous mutant mouse brains did not exhibit any obvious abnormalities at the macroscopic or microscopic level.

Because *Sez6* was isolated as a gene whose expression is increased by pentylenetetrazol (PTZ), a convulsant drug, the effect of PTZ on homozygous mutant mice was examined. SEZ-6$^{-/-}$ mice showed increased sensitivity to PTZ compared to wild-type and heterozygous mice.

Comments

These results suggest that SEZ-6 gene products are involved in the response cascade for seizure induced by PTZ.

Acknowledgements
Shigeru Noguchi
Department of Molecular Biology, Meiji Institute of Health Science, Odawara, Kanagawa, Japan

References

1. The SEZ-6$^{-/-}$ mouse is unpublished.
2. Herbst, R. and Nicklin, M.J.H. (1997) Mol. Brain Res. 44, 309–322.
3. Shimizu-Nishikawa, K. et al. (1995) Mol. Brain Res. 28, 201–210.
4. Shimizu-Nishikawa, K. et al. (1995) Biochem. Biophys. Res. Commun. 216, 382–389.

Other names
Steroidogenic factor 1, Fushi tarazu factor 1 (Ftz-F1), adrenal 4 binding protein (Ad4BP), embryonic long terminal repeat binding protein (ELP)

Gene symbol
Ftzf1

Accession number
MGI: 95591

Area of impact
Hormones

General description

SF-1 is an orphan nuclear receptor that is an upstream regulator of the cytochrome P450 steroid hydroxylases, Müllerian inhibiting substance, and markers of pituitary gonadotropes including the α subunit of glycoprotein hormones and the luteinizing hormone β subunit. SF-1 and ELP arise from the same gene by alternative promoter usage and splicing.

KO strain 1 construction[1]

A positive–negative selection scheme was used to insert *neo* (pMC1*neo*polyA) within exon 4 encoding the zinc-finger domain required for DNA binding and HSV-*tk* was added at the 3′ end. This strategy deleted both SF-1 and ELP. E14TG2a ES cells were targeted. The knockout allele is currently carried in a genetically heterogeneous line with contributions from strains 129, C57BL/6 and DBA/2 mice. The mutation is being backcrossed into a C57BL/6 background.

KO strain 2 construction[2]

A 100 nucleotide region spanning the region between the first and second zinc fingers of the DNA-binding domain of SF-1 was replaced by PGK*neo*. RW4 (129Sv) ES cells were electroporated and injected into C57BL/6 blastocysts.

Phenotype

The knockout mice were born without adrenal glands and gonads, and exhibited male-to-female sex reversal of their external and internal genitalia. They died within 3–8 days after birth due to adrenocortical insufficiency, although they could be kept alive by steroid replacement. Expression of several proteins specific for pituitary gonadotropes was impaired. Mutant mice also lacked the ventromedial hypothalamic nucleus.

Acknowledgements
Keith L. Parker
Duke University Medical Center, Durham, NC, USA

References

1 Luo, X. et al. (1994) Cell 77, 481–490.
2 Sadovsky, Y. et al. (1995) Proc. Natl Acad. Sci. USA 92, 10939–10943
3 Ingraham, H.A. et al. (1994) Genes Dev. 8, 2302–2312.
4 Ikeda, Y. et al. (1995) Mol. Endocrinol. 9, 478–486.

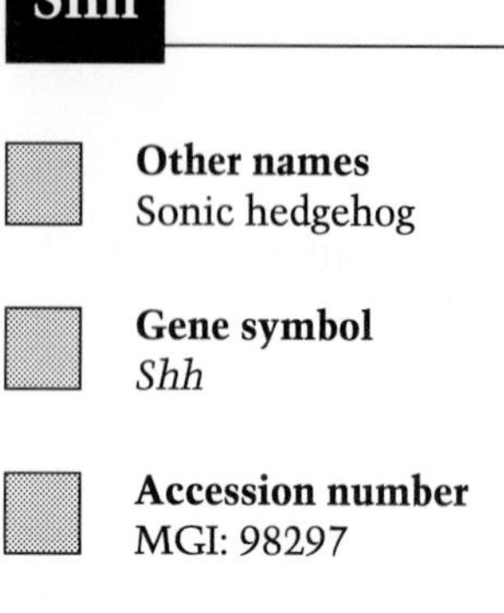

Other names
Sonic hedgehog

Gene symbol
Shh

Accession number
MGI: 98297

Area of impact
Development

General description

Sonic hedgehog is a member of the vertebrate hedgehog family of secreted proteins, related to the *Drosophila hedgehog* gene, which is involved in several patterning events in embryogenesis. Expression of Shh in vertebrates is associated with several areas of known important signaling function, including the node, the notochord, the floor plate and the zone of polarizing activity in the developing limb bud. Ectopic expression of Shh in the neural tube mimics the effect of ectopic notochord grafts in inducing floorplate and vental markers, while in chick, ectopic grafts of Shh-expressing cells to the anterior margin of the limb bud induces mirror-image duplication of digit pattern, as occurs when the ZPA is grafted ectopically.

KO strain construction

A PGK*neo* cassette replaced exon 2 and portions of the flanking introns, leading to a deletion of 97 of the 198 residues within the N-terminal product of Shh autocatalytic processing. This is the portion of the protein necessary for all Shh signaling functions. The vector was introduced into R1 129 ES cells, which were injected into C57BL/6 blastocysts. The phenotype was examined on a 129 × B6 background.

Phenotype

Most homozygotes die at or just before birth. The earliest abnormalities are seen at E9.5, with fusion of the optic vesicles in the midline. As development proceeds, the anterior region of the embryo shows defects reminiscent of holoprosencephaly, with no eyes, or reduced midline eye-like structures and loss of facial structures. Along the body axis, defects in dorsal-ventral patterning are detected. No floorplate is observed in the spinal cord and sclerotome structures are almost entirely absent. Markers of dorsal-ventral patterning are displaced or absent. Notochord is initially formed but fails to be maintained. In the limbs, distal structures are most severely affected with complete absence of digits and fusion of the paired distal limb bones. All of these defects are consistent with known expansion patterns and function of Shh.

Comments

One genetic cause for holoprosencephaly in humans has been shown to be mutations in the human *SHH* gene.

Reference

[1] Chiang, C. et al. (1996) Nature 383, 407–413.

Other names
SH-2-containing phosphatase, Syp, SH-PTP2, PTP1D, PTP2C, SAP-2, SH-PTP3

Gene symbol
Ptpn11

Accession number
MGI: 99511

Area of impact
Signal transduction

General description

Shp-2, Shp-1 and corkscrew comprise a small family of cytoplasmic tyrosine phosphatases that possess two tandem SH2 domains. Shp-2 is widely expressed and becomes tyrosine-phosphorylated after stimulation of cells with myriad growth factors and cytokines. Shp-2 can function as either a positive or negative regulator of signal transduction, depending on the specific receptor pathway stimulated.

KO strain construction

Part of the first SH2 domain and the inter-SH2 region (amino acids 46–110) of the Shp-2 gene (129Sv) were deleted and replaced by a neomycin-resistance cassette. The targeting vector was electroporated into R1 ES cells. Germ line chimeras were obtained by aggregating ES cells with CD1 morulas. No observable differences in phenotype between inbred (129Sv), mixed (129Sv × CD1) or outbred backgrounds (>95% CD1) were detected.

Phenotype

In the absence of wild-type Shp-2 protein, the production of axial mesoderm was severely perturbed, as indicated by abnormalities in the node, notochord and anterior–posterior axis. Shp-$2^{-/-}$ embryos also exhibited poorly developed somites and kinked or unclosed neural tubes. Extra-embryonic mesoderm also requires functional Shp-2 protein, as endothelial cells within the yolk sac of Shp-2 mutant embryos remained in the primitive honeycombed pattern rather than reorganizing into a highly vascularized network. Finally, the allantoic mesoderm was underdeveloped and failed to fuse with the maternal circulation, leading to death and resorption of these mutant embryos between E10 and E11 of gestation.

Comments

Disruption of the murine *Fgfr1* gene leads to a recessive lethal phenotype that is similar to the Shp-$2^{-/-}$ phenotype. Shp-2 is required for the full and sustained activation of MAP kinase following cell stimulation with FGFs, raising the possibility that the phenotype of Shp-2 mutant embryos results from a defect in FGF-receptor signaling.

Acknowledgements
Tracy Saxton
Samuel Lunenfeld Research Institute, Mount Sinai Hospital, Toronto, ON, Canada

References

[1] Saxton, T.M. et al. (1997) EMBO J. 16, 2352–2364.
[2] Tang, T.L. et al. (1995) Cell 80, 473–483.
[3] Bennett, A.M. et al. (1996) Mol. Cell Biol. 16, 1189–1202.
[4] Freeman, R.M. et al. (1992) Proc. Natl Acad. Sci. USA 89, 11239–11243.
[5] Pawson, T. (1995) Nature 373, 573–579.

Other names
DPC4, MAD homolog 4 (*Drosophila*)

Gene symbol
Madh4

Accession number
MGI: 894293

Area of impact
Development, oncogenesis

General description

Smad4 is a key signal transducer of the TGFβ-related pathways. It is ubiquitously expressed during embryogenesis and in adult tissues. It heterodimerizes with other members of the *Smad* gene family upon ligand stimulation and translocates to the nucleus. Smad4 has transcriptional activating potential and constitutes part of a functional transcription complex *in vivo*.

KO strain construction

The targeting vector replaced 5 kb of a genomic fragment comprising exon 8 and part of exon 9 with the neomycin-resistance gene. The *neo* gene was in the opposite transcriptional orientation relative to Smad4 and was flanked by *loxP* sites for future excision of the *neo* cassette. The targeting construct was electroporated into E14K ES cells derived from 129J/Ola. Positive clones were injected into E3.5 C57BL/6 blastocysts. Most analyses were performed with second and third generation offspring in a C57BL/6 or a CD1 background. The null mutation was confirmed by Western blot analysis of homozygous mutant ES cells.

Phenotype

Homozygous Smad4$^{-/-}$ mutant mice died before day 7.5 of embryogenesis. Mutant embryos were of reduced size, failed to gastrulate or express a mesodermal marker, and showed abnormal visceral endoderm development. Growth retardation of the Smad4-deficient embryos resulted from reduced cell proliferation rather than increased apoptosis. Aggregation of Smad4$^{-/-}$ ES cells with wild-type tetraploid morulae rescued the gastrulation defect. These results indicate that Smad4 is initially required for the differentiation of the visceral endoderm and that the gastrulation defect in the epiblast is secondary and non-cell autonomous. Rescued embryos showed severe anterior truncations, indicating a second important role for Smad4 in anterior patterning during embryogenesis.

Comments

The primary defect in the Smad4$^{-/-}$ mutant embryos demonstrates that TGFβ-related signaling is important in normal visceral endoderm development. The secondary defect affecting the anterior structures implies that Smad4 could be involved in the nodal pathway.

Acknowledgements
Christian Sirard and Tak W. Mak
Ontario Cancer Institute and Amgen Institute, Toronto, ON, Canada

Reference

1. Sirard, C. et al. (1998) Genes Dev. 12, 107–119.

Other names
Copper-zinc superoxide dismutase (CuZnSOD)

Gene symbol
Sod1

Accession number
MGI: 98351

Area of impact
Metabolism

General description

Sod1 encodes the enzyme copper-zinc superoxide dismutase (CuZnSOD). This is a housekeeping cytosolic and nuclear enzyme functioning in the catalysis of the dismutation reaction of superoxide radical anions (O_2^-) to form hydrogen peroxide (H_2O_2) and oxygen (O_2). SOD1 is often described as the first line of defense against free radical or oxidative damage which naturally occurs in all cells. Superoxide is constantly being produced as a byproduct of mitochondrial respiration as well as by a number of processes in the cytosol. A number of missense mutations in this gene are known to be responsible for approximately 20% of familial amyotrophic lateral sclerosis (ALS).

KO strain 1 construction[1]

Exon 5 of the mouse *Sod1* gene was replaced by a neomycin-resistance cassette. The targeting vector was constructed using plasmid pPNT. The knockout mice were generated on C57BL/6 and 129Sv hybrid backgrounds as well as on the 129SvEv inbred background.

KO strain 2 construction[2]

A standard replacement-type targeting vector was used which included positive–negative selection. A modified version of the vector pPNT was employed. Arms of homology included 4.9 kb on the 5′ side and 3.6 kb on the 3′ side. R1 ES cells were targeted. Homologous recombination resulted in the deletion of the entire coding sequence. The mutation has been studied primarily in a CD-1/129 mixed background as well as CD1/129 wild-type controls. Mice of inbred 129 and inbred C57BL/6 genetic background have also been examined.

Phenotype

Both homozygous male and female KO strain 1 $SOD1^{-/-}$ mice were healthy up to 12 months of age. No apparent pathologic changes were found in these knockout mice. However, these mice were extremely sensitive to paraquat. The female knockout mice also showed markedly reduced fertility.

The lifespan of KO strain 2 $SOD1^{-/-}$ animals is dramatically reduced. Half of the mutant mice died by about 630 days whereas only 25% of the wild-type littermates had died by 850 days of age. Mutant females had normal fertility when crossed to SOD1$^{+/+}$ males but markedly reduced fertility when crossed to SOD1$^{-/-}$ males. At 6 months, the mutant animals developed significant motor axonopathy which manifested as subtle motor behavior deficits at around 1 year of age. Cultures of primary embryonic fibroblasts from SOD1$^{-/-}$ mice did not survive for more than 48 hours whereas cultures from their wild-type littermates grew normally. Under stress conditions associated with oxidative injury, such as cerebral ischemia and facial motor axotomy, KO strain 2 SOD-deficient animals were more vulnerable than SOD$^{+/-}$ mice, which in turn were more vulnerable than the wild type[2]. Over 80% of KO strain 2 SOD1$^{-/-}$ mice developed very progressive liver tumors by 15 months in age (observed in outbred as well as 129 inbred lines)[3].

Comments

The phenotypes described to date in KO strain 2 SOD1$^{-/-}$ mice are consistent with "the free radical hypothesis of aging" in which cumulative oxidative damage is implicated in driving the aging process. In addition, increased vulnerability seen in the lesion models is consistent with oxidative damage being a component of certain stress regimens.

Acknowledgements
Andrew G. Reaume
Cephalon Inc., West Chester, PA, USA
Ye-Shih Ho
Institute of Chemical Toxicology, Wayne State University, Detroit, MI, USA

References

[1] Ho, Y.-S. et al. (1998) J. Biol. Chem. 273, 7765–7769.
[2] Reaume, A.G. et al. (1996) Nature Genet. 13, 43–47.
[3] Reaume, A.G. et al. (unpublished results).

Other names
Manganese superoxide dismutase (MnSOD)

Gene symbol
Sod2

Accession number
MGI: 98352

Area of impact
Metabolism

General description

Superoxide dismutases (SOD) are enzymes that catalyze the dismutation reaction of superoxide radical anions (O_2^-) to form hydrogen peroxide (H_2O_2) and oxygen (O_2). Three SOD isoenzymes exist: cytosolic and nuclear SOD1 (CuZnSOD), mitochondrial SOD2 (MnSOD), and SOD3 (extracellular SOD, EC-SOD). SODs protect cells and the extracellular matrix from free radical or oxidative damage which naturally occurs in all cells and tissues. SOD2 is localized in mitochondria and scavenges free radicals produced by oxidative phosphorylation. Impaired SOD2 activities have been correlated with ovarian cancer and type I diabetes in humans.

KO strain 1 construction[1]

Exon 3 of *Sod2* which contains the homodimerization domain, tetramer formation and manganese-binding regions, was inactivated by replacement with a neomycin-resistance gene. HSV-*tk* was placed at the 5′ end of the construct. CB1–4 ES cells derived from C57BL/6J mice were used and targeted, and injected into CD1 blastocysts.

KO strain 2 construction[2]

A 1.5 kb fragment containing exons 1 and 2 (including the transcriptional and translational start sites) of *Sod2* was replaced with an *Hprt* minigene.

Phenotype

KO strain 1 SOD2$^{-/-}$ mice died within 10 days after birth and exhibited dilated cardiomyopathy, lipid accumulation in skeletal muscle and liver, and acidosis. The activities of the mitochondrial enzymes succinate dehydrogenase and acetinase in the hearts were severely impaired. The ultrastructure of mitochondria and the activity of the cytochrome C oxidase were normal. These mice did not develop any pathological changes in the nervous system.

KO strain 2 SOD2$^{-/-}$ mice survived up to 3 weeks. These mice had severe anemia, neurodegeneration and progressive motor disturbances (weakness, rapid fatigue and circling behavior). Mice older than 7 days showed severe mitochondrial degeneration in neurons and myocytes. Only 10% of these mice developed dilated cardiomyopathies.

Comments

SOD2 appears to maintain the function of mitochondrial enzymes and protects them from inactivation by superoxides.

References

1 Li, Y. et al. (1995) Nature Genet. 11, 376–381.
2 Lebovitz, R.M. et al. (1996) Proc. Natl Acad. Sci. USA 93, 9782–9787.

Other names
Extracellular superoxide dismutase, EC-SOD, superoxide : superoxide oxidoreductase, EC1.15.1.1

Gene symbol
Sod3

Accession number
MGI: 103181

Area of impact
Metabolism

General description

Superoxide dismutases (SOD) catalyze the dismutation reaction of superoxide radical anions (O_2^-) to form hydrogen peroxide (H_2O_2) and oxygen (O_2). Three SOD isoenzymes exist: cytosolic and nuclear SOD1 (CuZnSOD), mitochondrial SOD2 (MnSOD), and SOD3 (extracellular SOD, EC-SOD). SODs protect cells and the extracellular matrix from free radical or oxidative damage which naturally occurs in all cells and tissues. SOD3 is a glycoprotein composed of 30 kDa subunits. Each subunit contains a copper and a zinc atom. The majority of SOD3 is bound to heparan sulfate proteoglycans in the extracellular matrix of tissues and the cell surface glycocalyx. A small amount of SOD3 can be also found in the serum, lymph and synovial fluid, and the cerebrospinal fluid.

KO strain construction

The coding sequence of *Sod3* is located in one exon. The active site of SOD3 was replaced by a neomycin-resistance gene. E14 ES cells were targeted and injected into C57BL/6 blastocysts.

Phenotype

$SOD3^{-/-}$ mice were viable and fertile and appeared normal up to 14 months after birth. Expression levels of SOD1 and SOD2 were normal, implying that there was no compensatory overexpression of these isoenzymes in $SOD3^{-/-}$ mutant mice. However, $SOD3^{-/-}$ mice were more susceptible to lung edema and died faster following exposure to high oxygen concentrations.

Comments

SOD3 appears to protect the extracellular spaces of tissues (particularly the lung) against superoxide radicals under stressful, but not physiological conditions.

Reference
[1] Carlsson, L.M. et al. (1995) Proc. Natl Acad. Sci. USA 92, 6264–6268.

Other names
Surfactant protein A

Gene symbol
Sftpa

Accession number
MGI: 109518

Area of impact
Metabolism

General description

SP-A is a phospholipid-associated protein of the pulmonary surfactant produced primarily by type II and bronchiolar cells in the lung epithelium. SP-A forms large complexes with other surfactant molecules such as tubular myelin. Functionally, SP-A may contribute to tubular myelin formation, increased surfactant spreading, the stabilization of phospholipid complexes, or inhibition of surfactant secretion. Moreover, SP-A can enhance the uptake of bacteria and viruses by alveolar macrophages.

KO strain construction

Portions of exons 3 and 4 and intron 3 were replaced by a neomycin-resistance cassette. HSV-*tk* was cloned 3′ to the inserted *neo* cassette. Targeted E14 ES cells were injected into C57BL/6 blastocysts.

Phenotype

Perinatal survival of SP-A$^{-/-}$ mice was normal. At high phospholipid concentrations, SP-A$^{-/-}$ mice produced the same surface tension as wild-type mice and had no apparent defects in lung morphology, type II cell morphology, lung compliance, phospholipid composition, or the surfactant proteins B, C and D. At low phospholipid concentrations, SP-A$^{-/-}$ mice displayed reduced surface activity. Tubular myelin figures were decreased, indicating that SP-A primarily regulates tubular myelin formation without changes in survival or lung functions. SP-A$^{-/-}$ mice were susceptible to group B streptoccocal infections, indicating that SP-A has a role in the clearance of streptococci from the lung.

References
[1] Korfhagen, T.R. et al. (1996) Proc. Natl Acad. Sci. USA 93, 9594–9599.
[2] LeVine, A.M. et al. (1997) J. Immunol. 158, 4336–4340.

Other names
Surfactant protein B

Gene symbol
Sftpb

Accession number
MGI: 109516

Area of impact
Metabolism

General description

SP-B is a 79 amino acid hydrophobic protein that enhances stability and spreading of of monolayers formed by surfactant phospholipids in alveoli. It also contributes to tubular myelin formation. Defects in SP-A and other surfactant proteins are the cause of the respiratory distress syndrome of premature infants and adult respiratory distress syndrome (ARDS).

KO strain construction

PGK-*neo* was inserted into exon 4 of the murine *Sftpb* gene. HSV-*tk* was placed at the 3′ terminal end of the construct. Targeted D3 ES cells were injected into C57BL/6 blastocysts.

Phenotype

SP-B$^{-/-}$ mice died immediately after birth due to respiratory failure. Lungs developed normally but remained atelectatic and lacked tubular myelin figures. Type II alveolar cells lacked fully formed lamellar bodies. Aberrant pro-SP-C was detectable whereas the fully processed SP-C molecule was decreased.

Comments

SP-B appears to affect storage, routing and function of surfactant phospholipids.

Reference
[1] Clark, J.C. et al. (1995) Proc. Natl Acad. Sci. USA 92, 7794–7798.

Spalt

Other names
mSal-1

Gene symbol
Spalt

Accession number
MGI: 109205

Area of impact
Development, transcription factors

General description

Spalt is a transcription factor of 1355 amino acids which contains nine zinc fingers. It is expressed in brain, spinal cord, kidneys, limb buds, genitals, palate and teeth. Spalt is first expressed at E7 in mesoderm. Expression in the limb buds is restricted to the progress zone and declines with age. Expression in kidney is late.

KO strain construction

Eight zinc fingers of the *Spalt* gene (from 129J) were replaced by the β-galactosidase/neomycin cassette from the pHM2 vector[2]. Targeted ES cells were injected into C57BL/6 blastocysts. Heterozygotes were crossed to strains C57BL/6 and 129SvEv.

Phenotype

Spalt$^{-/-}$ mice died at birth, probably from a feeding defect. A small cleft palate was visible. The tongue was shorter and the animals did not feed. Brain, spinal cord, limbs and kidneys were normal.

Comments

The phenotypic changes observed in Spalt$^{-/-}$ mice probably result from compensation by other *Spalt*-related genes. There are at least two additional such genes in mice and the expression pattern of one is very similar to that of *Spalt*.

Acknowledgements
Günther Schütz
German Cancer Research Center, Heidelberg, Germany

References
[1] The Spalt$^{-/-}$ mouse is not yet published.
[2] Kaestner, K.H. et al. (1994) Gene 148, 67–70.
[3] Ott, T. et al. (1996) Mech. Dev. 56, 117–128.

Srd5a1

Other names
Steroid 5α-reductase type 1, steroid 5α-reductase 1

Gene symbol
Srd5a1

Accession number
MGI: 98400

Area of impact
Hormones

General description

Steroid 5α-reductase is one of two isozymes that catalyze the 5α-reduction of steroid hormones that have a 3-oxo, δ 4,5 stereochemistry. For example, these isozymes convert testosterone to dihydrotestosterone, which is a more potent androgen. In mice, the type 1 isozyme is expressed in many different tissues, including the liver, kidney and reproductive tissues.

KO strain construction

A replacement construct leading to the deletion of 340 bp of 5′ flanking DNA and 110 bp of exon 1 of the gene and insertion of a neomycin gene was used to create a null allele at the *Srd5a1* gene on mouse chromosome 13. JH-1 ES cells (derived from 129Sv mice) were injected into C57BL/6 blastocysts. The targeted mutation was maintained on an isogenic 129Sv background and in mixed-strain (C57BL/6J/129Sv) animals.

Phenotype

Males were normal. Females exhibited parturition and fecundity defects. About 70% of pregnant females failed to deliver at term and most died in distressed labor. This defect was reversed by administration of 5α-reduced androgens. The litter size of homozygous KO females was 3.0 pups versus 8.0 in wild-type controls. Fetal death occurred between gestational days 10.75 and 11.0 due to estrogen excess and could be prevented by administration of estrogen receptor antagonists or inhibitors of estrogen biosynthesis. Both the parturition and fecundity defects were maternal in origin.

Comments

The phenotype of Srd5a1$^{-/-}$ mice reveals a role for 5α-reduced androgens in female reproduction and fetal loss in midgestation due to estrogen excess. Parturition defects in rodents are rare.

Acknowledgements
David Russell
University of Texas Southwestern Medical Center, Dallas, TX, USA

References
[1] Mahendroo, M.S. et al. (1996) Mol. Endocrinol. 10, 380–392.
[2] Mahendroo, M.S. et al. (1997) Mol. Endocrinol. 11, 917–927.

Srm

Other names
Src-related kinase lacking C-terminal regulatory tyrosine and N-terminal myristylation sites

Gene symbol
Srms

Accession number
MGI: 101865

Area of impact
Oncogenes

General description

Srm is a non-receptor protein tyrosine kinase which contains SH2, SH2′ and SH3 domains and a tyrosine residue for autophosphorylation in the kinase domain. However, Srm lacks an N-terminal glycine for myristylation and the regulatory C-terminal tyrosine. Although this structure resembles that of the Tec family of non-receptor tyrosine kinases, the N-terminal domain of Srm is unique and most closely resembles Src in the SH region. Two Srm transcripts are expressed almost ubiquitously and are influenced by tissue type and developmental stage.

KO strain construction

neo was inserted into the kinase domain to create a deletion construct for positive–negative selection. Homologous recombinants in the ES cell line TT2 (C57BL/6 × CBA/J) were aggregated with ICR. Chimeras were mated with C57BL/6.

Phenotype

Mutant mice displayed no apparent abnormalities, similar to mice carrying mutated alleles of the Src family kinases.

Comments

Srm may represent a new family of non-receptor tyrosine kinases which are redundant in function.

Reference
[1] Kohmura, N. et al. (1994) Mol. Cell. Biol. 14(10), 6915–6925.

Other names
Steroidogenic acute regulatory protein

Gene symbol
Star

Accession number
MGI: 102760

Area of impact
Hormones

General description

Star encodes the protein which regulates the acute phase of steroidogenesis. By facilitating the transport of cholesterol across the mitochondrial membranes, StAR acts as the rate-limiting step of steroidogenesis. It is expressed in the adrenal cortex and gonads.

KO strain construction

A standard positive–negative selection scheme was used to insert *neo* into exon 2. Insertion of *neo* also deleted exon 3 of the gene. The knockout allele is currently carried in a genetically heterogeneous line with contributions from strains 129, C57BL/6 and DBA/2 mice.

Phenotype

$StAR^{-/-}$ mice appeared normal at birth and fed well, but failed to thrive and died around post-natal day 14 due to adrenocortical insufficiency. Knockout animals could be rescued with hormone replacement therapy. There was a marked impairment of all steroid production. $StAR^{-/-}$ animals appeared externally female, irrespective of their genetic sex. Large amounts of lipid deposits were found in the degenerating adrenal cortex and gonads.

Comments

These mice serve as a murine model for an autosomal recessive steroid disorder called lipoid congenital adrenal hyperplasia.

Acknowledgements
Keith L. Parker
Duke University Medical Center, Durham, NC, USA

References

[1] The $StAR^{-/-}$ mouse is unpublished.
[2] Caron, K. M. et al. (1997) Mol. Endocrinol. 11, 138–147.
[3] Clark, B. J. et al. (1994) J. Biol. Chem. 269, 28314–28322.

Stat1

Other names
Signal transducer and activator of transcription 1, ISGF3 p91

Gene symbol
Stat1

Accession number
MGI: 103063

Area of impact
Immunology and inflammation, signal transduction

General description

The Stat1 protein is a transcription factor that is inactive until tyrosine-phosphorylated, a modification that occurs in response to cytokine and growth factor stimulation of cells. Stat1 is activated by IFNα, IFNβ, and IFNγ and is required for cellular responses to these cytokines. It is also activated in response to other cytokines and growth factors, but its role in responses to these agents is currently unknown. Once activated, Stat1 multimerizes either as a homodimer or as a heteromultimer with other transcription factors and translocates to the nucleus, where it binds to DNA. Stat1 contains a single site for tyrosine phosphorylation, an SH2 domain involved in receptor recognition and multimerization, a DNA-binding domain, a tetramerization domain, and a putative SH3 domain.

KO strain 1 construction[1]

The *Stat1* gene was disrupted by the replacement of about 3.5 exons (encoding 144 amino acids of the DNA-binding domain) with the neomycin-resistance gene in the opposite transcriptional orientation. Gene targeting was done in CCE ES cells that were injected into C57BL/6 blastocysts. Transmission of the deleted allele was maintained in 129J, C57BL/6J, BALB/c and CD1 mice.

KO strain 2 construction[2]

The first three translated exons of the *Stat1* gene and 0.7 kb of 5′ upstream sequence were replaced by a neomycin resistance cassette. HSV-*tk* was placed at the 3′ end of the construct. GS-1 ES cells were electroporated and targeted ES cells were injected into C57BL/6 blastocysts.

Phenotype

Stat1$^{-/-}$ mice were viable and had no gross developmental defects. However, mutant animals failed to thrive, were very susceptible to viral and microbial pathogens, and exhibited altered regulation of terminal T lymphocyte differentiation in the presence of infectious agents. Homozygous null animals showed no induction of a panel of IFN-inducible genes in response to treatment with IFN-α or IFN-γ. However, cells and tissues from Stat1$^{-/-}$ mice remained responsive to all other cytokines tested.

Comments

The phenotype of Stat1$^{-/-}$ mice demonstrates the essential role of IFN in innate responses to viral and microbial pathogens, and the interaction between the innate and adaptive immune responses. It also demonstrates the high degree of specificity of the Stat1 transcription factor for IFN signaling, despite Stat1 activation by other cytokines and growth factors.

Acknowledgements

David E. Levy
Department of Pathology and Kaplan Cancer Center, New York University School of Medicine, New York, USA

References

[1] Durbin, J.E. et al. (1996) Cell 84, 443–450.
[2] Meraz, M.A. et al. (1996) Cell 84, 431–442.

Other names
Signal transducer and activator of transcription 3, acute phase response factor (APRF)

Gene symbol
Stat3

Accession number
MGI: 103038

Area of impact
Signal transduction, development

General description

Stat3 is widely expressed in adult tissues and in the visceral endoderm and deciduum during embryogenesis. It is activated by phosphorylation of a tyrosine residue located in its C-terminal portion. This phosphorylation can be induced by a variety of cytokines. Stat3-linked receptors include receptors for the IL-6 family of cytokines (IL-6, leukemia inhibitory factor (LIF), ciliary neurotropic factor (CNTF) and IL-11), leptin, granulocyte colony-stimulating factor (G-CSF), and epidermal growth factor (EGF). Once activated, Stat3 forms a homodimer, rapidly translocates to the nucleus, and induces the transactivation of several target genes.

KO strain construction

The targeting vector was constructed by replacing a 3.0 kb fragment of genomic DNA containing three exons (encoding the SH2 domain and the tyrosine residue crucial for Stat activation) with a neomycin-resistance gene (*neo*). *neo* was flanked by an upstream 0.9 kb short arm and a downstream 5.0 kb long arm fragment. HSV-*tk* was placed at the 3′ end of the construct. E14.1 ES cells were targeted and positive clones were injected into C57BL/6 blastocysts.

Phenotype

Stat3$^{-/-}$ embryos developed normally until the egg cylinder stage at E6.0 but showed rapid degeneration between E6.5 and E7.5. Stat3-deficient blastocysts were able to exhibit outgrowth of the inner cellular mass in culture *in vitro*, indicating that Stat3-deficiency does not affect embryonic development during the early post-implantation period. During early embryogenesis, Stat3 mRNA was exclusively expressed in the visceral endoderm. The early embryonic lethality of the Stat3-deficient embryos may be the result of a functional defect of the visceral endoderm, which is important for metabolic exchange from maternal blood.

Comments

Stat3 is essential for early embryonic development in the mouse.

Acknowledgements
Shizuo Akira
Department of Biochemistry, Hyogo College of Medicine, Hyogo, Japan

References

[1] Takeda, K. et al. (1997) Proc. Natl Acad. Sci. USA 94, 3801–3804.
[2] Akira, S. et al. (1994) Cell 77, 63–71.
[3] Zhong, Z. et al. (1994) Science 264, 95–98.

Gene symbol
Stat4

Accession number
MGI: 103062

Area of impact
Transcription factor, signal transduction

General description

Stat 4 is a member of the signal transducers and activators of transcripiton family of transcription factors. It is inducibly tyrosine phosphorylated and activated by the cytokine IL-12. Following activation it is speculated to regulate a number of genes that are activated by IL-12.

KO strain 1 construction[1]

The targeting construct was an insertion construct containing 2.0 and 4.0 kb of 5′ and 3′ homology respectively. The construct was inserted into the second coding exon (exon 3) and contained the hygromycin selectable gene. In addition, the vector also contained the thymidine kinase gene for negative selection. The targeting was done in the 129-derived E14 ES cells, and chimeric mice were generated by injection into C57BL/6 blastocysts.

KO strain 2 construction[2]

The targeting vector was an insertion construct containing the neomycin-resistance gene for positive selection of homologous recombinants, and the thymidine kinase gene for negative selection of random integrants. The neomycin resistance gene was cloned into a *Bam*HI restriction site such that it interrupted the exon encoding amino acids 262–314 of Stat 4. D3 (strain 129) ES cells were used, and Stat $4^{-/-}$ cells were injected into BALB/c blastocysts.

Phenotype

The mice were viable and fertile with no detectable defects in hematopoiesis. However, all IL-12 functions tested were disrupted, and indeed the mice were phenotypically identical to the IL-12 knockout mice. IL-12 induction of IFN-γ by spleen cells was impaired in the KO mice. In addition, the IL-12-induced enhancement of NK cytolytic function was impaired, as was the IL-12-supported *in vitro* development of Th1 cells. The development of Th1 cells in response to the pathogen *Listeria monocytogenes* was also reduced in Stat 4-deficient mice, and the mice exhibited a general propensity towards the development of Th2 responses.

Comments

The Stat 4-deficient mice demonstrate the obligatory role of Stat 4 signal transduction in the response of T cells to IL-12, a Th1-inducing cytokine. The induction by IL-12 of IFNγ production is dependent upon Stat 4 signaling. There was no evidence from the Stat 4-deficient mice to indicate an obligatory role for Stat 4 in spermatogenesis or hematopoiesis.

Acknowledgements
Michael Grusby
Harvard School of Public Health, Department of Cancer Biology, Boston, MA, USA
James Ihle
Howard Hughes Medical Institute, Department of Biochemistry, St. Jude's Children's Research Hospital, Memphis, TN, USA

References

[1] Thierfelder, W.E. et al. (1996) Nature 382, 171–174.
[2] Kaplan, M.H. et al. (1996) Nature 382, 174–177.

Stat5a

Other names
Signal transducer and activator of transcription 5A, mammary gland factor (MGF)

Gene symbol
Stat5a

Accession number
MGI: 103036

Area of impact
Signal transduction

General description

Prolactin (PRL) induces mammary gland development (defined as mammopoiesis) and lactogenesis. Binding of PRL to its receptor leads to the phosphorylation and activation of Stat (signal transducers and activators of transcription) proteins, which in turn promote the expression of specific genes. The activity pattern of two Stat proteins, Stat5a and Stat5b, in mammary tissue during pregnancy suggests an active role for these transcription factors in epithelial cell differentiation and milk protein gene expression.

KO strain construction

A deletion vector was constructed to remove exons 1–3 by *neo*^{r} replacement. The vector was transfected into 129 ES cells and chimeric mice were mated to 129 mice.

Phenotype

Stat5a-deficient mice developed normally and were indistinguishable from hemizygous and wild-type littermates in size, weight and fertility. However, mammary lobuloalveolar outgrowth during pregnancy was curtailed, and females failed to lactate after parturition due to a failure of terminal differentiation. Although Stat5b shows 96% similarity to Stat5a and a superimposable expression pattern during mammary gland development, it failed to compensate for the absence of Stat5a.

Comments

These results document that Stat5a is the principal and an obligate mediator of mammopoietic and lactogenic signaling.

Acknowledgements
Lothar Hennighausen
National Institutes of Health, Bethesda, MD, USA

References

1. Liu, X. et al. (1997) Genes Dev. 11, 179–186.
2. Liu, X. et al. (1996) Mol. Endocrinol. 10, 1496–1506.
3. Liu, X.W. et al. (1995) Proc. Natl Acad. Sci. USA 92, 8831–8835.
4. Li, M. et al. (1997) Proc. Natl Acad. Sci. USA 94, 3425–3430.
5. Hennighausen, L. et al. (1997) J. Biol. Chem. 272, 7567–7569.

Other names
IL-4-Stat

Gene symbol
Stat6

Accession number
MGI: 103034

Area of impact
Immunity and inflammation, signal transduction

General description

Stat 6 is a member of the Stat ("signal transducers and activators of transcription") family of transcription factors. It is widely expressed in adult tissues, and is inducibly tyrosine phosphorylated and activated by the cytokines IL-4 and IL-13. Following tyrosine phosphorylation, Stat6 homodimerizes, acquires the ability to bind DNA, translocates to the nucleus and activates a number of genes involved in IL-4 responsiveness.

KO strain 1 construction[1]

The exons encoding amino acids 505–584 of Stat 6 were replaced with a neomycin-resistance gene/polyA cassette. In addition, the vector contained an HSV-*tk* cassette for negative selection of non-homologous recombination events. The targeting vector was electroporated into strain 129 ES cells (D3), and Stat $6^{-/-}$ clones were injected into BALB/c blastocyts. The null mutation was confirmed using Southern blotting and immunoprecipitation analyses.

KO strain 2 construction[2]

The targeting vector was constructed by replacing a 2.2 kb fragment of genomic DNA containing two exons encoding the SH2 domain of Stat 6 with the neomycin resistance gene (*neo*). *neo* was flanked by 7.0 kb of *Stat6* homology upstream and 1.2 kb of homology downstream. The vector was transfected into the E14.1 ES cell line, and chimeric mice were backcrossed into a C57BL/6 genetic background. The null mutation was confirmed by Northern and Southern blot analyses.

KO strain 3 construction[3]

The targeting construct was a replacement targeting vector for the 5′ region of the gene which contains the first coding exon. The vector contained 3.1 and 3.6 kb of 5′ and 3′ homology, respectively. The *neo* gene was introduced for positive selection. The targeting vector was used to disrupt the gene in RW4 ES cells, derived from 129 mice. The cells were injected into C57BL/6 blastocysts, and chimeric mice were backcrossed with C57BL/6 mice. The null mutation was confirmed by Southern blotting, Western blotting and immunoprecipitation.

Phenotype

Mutant mice had specific immunological deficiencies associated with responses to IL-4, and were phenotypically similar to IL-4 knockout mice. B cells exhibited an inability to undergo class switching to IgE in response to anti-IgD, and to upregulate MHC class II and CD23 expression upon stimulation with IL-4. B cell proliferation in response to IL-4 was impaired, as was IL-4-induced T cell proliferation. Development of Th2 responses was also impaired, and the production of IgE and IgG1 in response to nematode infection was profoundly reduced. In addition, the response of macrophages to IL-13 was impaired, indicating that IL-13 also makes use of Stat 6 signaling in macrophages.

Comments

Stat 6 plays an obligatory role in signaling through both the IL-4 and the IL-13 receptors in response to their respective ligands. The important role of IL-4 in the initiation and maintenance of Th2 responses is highlighted by the reduced Th2 phenotype of Stat $6^{-/-}$ mice, and its role in IgE class switching is also demonstrated.

Acknowledgements
Shizuo Akira
Department of Biochemistry, Hyogo College of Medicine, Hyogo, Japan
Michael Grusby
Harvard School of Public Health, Department of Cancer Biology, Boston, MA, USA
James Ihle
Howard Hughes Medical Institute, Department of Biochemistry, St. Jude's Children's Research Hospital, Memphis, TN, USA

References

[1] Kaplan, M.H. et al. (1996) Immunity 4, 313–319.
[2] Takeda, K. et al. (1996) Nature 380, 627–630.
[3] Shimoda, K. et al. (1996) Nature 380, 630–633.
[4] Takeda, K. et al. (1996) J. Immunol. 157, 3220–3222.

Other names
RON

Gene symbol
Ron

Accession number
MGI: 99614

Area of impact
Immunity and inflammation

General description

STK, the murine homolog of human RON, is a receptor tyrosine kinase belonging to the Met family. STK is expressed in the CNS (ganglia, hippocampus, hypothalamus), keratinocytes and other epithelial cells (gut, kidney, bronchus, nasal cavity, oviduct and testis), the endocrine system (pancreatic islets, parathyroid, adrenal medulla), megakaryocytes and some tissue-resident macrophages. The ligand for STK, MSP (macrophage-stimulating protein), was isolated from serum via its ability to induce chemotaxis of murine peritoneal macrophages *in vitro*. MSP is activated by members of the coagulation cascade in response to tissue damage. In addition to its potential to induce chemotaxis and phagocytosis of C3bi-coated erythrocytes, MSP has an inhibitory effect on the production of nitric oxide (NO) by activated peritoneal macrophages *in vitro*. NO is a principal mediator of many of the cytokine-inducible macrophage activities observed during a normal cell-mediated immune response.

KO strain construction

The STK targeting construct was designed to replace a 850 bp region within the first translated exon of the *Ron* gene (including the translational start site) with a *lacZ* reporter gene under the transcriptional control of the *Ron* promoter; translation was from the *lacZ* ATG. Homologous recombination between the endogenous *Ron* gene and the targeting construct was expected to abrogate translation of full-length STK while permitting visualization of *Ron* expression. The targeting vector was introduced into R1 ES cells (129Sv). $STK^{+/-}$ ES cells were aggregated with blastocysts from CD-1 mice, and the resulting chimeras were mated with CD-1 females.

Phenotype

Peritoneal macrophages from mice lacking STK produced elevated levels of NO in response to IFNγ in a dose-dependent manner, without the need for a costimulus. However, production of pro-inflammatory cytokines by activated macrophages from $STK^{-/-}$ mice was unaltered. *In vivo*, $STK^{-/-}$ mice exhibited increased inflammation in an IFNγ-mediated delayed-type hypersensitivity (DTH) reaction and increased susceptibility to endotoxic shock. Furthermore, the levels of NO in the serum of mice injected with LPS were significantly higher than those in control littermates. Nevertheless, the serum levels of IFN and the intermediate cytokines generated by the inflammatory response, which have previously been shown to play a critical role in septicemic shock, did not differ significantly from controls. These data suggest that the STK receptor suppresses NO production through negative regulation of IFN-activated macrophages, thereby ameliorating the potentially tissue-damaging effects of a cell-mediated immune response.

Comments

The $STK^{-/-}$ mice will be useful in defining the intracellular signaling pathways that regulate the response to IFNγ through the STK receptor. In addition, these mutant animals may be valuable models in which to study inflammatory disorders and provide a unique *in vivo* model to investigate the role of a receptor tyrosine kinase in the pathogenesis of infectious and inflammatory diseases.

Acknowledgements
Pamela Correll
Pennsylvania State University, Department of Veterinary Science, University Park, PA, USA

Reference

[1] Correll, P.H. et al. (1997) Genes Function 1, 69–83.

Stromelysin

Other names
Stromelysin-1 (SLN1), matrix metalloproteinase 3 (MMP-3), transin-1

Gene symbol
Mmp3

Accession number
MGI: 97010

Area of impact
Immunity and inflammation

General description

Stromelysin (SLN1) is a member of the matrix metalloproteinase (MMP) family implicated in tissue remodeling during developmental and pathophysiological processes. The gene product is able to cleave a number of matrix substrates, as well as mediators of inflammatory responses. In addition, it has been proposed that SLN1 contributes directly to the degradation of articular cartilage matrix components during naturally occurring and experimentally induced arthritis. SLN1 has been reported to enhance the activation of other MMPs.

KO strain construction

The targeting vector pSLNRV1.1 was designed to delete part of exon 3 (from the internal *Hin*dIII restriction site) and exon 4, which encode the SLN1 Zn-binding site. This genomic material was replaced with a neomycin-expression cassette in the anti-sense direction. The targeting vector was introduced into AB2.1 ES cells (129SvEv) and was derived on 129SvEv, C57BL/6 and B10.RIIII backgrounds. Northern and Western analyses confirmed that there was no detectable SLN1 mRNA or protein in cells or tissues of the knockout mouse.

Phenotype

The fertility and lifespan of the SLN1$^{-/-}$ knockout mouse were normal. No gross or histologic changes were observed in tissues from mutant mice compared with those from wild type mice at 8 weeks and 14 months of age. Likewise, no clinically significant changes in the serum chemistry or hematology were observed. SLN1$^{-/-}$ mice were susceptible to collagen-induced arthritis, and arthritic paws had levels of cartilage degradation equal to those observed in wild-type mice. SLN1$^{-/-}$ mice demonstrated delayed repair of 6–12 mm dermal wounds, but had similar tensile strengths of repaired incisional wounds. The mutant mice also displayed subtle differences in the involution of mammary tissue.

Comments

SLN1$^{-/-}$ mice have been reported to be less healthy in environments which challenge the mice with mites or dermatological abrasions, possibly because of the observed delay in wound repair observed in these mutants.

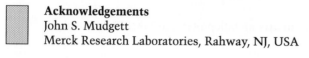

Acknowledgements
John S. Mudgett
Merck Research Laboratories, Rahway, NJ, USA

References

[1] Mudgett, J.S. et al. (1998) Arthritis Rheum. 41, 110–121.
[2] Rudolph, L.A. et al. (1997) Endocrinology 138, 4902–4911.

5′ Switch γ1 region

Other names
$5's_{\gamma}1$

Area of impact
Immunity and inflammation, B cell

General description

Immunoglobulin class switch recombination (CSR) is directed by cytokines to distinct immunoglobulin classes via specific activation of repetitive recombinogenic switch regions, located upstream of each C_H gene segment. Prior to recombination, flanking sequences 5′ of the switch region are subject to cytokine-induced activation of their promoter elements, leading to transcription of unrearranged (germ line) switch regions and their associated C_H genes. Based on correlative evidence, germ line transcription and processed, structurally conserved germ line transcripts have been implicated in CSR control.

KO strain construction

The long homology arm (8 kb) and short homology arm (0.8 kb) of targeting vectors were derived from BALB/c genomic DNA. A 1.6 kb fragment of the 5′ switch γ1 region was replaced by the neomycin-resistance gene flanked by two FRT recombination signal sequences. The KO vector was transfected into ES cells from 129Ola mice. The neo^r gene was subsequently deleted by transient transfection of the G418-resistant ES cells with a Flp recombinase expression vector. In this way, ES cells were created which had a targeted deletion of the flanking sequences 5′ of the switch γ1 region, but which did not have the influence of a heterologous promoter and enhancer from the neo^r gene. Chimeric mice were backcrossed with C57BL/6 and CB.20 strain mice, and the presence of the mutated allele in heterozygous and homozygous mice was confirmed by Southern blot analysis.

Phenotype

Mutant mice lacking the 5′ sγ1 region developed normally, and were grossly normal in appearance. However, the mice exhibited a selective IgG1 agammaglobulinemia, and failed to produce IgG1 in response to immunization. Production of the other immunoglobulin isotypes was unaffected. This phenotype established that a switch region lacking its 5′ flanking sequence is severely impaired as a partner for CSR. Mutant mice with impaired class switch to IgG1 were also used to show that, although the majority of murine IgE-expressing B cells arises from sequential switching, frequency of IgE CSR is autonomously determined and independent of prior class switch to other classes.

In addition to the 5′ switch-deleted strain described above, mice were constructed which had a heterologous, inducible promoter inserted in place of the 5′ switch region flanking sequences, either with or without the endogenous splice donor site. Analysis of these mice showed that transcription of the switch region and the adjacent C_H gene segment, as well as correct processing of the resulting transcript, were required to efficiently direct CSR to the adjacent switch region.

Comments

The flanking sequences 5′ of the switch γ1 region are necessary to direct efficient class switching to the adjacent constant region gene segment. Transcription of the switch region and the flanking C_H gene segment is required for efficient class switching, and the correctly processed transcript may also be required. The strictly *cis*-acting nature of the switch recombination control element is consistent with a model attributing directed CSR to modulation of accessibility of individual switch regions. Class switching to IgE occurs sequentially, but is not affected by the deletion of the $5's_{\gamma}1$ region.

Acknowledgements
Andreas Radbruch
Deutsches Rheum-Porschungszentrum, Berlin, Germany

References

[1] Jung, S. et al. (1993) Science 259, 984–987.
[2] Coffman, R.L. et al. (1993) Adv. Immunol. 54, 229–270.
[3] Zhang, J. et al. (1993) EMBO J. 12, 3529–3537.
[4] Lorenz, M.S. et al. (1995) Science 267, 1825–1828.
[5] Jung, S. et al. (1994) J. Exp. Med. 179, 2023–2026.

Gene symbol
Syk

Accession number
MGI: 99515

Area of impact
Immunity and inflammation, signal transduction

General description

Syk is a cytoplasmic protein tyrosine kinase expressed in many hematopoietic cells. It contains two N-terminal SH2 domains and a C-terminal kinase domain. Syk is closely related structurally to ZAP-70 kinase, shown to be crucial for TCR signaling and T cell development. Syk has been found in association with Fc receptors, the BCR and the TCR, suggesting that it plays a role in transducing signals between the antigen and Fc receptors to downstream molecules. Syk becomes associated with the BCR on both immature and mature B cells. The tandem SH2 domains in Syk bind to a pair pf phosphotyrosine sites in the ITAM motifs of the Igα/Igβ receptor signaling chains, leading to the activation of Syk. Syk has also been implicated in signaling through the receptors for IL-2, G-CSF and several platelet agonists.

KO strain 1 construction[1]

The targeting vector was designed such that the *Syk* gene (derived from 129Sv) was disrupted at an *Nco*I site located in the exon containing subdomain I of the kinase domain. PGK*neo* was inserted at this site in the opposite transcriptional orientation. The targeting vector was electroporated into D3 ES cells and positive clones were injected into C57BL/6 blastocysts. Chimeric mice were bred with (B6D2) F1 mice to obtain heterozygotes, which were intercrossed to obtain homozygous Syk$^{-/-}$ mice. The null mutation was confirmed by immunoblot of fetal liver lysates, using antibodies directed against several regions of the protein. No full-length or truncated forms of Syk protein were detected in mutants.

KO strain 2 construction[2]

The targeting vector was designed to delete 3 kb of the *Syk* gene (derived from 129Sv) which included the conserved exon encoding 41 residues located in subdomain VI of the Syk kinase domain. This region was replaced by PGK*neo*, and PGK*tk* was added to the 3′ end of the genomic sequences. The targeting vector was electroporated into R1 ES cells and positive clones were isolated and used to generate chimeric mice, from which homozygous Syk$^{-/-}$ mice were obtained. No Syk protein was detected by immunoblot analysis of mutant fetal liver cells.

Phenotype

$Syk^{-/-}$ mice died shortly after birth, exhibiting swollen red footpads. Analysis of embryos of heterozygous intercrosses showed the expected numbers of $Syk^{-/-}$ fetuses at days 16.5 and 18.5 of gestation. There was extensive hemorrhaging of mutant embryos only between E14 and 18: hemorrhaging was reduced in live-born pups. Mutant pups accumulated large amounts of chylous ascites after feeding. Hematopoiesis was not grossly affected. Early stages of B cell development were normal, but there was a block in later B cell development at the pro-B to pre-B cell transition. $Syk^{-/-}$ radiation chimeras showed low numbers of immature B cells in the periphery, but no mature B cells. Dμ signaling was blocked, a phenomenon also observed in λ5-deficient B cells. αβ T cell development was normal, but the development of $V\gamma3^{+}$ thymocytes (which become dendritic epidermal T cells) was impaired. Signaling through the G-CSF and IL-2 receptors was normal. $RAG\text{-}2^{-/-}$ mice reconstituted with $Syk^{-/-}$ lymphocytes showed a signficant decrease in intra-epithelial γδ T cells in the skin and gut. The development of these cells was arrested at the stage following TCR gene rearrangement[3]. Vγ/Vδ gene rearrangements associated with specific epithelia were also disrupted.

Comments

Syk is essential for normal vascular integrity in the embryo and for several steps in the differentiation of B cells. Syk appears to be particularly necessary for the transduction of signals from the pre-BCR permitting clonal expansion and further maturation of pre-B cells.

References

1. Turner, M. et al. (1995) Nature 378, 298–302.
2. Cheng, A.M. et al. (1995) Nature 378, 303–306.
3. Mallick-Wood, C.A. et al. (1996) Proc. Natl Acad. Sci. USA, 93, 9704–9709.

Other names
Synaptotagamin 1, vesicle docking protein

Gene symbol
Syt1

Accession number
MGI: 99667

Area of impact
Neurology

General Description

SYT1 associates with the neurotransmitter vesicle and binds Ca^{2+}. It interacts with other synaptic vesicle proteins such as syntaxin and neurexin, and may play a role in vesicle docking.

KO strain construction

neo^r was inserted into an exon encoding amino acids 270–308 in the *Syt1* gene with *tk* on the 3′ flank. The vector was electroporated into AB1 (129) ES cells, and homologous recombinants were injected into blastocysts to generate chimeric mice.

Phenotype

Animals homozygous for a null mutation in the *Syt1* gene showed normal embryonic development and were indistinguishable immediately after birth. However these animals died within 48 hours of birth. No gross morphological abnormalities were observed in these animals, although they apparently failed to suckle. Electron microscopy of synapses of SYT1$^{-/-}$ animals also showed no major abnormalities, suggesting that synaptogenesis *per se* was normal.

Analysis of hippocampal slice cultures (in which pyramidal neurons had completed their synaptic connections) showed that the synchronous, fast component of calcium-dependent neurotransmitter release was decreased (7.6 ± 1.4 in SYT1$^{-/-}$ mice vs. 23 ± 0.5 in SYT1$^{+/+}$ mice; mEPSC frequency per single bouton) whereas asynchronous release processes, including spontaneous synaptic activity and release triggered by hypertonic solution or α-latrotoxin, were unaffected. The data support the hypothesis that SYT1 is involved in the final stage of synaptic transmission, vesicle fusion with the plasma membrane. Heterozygotes showed no phenotypic abnormalities.

Comments

SYT1 is probably a Ca^{2+} sensor necessary for synchronous neurotransmitter release.

Reference

[1] Geppert, M. et al. (1994) Cell, 79, 717–727.

Other names
c-erbA-alpha, TR-alpha, thyroid hormone receptor alpha

Gene symbol
Thra

Accession number
MGI: 98742

Area of impact
Hormone

General description

Thyroid hormones are thought to mediate their effects by binding to two nuclear receptors, T3Rα1 and T3Rβ. The *Thra* gene encodes the T3Rα1 receptor isoform. Through alternative splicing, it also encodes the α2 isoform which does not bind thyroid hormone and which is thought to be a trans-dominant inhibitor of T3Rα1.

KO strain construction

Exon 2 was disrupted by the insertion of a *lacZ* cassette together with a neo^r gene. ES cells were of the 129 line. Homozygous mutants were produced in either an inbred C57BL/6 background or an outbred OF1 background.

Phenotype

Homozygous mutants showed normal embryonic development and were born normally. However one week after birth, they stopped growing entirely and died at 2–4 weeks. Mutant mice showed a delay in the development of the small intestine and bone, and hypothyroidia.

Acknowledgements
Jacques Samarut
Ecole Normale Superior Lyon, Lyon, France

Reference

[1] Fraichard, A. et al. (1997) EMBO J. 16, 4412–4420.

Other names
T cell acute lymphocytic leukemia 1, SCL, TCL 5, Tal1Scl, T cell leukemia oncoprotein

Gene symbol
Tal1

Accession number
MGI: 98480

Area of impact
Hematopoiesis, transcription factors

General description

tal1/scl is a member of the basic helix-loop-helix (bHLH) family of transcription factors. It is expressed in embryonic and adult early blood cells, in vascular endothelium, and has a restricted expression in CNS. It is frequently disrupted in human T cell leukemias.

KO strain 1 construction

The targeting vector was a typical positive-negative *neo*-selection construct. The entire coding region of the *Tal1* gene (originally derived from strain 129) was deleted and replaced with the *neo*-resistance cassette. The HSV-*tk* cassette was present outside the area of homology. The targeting vector was electroporated into J1 ES cells. Positive clones were injected into C57BL/6 blastocysts to generate chimeras. The null mutation was confirmed by Southern analysis and RT-PCR of RNA from yolk sac tissue.

KO strain 2 construction

The portion of the exon coding for the basic binding region and HLH domain was replaced by PGK*neo*polyA. The PGK-*tk* negative selection cassette was ligated into the 5′ end of the construct. The targeting vector was electroporated into W9.5 ES cells (129Sv). Positive clones were injected into (C57BL/6 × C57BL/10) F2 blastocysts. Chimeras were mated to C57BL/6 mice to generate heterozygotes. The null mutation was confirmed by Southern analysis.

Phenotype

tal1/scl$^{-/-}$ embryos died around E9.5–E10. They were pale, edematous and markedly growth-retarded. Histological studies showed the complete absence of recognizable hematopoiesis in the yolk sac of these embryos. Early organogenesis appeared to be otherwise normal. Culture of yolk sac cells of wild-type, heterozygous and homozygous littermates confirmed the absence of hematopoietic cells in *tal1/scl*$^{-/-}$ yolk sacs, although yolk sac vascular cells were present. RT-PCR was used to examine the transcription of several genes implicated in early hematopoiesis. Transcription of GATA-1 and PU.1 transcription factors was absent when RNA from *tal1/scl*$^{-/-}$ yolk sacs and embryos was analyzed.

The hematopoietic defect was shown to be cell autonomous by analysis of the contribution of *tal1/scl*$^{-/-}$ ES cells to hematopoiesis in chimeric mice. Study of *tal1/scl*$^{-/-}$ ES cells showed no adult hematopoiesis of any lineage, either *in vivo* or *in vitro*.

Comments

tal1/scl is an essential regulator of the earliest steps in hematopoietic development and thus is required for the formation of embryonic blood *in vivo*. It is dispensable for the development of vascular cells. The phenotype of *tal1/scl* mutants resembles those following the loss of the erythroid transcription factor GATA-1 or the LIM protein Rbtn2.

Acknowledgements
Stuart Orkin
Divisions of Hematology/Oncology and Pediatric Oncology, Children's Hospital and Dana-Farber Cancer Institute, Harvard Medical School, Boston, MA, USA
Lorraine Robb
The Walter and Eliza Hall Institute of Medical Research, The Royal Melbourne Hospital, Victoria, Australia

References
1. Shivdasani, R.S. et al. (1995) Nature 373, 432–434.
2. Robb, L. et al. (1995) Proc. Natl. Acad. Sci. USA 92, 7075–7079.
3. Robb, L. et al. (1996) EMBO J. 15, 4123–4129.
4. Elefanty, A.G. et al. (1997) Blood 90, 1435–1447.
5. Porcher, C. et al. (1996) Cell 86, 47–57.

TAP1

Other names
Transporter associated with antigen processing 1

Gene symbol
Tap1

Accession number
MGI: 98483

Area of impact
Immunity and inflammation, antigen presentation

General description

Tap1 encodes a subunit for a peptide transporter termed TAP1 (transporter associated with antigen processing), which translocates antigenic MHC class I-binding peptides from the cytoplasm of the cell to the lumen of the ER, for association with MHC class I molecules. *Tap1* is expressed in all nucleated cells and its expression is inducible by IFNγ.

KO strain construction

The construct was designed to delete a 7 kb fragment from the *Tap1* gene, encompassing most of the protein-encoding region. The construct contained a 6.5 kb fragment from the 5′ end and a 2.5 kb fragment from the 3′ end of the *Tap1* gene (from strain C57BL/6), interrupted by a neomycin-resistance gene. The vector was transfected into ES cells from strain 129Sv, and chimeric mice were bred with C57BL/6 mice. The gene knockout was confirmed by Southern blotting.

Phenotype

TAP1 mutant mice were not able to transport cytosolic peptides into the ER for association with class I MHC molecules. As a result, the mice were defective in the stable assembly and intracellular transport of MHC class I molecules, and consequently showed severely reduced levels of class I molecules at the cell surface. Expression of class I molecules at the surface of these cells could be partially restored by treatment with presentable peptides or by culture at reduced temperature. These cells were unable to present most cytosolic antigens to class I-restricted cytotoxic T lymphocytes. TAP1$^{-/-}$ mutant mice had very few $CD4^-CD8^+$ T cells, although those that did reach the periphery seemed relatively normal. NK cells in these mice were normal in number, but had altered specificities for killing target cells.

Comments

The TAP1 knockout mice demonstrate the crucial role of peptide transport in the assembly of mature class I MHC molecules in the ER. In addition, the results show that class I MHC expression is required for the development of normal numbers of CD8$^+$ T cells in the thymus, as well as for normal NK cell function. TAP1 knockout mice are useful for studies on MHC class I-restricted antigen presentation, T cell repertoire selection, and allograft rejection.

Acknowledgements
Luc Van Kaer
Howard Hughes Medical Institute, Department of Microbiology and Immunology, Vanderbilt University School of Medicine, Nashville, TN, USA

References

1 Van Kaer, L. et al. (1992) Cell 71, 1205–1214.
2 Ashton-Rickardt, P.G. et al. (1993) Cell 73, 1041–1049.
3 Ljunggren, H.G. et al. (1994) Proc. Natl Acad. Sci. USA 91, 6520–6524.

Tau

Other names
Microtubule-binding protein

Gene symbol
Mtapt

Accession number
MGI: 97180

Area of impact
Neurology

General description

Tau protein binds to microtubules and can induce tubulin polymerization *in vitro*. It has been proposed that Tau makes crossbridges between the microtubules in axons and plays a role in axon extension. This was shown by over- or underexpression experiments in transfected cell lines.

KO strain construction

neo^r was inserted into exon 1 of the *Mtapt* gene. *tk* was spliced to the 5′ end. The linearized vector was electroporated into J1 ES cells (129Svter). Selected ES clones showing homologous recombination were injected into C57BL/6 blastocysts. Chimeras were interbred to produce offspring for analysis.

Phenotype

No gross abnormalities were observed in Tau$^{-/-}$ mice and no developmental deficits were revealed by histological examination. The relative levels of various microtubule-associated proteins, neurofilament proteins, synapsin 1, and various tubulin isoforms were similar to those in control littermates. Only the amount of MAP1A was increased about 2-, 1.3- and 1.3-fold in day 7, 14 and adult Tau$^{-/-}$ mice, respectively.

The microtubule (MT) number and density of parallel fiber axons in the cerebellar tissue was reduced in Tau$^{-/-}$ mice but the MT density in Purkinje cell dendrites was not significantly affected. MT density also did not differ significantly in axons with larger diameters (optic nerve and sciatic nerve), suggesting that Tau affects the stability of MTs in small-caliber axons. The frequency of crossbridges between MTs was greatly reduced in Tau$^{-/-}$ axons but filamentous crossbridges between MT and axonal plasma membrane occurred in both Tau$^{-/-}$ and Tau$^{+/+}$ axons.

Comments

Tau is important for the stabilization of microtubules in small-caliber axons, rather than dendrites. Large-caliber axons were not affected in Tau$^{-/-}$ mice, possibly due to compensation by MAP1A. Tau is not necessary for axon elongation.

Reference

[1] Harada, A. et al. (1994) Nature 369, 488–491.

Other names
T cell factor 1, *Tcf-7* (on genomic maps: mouse chromosome 11)

Gene symbol
Tcf1

Accession number
MGI: 98504

Area of impact
Immunity and inflammation, transcription factors

General description

Tcf-1 is a DNA-binding protein of the high mobility group (HMG) protein family. It was initially identified as a transcription factor binding to a sequence motif in the enhancer elements of a number of genes, including CD3ε and TCRα. Tcf-1 enables β-catenin to transduce Wnt signals by association with its monomer. It is closely related to lymphoid enhancer factor (LEF-1). The expression pattern of both genes is complex and largely overlapping during embryogenesis, but expression of both is restricted to lymphocytes postnatally.

KO strain construction

Using the genomic *Tcf1* gene derived from a C57BL/6 × C3H EMBL 3 library (Stratagene), two replacement type targeting vectors were created as follows:

1. TCF-$1^{\Delta VII}$: In a 12 kb *Sal*I/*Bgl*II genomic fragment, a 67 bp *Xho*I/*Sma*I fragment containing part of exon VII (which encodes the N-terminus of the HMG box) was replaced by a PGK*neo*^{r} cassette.
2. TCF-$1^{\Delta V}$: In a 9 kb *Sac*II/*Xho*I genomic fragment, a 350 bp *Nde*I fragment containing part of intron IV and the evolutionarily conserved exon V (function unknown) was replaced by a PGK*neo*^{r} cassette.

The targeting vectors were electroporated into the 129Ola-derived E14 ES cell line. Positive clones were injected into C57BL/6 blastocysts. The chimeras and F1s were crossed with C57BL/6. Homozygous Tcf-$1^{-/-}$ mouse strains were established by intercrossing the heterozygous F2s. The null mutations were confirmed by Southern blot analysis.

Phenotype

Mice heterozygous for either mutation had a wild-type phenotype. Both Tcf-$1^{\Delta VII}$ and Tcf-$1^{\Delta V}$ knockout mice were healthy and fertile and had normal lifespans. Apart from the lymphoid system, internal organs showed no gross morphological abnormalities. The thymus of Tcf-$1^{\Delta VII}$ mutant mice was greatly reduced in size. Histological analysis revealed that the thymus was organized into a cortex and medulla, but that the cortex was narrow and pale-staining, indicative of hypocellularity. Total thymocyte numbers were reduced 10–100-fold, a difference that became more dramatic with age. The thymus of Tcf-$1^{\Delta V}$ mutant mice showed no histological aberrations apart from a moderate reduction in size. Total thymocyte numbers were reduced 2–5-fold at all ages.

In the thymus of Tcf-$1^{\Delta VII}$ mutant mice, the following were observed: (1) the $CD4^-CD8^-$ precursor population was unaffected; (2) the absolute numbers of $CD4^+CD8^+$ and mature $CD4^+$ and $CD8^+$ single positive cells were greatly reduced; (3) the percentage of immature $CD8^+CD3^-CD4^-$ thymocytes was increased. Thus, thymocyte development was blocked at the transition from the $CD8^+$ immature single positive to the $CD4^+/CD8^+$ double positive stage. Most of the immature single positive cells were at the resting stage of the cell cycle. Phenotypic changes in the Tcf-$1^{\Delta V}$ mutant mice were qualitatively similar but less striking.

In lymph nodes and spleens of Tcf-$1^{\Delta VII}$ mutant mice, T cell numbers were decreased 10-fold and 3-fold respectively. The CD4:CD8 and TCRαβ : TCRγδ ratios were unaffected. B cell and NK cell numbers were normal. The development of two independent and putatively extrathymic T cell lineages, intestinal $TCR\gamma\delta^+$ cells and liver $NK1^+TCR\alpha\beta^+$ cells, was selectively impaired. However, $TCR\alpha\beta^+CD4^-8\alpha^+\beta^-$ intra-epithelial lymphocytes, of extrathymic origin, and $NK1^-TCR\alpha\beta^+$ liver T cells, of unknown origin, both developed normally. In Tcf-$1^{\Delta V}$ mutant mice, there was no significant decrease in T or B cell numbers. Splenic T cells were functional in proliferation assays in response to ConA and alloantigen, as well as in allogeneic cytotoxicity assays. Total IgM and IgG were normal, as were specific immune responses to foreign protein antigens.

Comments

While Tcf-1 controls an essential step in thymocyte differentiation, it can be concluded that mature T lymphocytes do not depend on a functional Tcf-1 gene. Tcf-1 is not required for the development of some extrathymic T cell subsets.

It is likely that Tcf-1 and LEF-1 perform redundant functions, at least in part, precluding the direct identification of target genes in the single Tcf-1 knockout mouse. A Tcf-1/LEF-1 double knockout mouse would likely give more insight into this question.

Acknowledgements
J.S. Verbeek and H.C. Clevers
University Hospital Utrecht, Utrecht, The Netherlands

References

1 Verbeek, S. et al. (1995) Nature 374, 70–74.
2 Ohteki, T. et al. (1995) Eur. J. Immunol. 26, 351–355.

Tcp10b^t

Gene symbol
Tcp10b

Accession number
MGI: 98542

Area of impact
Development

General description

The Tcr^t locus (t complex responder) is essential for the phenomenon of transmission ratio distortion (TRD) associated with certain mouse t complex haplotypes. In TRD, heterozygous +/− males transmit the t-bearing chromosome to nearly all their offspring. The Tcp10b^t gene, being located within the genetically defined Tcr^t region, was considered to be a candidate gene for the Tcr^t. Tcp10b^t expression is restricted to male germ cells from pachytene spermatocytes onwards, and exhibits a unique post-meiotic alternative splicing pattern.

KO strain construction

The targeting construct was designed to replace most of *Tcp10b* exon 6 and all of exons 7–9 with a neomycin cassette, deleting the region containing two unique alternative post-meiotic splicing events. The reading frame was altered such that 145 amino acids of the C-terminus were eliminated and replaced by sequences potentially encoding 45 different amino acids. The neomycin gene was under the control of the murine β-actin promoter. HSV-*tk* was included at the 3′ end of the genomic sequences. The targeting vector was introduced into ES cells which were specially derived from mice homozygous for a t haplotype, t^{h49}. Positive clones were injected into C57BL/6J blastocysts. Since the t^{h49} ES cells were homozygous for the wild-type agouti allele, chimeras were detected by coat color.

Phenotype

The Tcp10b$^{t-/-}$ mice appeared to be entirely normal and the targeted chromosome retained full responder activity.

Comments

Tcp10b^t does not appear to be involved in the transmission ratio distortion of mouse t haplotypes and therefore is not Tcr^t. The function of Tcp10b^t remains unknown, as homologs appear to provide functional redundancy.

Acknowledgements
John Schimenti
The Jackson Laboratory, Bar Harbor, ME, USA

Reference

[1] Ewulonu, U.K. et al. (1996) Genetics 144, 785–792.

Other names
T cell receptor alpha, TCRA

Gene symbol
Tcra

Accession number
MGI: 98553

Area of impact
Immunity and inflammation, T cell

General description

Tcra encodes one of the components of the T cell receptor αβ heterodimer which recognizes MHC class I or class II antigens complexed with peptide. TCRα is expressed in association with the TCRβ chain and the CD3 complex on >95% of peripheral T cells, and on some double positive (CD, $CD4^+8^+$) and all mature single positive (CD, $CD4^+CD8^-$ or $CD4^-CD8^+$) thymocytes. The αβTCR is considered to be a key regulator of the immune response, involved in the process of positive selection and the double positive to single positive transition. The TCRα gene is rearranged after the TCRβ gene in the developing fetal thymus.

KO strain 1 construction[1]

The transposon Tn5 neomycin-resistance gene (*neo*) driven by a eukaryotic promoter was cloned into an *Eco*RV site within the first constant region exon of the murine *Tcra* gene. To permit negative selection, the HSV-*tk* gene was cloned into an *Eco*RV site upstream of the construct region exon, deleting a 0.5 kb *Eco*RV fragment. ES cells of the GK129 line (derived from a 129Ola/Hsd mouse strain) were electroporated with the targeting construct. Positive ES clones were injected into BALB/c blastocysts. Chimeras were crossed with BALB/c females. FACS staining showed the complete elimination of $\alpha\beta^+$ T cells from the thymus, spleen and Peyer's patches.

KO strain 2 construction[2]

The targeting vector was designed to disrupt the unique TCR-Cα gene segment by inserting the PGK*neo* neomycin-resistance cassette into its first exon, which encodes the extracellular domain. A 3.9 kb fragment of genomic BALB/c sequence was used in the construct. The targeting vector was electroporated into D3 ES cells and chimeric mice were generated by standard procedures. Intercrossing of mice heterozygous for the mutated allele gave rise to $TCR\alpha^{-/-}$ homozygous mutant mice at the expected frequency. The null mutation was confirmed by Southern blotting.

Phenotype

TCR$\alpha^{-/-}$ mice were grossly normal in appearance. Size and cell numbers of the thymus and spleen were approximately normal. However, the Peyer's patches were small and shriveled in appearance. In addition, TCR$\alpha^{-/-}$ thymi lacked a medulla and had instead a network of cortical epithelial cells that extended throughout the organ. Normal numbers of thymic $\gamma\delta$ T cells of a normal surface marker profile were present, but $\alpha\beta^+$ T cells were missing and thymocyte development was arrested at the double positive to single positive thymocyte checkpoint. All earlier thymocyte developmental stages were unaffected. The rearrangement and transcription of TCRβ was normal. Surface expression of TCRβ chains was detected on TCR$\alpha^{-/-}$ thymocytes.

In the periphery, B and $\gamma\delta^+$T cells were expanded in the secondary lymphoid organs. B cells secreted copious amounts of antibodies of T-dependent isotypes which were reactive to self-antigens. Production and class switch of antibodies, especially IgG1 and IgE, occurred efficiently. Germinal centers developed spontaneously in mice lacking TCRα genes despite the absence of αβ T cells. In addition to B cells, these germinal centers contained CD4$^+$ $\gamma\delta$ T cells and a novel population of CD4$^+\beta^+\alpha^+$ T cells, the expansion of which contributed to the increased size of some lymphoid organs. The formation of germinal centres in TCR$\alpha^{-/-}$ mice was pathogen-driven, since these structures were not observed in a germ-free colony of these mice. TCR$\alpha^{-/-}$ mice developed spontaneous intestinal inflammation, the penetrance of which varied with genetic background and which appeared to be initiated by organisms present in the gut flora. The appendix lymphoid follicle (ALF) showed an increase of IgA-, IgG1- and IgG2a- but not IgM-secreting B cells in TCR$\alpha^{-/-}$ ALF. CD3-induced apoptosis of CD4$^+$8$^+$ thymocytes can still occur in TCR$\alpha^{-/-}$ mice.

Comments

TCRα expression is not necessary for TCRβ rearrangement and transcription. TCRα is required for the "development clock" that directs double positive thymocytes to mature into single positive cells. TCRα is also required for the development of the thymic medulla. B cells and $\gamma\delta^+$ T cells develop independently of $\alpha\beta^+$ T cells.

Acknowledgements
Michael Owen
Imperial Cancer Research Fund, Lincoln's Inn Fields, UK

References

[1] Philpott, K.L. et al. (1992) Science 256, 1448–1452.
[2] Mombaerts, P. et al. (1992) Nature 360, 225–231.
[3] Palmer, D.B. et al. (1993) Dev. Immunol. 3, 175–179.
[4] Wen, L. et al. (1994) Nature 369, 654–658.
[5] Dianda, L. et al. (1996) Eur. J. Immunol. 26, 1603–1607.
[6] Dianda, L. et al. (1997) Am. J. Pathol. 150, 91–97.
[7] Misoguchi, A. et al. (1996) J. Exp. Med. 184, 707–715.
[8] Tokoro, Y. et al. (1996) Eur. J. Immunol. 26, 1012–1017.

Other names
T cell receptor beta, TCRB

Gene symbol
Tcrb

Accession number
MGI: 98578

Area of impact
Immunity and inflammation, T cell

General description

Tcrb encodes one of the components of the T cell receptor αβ heterodimer which recognizes MHC class I or class II antigens complexed with peptide. TCRβ is expressed in association with the TCRα chain and the CD3 complex on >95% of peripheral T cells, and on some double positive (CD, $CD4^+CD8^+$) and all mature single positive (CD, $CD4^+CD8^-$ or $CD4^-CD8^+$) thymocytes. The αβTCR is considered to be a key regulator of the immune response, involved in the process of positive selection and the DP to SP transition. The TCRβ gene is rearranged before the TCRα gene in the developing fetal thymus.

KO strain construction

The targeting vector was designed to delete a 15 kb fragment of the *Tcrb* gene containing one Dβ, 10 Jβ and both Cβ gene segments. The 5′ deletion breakpoint was between gene segments Jβ1.2 and Jβ1.3. This region was replaced by PGK*neo*. HSV-*tk* expression cassettes were introduced 5′ and 3′ of the genomic sequences. The targeting vector was electroporated into 129Ola-derived ES cells, and germ line chimeras were obtained using standard protocols. Intercrossing of mice heterozygous for the mutated allele gave rise to homozygous $TCR\beta^{-/-}$ mice at the expected frequency. The null mutation was confirmed by Southern blotting.

Phenotype

$TCR\beta^{-/-}$ mice had no gross phenotype. Thymi of mutant mice showed a 6–60-fold reduction in total cell number. No $TCR\beta^+$ cells were detected and only about 50% (on average) of thymocytes were double positive (representing only 6% of the wild-type number of double positive cells). As in the wild type, these double positive cells were small and IL-$2R^-$. Due to the reduction in double positive cells, the proportion of large double negative cells was increased to 50% of the total cell number, but appeared to be of the normal subpopulation composition: immature IL-$2R^+$ TCR^- cells, and mature IL-$2R^-$ $TCR\gamma\delta^+$ cells. A few $CD4^+$ and $CD8^+$ single positive cells were present, of which half to two-thirds were γδ T cells (the remainder were considered to cells in transition to the double positive stage). Loss of TCRβ did not affect γδ thymocyte differentiation or γδ T cell number in lymphoid organs. A significant level of gene rearrangement was able to occur in the thymi of mutant mice.

Double TCRα$^{-/-}$TCRβ$^{-/-}$ knockout mice[2]

Thymi of mice doubly mutated to lack both TCRα and TCRβ were similar in phenotype to those of TCRβ$^{-/-}$ single mutant mice.

Double TCRβ$^{-/-}$TCRδ$^{-/-}$ knockout mice[2]

Thymi of mice lacking both TCRβ and TCRδ had almost no double positive T cells, and no single positive cells at all.

Comments

TCRβ is important in the generation of normal cell numbers in the thymus. The rearrangement or expression of a single productively rearranged TCRβ gene is necessary and sufficient for the double negative to double positive transition and the expansion of double positive cells. Mutation of TCRβ blocks αβ thymocyte differentiation at an earlier stage than does mutation of TCRα, but does not affect γδ thymocyte differentiation. TCRβ rearrangement is not necessary for TCRα rearrangement.

References

1 Mombaerts, P. et al. (1991) Proc. Natl Acad. Sci. USA 88, 3084–3087.
2 Mombaerts, P. et al. (1992) Nature 360, 225–231.

Other names
T cell receptor delta chain, TCRD

Gene symbol
Tcrd

Accession number
MGI: 98611

Area of impact
Immunity and inflammation, T cell

General description

The *Tcrd* gene is located within the *Tcra* gene on mouse chromosome 14, between the Vα and Jα regions. TCRδ rearrangements precede TCRα rearrangements. TCRδ forms heterodimers with TCRγ to form the TCRγδ. T cells expressing TCRγδ appear to be a lineage distinct from TCRαβ T cells. TCRγδ T cells preferentially localize to epithelial surfaces, suggesting a unique role for γδ T cells in epithelial environments, but their precise function remains unknown. Mutations in either TCRα or TCRβ do not affect the differentiation of TCRγδ thymocytes.

KO strain construction

The sequences encoding the constant region of the TCRδ chain were replaced with a neomycin-resistance cassette. Variable (V), diversity (D), and joining (J) regions of the TCRδ locus were left intact. E14 and D3 ES cells were used for the gene targeting. The chimeras were originally crossed to either C57BL/6 or BALB/c, followed by backcrossing to these strains, respectively. The targeting of the TCRδ locus was confirmed by Southern blotting.

Phenotype

The mutation caused the complete loss of T cells bearing TCRγδ chains, but had little or no effect on the development of T cells bearing TCRαβ chains. The lack of expression of TCRγδ did not affect the pattern of TCRγδ V(D)J rearrangement, V gene usage, or level of diversification normally observed during fetal and adult γδ T cell development. Macrophages from the mutant mice produced only small amounts of TNF-α in response to LPS and showed a reduced level of expression of CD14. As a consequence, the mutant mice became resistant to septic shock by *Salmonella choleraesuis* infection. Although the mutant mice resolved *Listeria monocytogenes* infections, extensive abscesses were formed. The existence of αβ T cells is required for this inflammatory reaction.

The absence of γδ T cells in TCRδ$^{-/-}$ mice was associated with a reduction in intestinal epithelial cell turnover and a downregulation of the expression of MHC class II molecules. The mutant mice produced much lower levels of IgA antibodies when immunized orally with a vaccine of tetanus toxoid plus cholera toxin as adjuvant. Mutant mice infected with *Eimeria vermiformis* displayed exaggerated intestinal damage, due to a failure to regulate the consequences of the αβ T cell response. The epidermis of normal mice that had spontaneously recovered from cutaneous graft-vs.-host disease (GVHD) induced by local injection of $CD4^+$ autoreactive T cells became resistant to subsequent attempts to induce GVHD in a site-restricted manner. However, resistance to the cutaneous GVHD was not induced in the epidermis of the mutant mice.

Comments

The unique pattern of V gene usage, diversification, and V(D)J recombination seen in γδ T cells of wild-type mice likely results from the control of intracellular mechanisms acting at the level of DNA rearrangement, rather than selective mechanisms in the thymic microenvironment selecting T cells expressing certain γδTCRs. In addition, the data suggest a role for γδ T cells in regulating αβ T cell function, as well as in regulating the physiology of the intestinal and other epithelial tissues.

Acknowledgements
Shigeyoshi Itohara
Kyoto University, Kyoto, Japan

References

[1] Itohara, S. et al. (1993) Cell 72, 337–348.
[2] Komano, H. et al. (1995) Proc. Natl Acad. Sci. USA. 92, 6147–6151.
[3] Fujihashi, K. et al. (1996) J. Exp. Med. 183, 1483–1489.
[4] Mombaerts, P. et al. (1993) Nature 365, 53–56.

TCR-Vγ5

Other names
TCR-Vγ5/3/GVISI

Gene symbol
Tcrg-V5

Accession number
MGI: 98635

Area of impact
Immunity and inflammation, T cells

General description

The TCR-Vγ5 gene encodes a chain of the $\gamma\delta$TCR that is very strongly associated with murine skin T cells. In most inbred strains, more than 90% of skin T cells express this TCR chain, as do the first fetal thymocytes to appear in ontogeny.

KO strain construction

The gene was targeted using a positive–negative selection replacement vector strategy. *neo* was driven by the PGK promotor, and was used to disrupt the coding region of the Ig-like domain of the gene. The KO mouse was of mixed 129 × B6 background. Backcrosses to B6 are in progress.

Phenotype

TCR-Vγ5$^{-/-}$ mice were outwardly healthy, with no obvious skin defects. The success of the gene disruption was indicated by the fact that none of the skin T cells were positive for 536, an antibody that reacts with cells expressing Vγ5. Nonetheless, a skin T cell repertoire developed that appeared normal by the criteria of cell numbers and dendritic morphology. In addition, all other $\gamma\delta$ T cell subsets appeared to develop largely normally, indicating that although Vγ5^{+} cells are the first T cells to appear, they do not play a critical role in the development of subsequent T cells. The skin T cell repertoire that developed in Vγ5-deficient mice expressed the epitope for the 17D1 antibody. In normal mice, this antibody reacts exclusively with the T cell receptor of skin T cells, causing it to be regarded as an anti-clonotypic antibody specific for skin $\gamma\delta$ T cells. This result suggests that the Vγ5 T cell receptor has an important function in the skin, and that in the absence of Vγ5, an alternative combination of genes encodes the relevant conformation.

Comments

The Vγ5 chain is not required for the development of an apparently normal-looking skin T cell repertoire. The TCR-Vγ5$^{-/-}$ mouse is one of a series of knockouts designed to disrupt every possible $\gamma\delta$TCR.

Acknowledgements
Adrian Hayday
Department of Biology, Yale University, New Haven, CT, USA

Reference

[1] Mallick-Wood, C. et al. (1998) Science 279, 1729–1733.

Other names
Terminal deoxynucleotidyl transferase

Gene symbol
Tdt

Accession number
MGI: 98659

Area of impact
Immunity and inflammation

General description

TdT is a unique, template-independent DNA polymerase which adds random nucleotides to the free 3′ ends of DNA molecules. Its only known physiologic function is to generate antigen receptor diversity in the immune system by adding template-independent "N" nucleotides at the junctions during the rearrangement of the variable, diversity and joining gene segments of the immunoglobulin heavy chain and T cell receptor. TdT expression is limited almost entirely to immature lymphocytes and has not been detected at significant levels early during ontogeny in most species examined.

KO strain 1 construction[1]

A 1.1 kb neomycin-resistance gene (pMC1*neo*polyA, Stratagene) was inserted without deletion into the *Eco*RV site in exon 4 of an 11 kb *Xho*I/*Kpn*I genomic fragment (isolated from an A/J genomic library). No *lox* sites were present. The HSV-*tk* gene was appended to the 3′ end to allow double selection. The targeting construct was originally electroporated into D3 (129Sv) ES cells. Chimeras were crossed to C57BL/6 mice and heterozygotes were intercrossed to produce TdT$^{-/-}$ mutants. The TdT-null mutation has since been fairly extensively backcrossed to C57BL/6 (12 times), and also onto B6, NOD (extensive), B10.Br, BALB/c, NZB and NZW strains. The null mutation was confirmed by Southern analysis, PCR amplification of thymus RNA, and *in situ* hybridization to thymic sections.

KO strain 2 construction[2]

The targeting vector was constructed such that the promoter and first exon of the *Tdt* gene were deleted and replaced with PGK*neo*. PGK-*tk* was present at the 5′ end of the vector. The construct was transfected into CCE ES cells. Positive clones were injected into blastocysts of RAG-2-deficient mice and transferred into B6CBAF1/J females. Neither TdT enzymatic activity nor TdT RNA transcripts were detected in mutant thymocytes.

Phenotype

Heterozygotes had a normal phenotype. TdT$^{-/-}$ homozygous mutant mice bred well and appeared healthy in a conventional animal facility. No anomalies other than those affecting the immune system were observed. There was no increased susceptibility to infections. No significant alterations in the major T and B cell compartments were detected. $TdT^{-/-}$ mice mounted normal T and B cell responses to complex antigens.

N nucleotides were essentially absent at the V(D)J junctions of rearranged Ig and TCR genes in $TdT^{-/-}$ mutant mice, so that the Ig and TCR repertoires of adult $TdT^{-/-}$ animals were less diverse than those of TdT^{+} animals. Repetitive, homology-directed V(D)J junctions were formed at relatively high frequencies in $TdT^{-/-}$ mice, further limiting the diversity, particularly of Ig heavy chains and γδTCRs. Thus, $TdT^{-/-}$ mice carried antigen receptor repertoires resembling those occurring normally in fetal and neonatal mice (prior to the onset of TdT expression) which show limited junctional diversity.

In αβTCRs, the TdT-null mutation affected the length and diversity of the CDR3 loops thought to be important in "directing" MHC/peptide recognition. Because CDR3 loops lacking N nucleotides (N^{-}) are shorter and less diverse, they appear to wield less influence than do their N nucleotide-containing (N^{+}) counterparts. Positive selection was more efficient in $TdT^{-/-}$ animals and the N^{-} peripheral repertoire was more polyreactive and less peptide-oriented than the N^{+} repertoire. However, this loss of specificity did not markedly diminish the response to specific peptides. Despite the lack of TdT-mediated junctional diversity, $TdT^{-/-}$ mice were quite robust and responded effectively to a wide variety of challenges to the immune system. Mutants did display perturbations of autoimmune phenomena.

Comments

TdT is responsible for N nucleotide addition and is essential for the creation of diversity in lymphocyte antigen receptor repertoires. TdT activity blocks homology-directed recombination, causing a greater expansion of repertoire diversity than was previously thought.

Acknowledgements
Susan Gilfillan
Basel Institute for Immunology, Basel, Switzerland
Christophe Benoist
CNRS INSERM Institut de Génétique et de Biologie Moléculaire et Cellulaire, Université Louis Pasteur, Strasbourg, France

References

1 Gilfillan, S. et al. (1993) Science 261, 1175–1178.
2 Komori, T. et al. (1993) Science 261, 1171–1175.
3 Gilfillan, S. et al. (1994) Int. Immunol. 6, 1681–1686.
4 Gilfillan, S. et al. (1995) Eur. J. Immunol. 25, 3115–3122.
5 Gavin, M.A. and Bevan, M.J. (1995) Immunity 3, 793–800.
6 Gilfillan, S. et al. (1995) Immunol Rev. 148, 201–219.

Other names
Tie2

Gene symbol
Tek

Accession number
MGI: 98664

Area of impact
Development

General description

Tek is a 140 kDa receptor tyrosine kinase expressed in endothelial cells and some early hematopoietic cells. It contains complex extracellular domains consisting of two Ig-like loops separated by three EGF-like repeats that are followed by three fibronectin type III-like repeats. Tek expression can be detected during mouse development in regions that overlap with Flk-1^+ extra-embryonic mesoderm at E7.5. It is thought that Tek may play a role in the vascularization of the embryo.

KO strain construction

The null allele of *Tek* was created by deleting the last 52 bp of exon 1 which encodes the first 17 amino acids of Tek, including the translation start site and the signal sequence. A positive–negative targeting vector was constructed in which PGK*neo* disrupted the *Tek* genomic gene isolated from strain 129Sv. The targeting vector was electroporated into R1 ES cells and positive clones were injected into C57BL/6J blastocysts. Intercrosses of heterozygous progeny were carried out either in an outbred (129SvJ × CD-1) or inbred (129SvJ) background. No differences in phenotype due to genetic background were observed.

Phenotype

$Tek^{-/-}$ embryos died by day 9.5–10.5 of embryogenesis. There were signs of vascular hemorrhaging and a loss of vascular integrity. The development of the heart was affected, probably due to a deficit in signaling between the endocardium and myocardium. There appeared to be a progressive loss of endothelial cells both in the embryo proper and in the extra-embryonic vasculature.

Comments

Tek signaling is not required for the initial appearance of early endothelial cells during embryogenesis but is crucial for their subsequent survival and/or proliferation. Tek is particularly necessary for endocardial/myocardial interactions during development.

Acknowledgements
Daniel Dumont
Amgen Institute, Toronto, ON, Canada

References

1. Dumont, D.J. et al. (1994) Genes Dev. 8, 1897–1909.
2. Dumont, D.J. et al. (1992) Oncogene 7, 1471–1480.
3. Dumont, D.J. et al. (1993) Oncogene 8, 1293–1301.
4. Sato, T.N. et al. (1995) Nature 376, 70–74.

Other names
ets variant gene 6 (TEL oncogene), ets family transcription factor

Gene symbol
Etv6

Accession number
MGI: 109336

Area of impact
Oncogenes, transcription factors

General description

TEL is an ets transcription factor first discovered at the site of a chromosomal translocation in leukemia. It is widely expressed in the embryo and adult.

KO strain construction

The DNA-binding domain was deleted using a typical positive–negative *neo*-selection construct. The mutation is on a 129 background.

Phenotype

The phenotype was embryonic lethal and $TEL^{-/-}$ mice died at E10–11 from defective angiogenesis in the yolk sac. There was apoptosis in selected regions within the embryo proper.

Comments

TEL is essential for normal yolk sac vascular development and for the prevention of apoptosis in selected tissues.

Acknowledgements
Stuart H. Orkin
Divisions of Hematology/Oncology and Pediatric Oncology, Children's Hospital and Dana-Farber Cancer Institute, Harvard Medical School, Boston, MA, USA

Reference
[1] Wang, L.C. et al. (1997) EMBO J. 16, 4374–4383.

Tensin

Gene symbol
Tns

Accession number
MGI: 104552

Area of impact
Development, nephrology

General description

Tensin is a 200 kDa focal contact phosphoprotein that contains three actin-binding domains and an SH2 domain. Tensin is tyrosine phosphorylated upon extracellular matrix stimulation or cell transformation by v-Src. Kinetic studies suggest that localization of tensin and focal adhesion kinase to sites of integrin clustering are early events in the formation of focal adhesions. Tensin is highly expressed in kidney and heart in adult mice and is expressed no later than day 7 in the mouse embryo.

KO strain construction

The 4 kb *Spe*I/*Eco*RI fragment of the mouse tensin gene was replaced with a 1.8 kb fragment containing a PGK1*neo* gene for positive selection. The PGK1*neo* gene was flanked 3′ with a 3 kb *Tns* fragment and 5′ with a 2.2 kb *Tns* fragment, and a PGK1*tk* gene for negative selection. The targeting vector was electroporated into R1 ES cells and positive clones were injected into C57BL/6 blastocysts. Chimeras were bred to obtain germ line transmission.

Phenotype

Tensin$^{-/-}$ mice appeared normal at birth. As judged by external appearance, body weight and general behavior, the developing post-natal mutant mice did not seem to differ from their wild-type or heterozygote littermates. Gross anatomical examination of the brain, liver, heart, spleen, stomach, skin, lung, eyes and intestine revealed no obvious defects. However, the kidneys exhibited significant and consistent bilateral abnormalities. They were pale, had a granular appearance on their surfaces, and were often slightly larger than age-matched wild-type kidneys. When hemisected, the wild-type kidneys always displayed a dark reddish brown cortex which was readily distinguishable from the lighter medulla. This distinction in coloring was notably less prominent in the mutant kidneys and the renal pelvis opening was markedly enlarged.

Histologic examination of sections of tensin$^{-/-}$ kidneys revealed various degrees of cystic defects, ranging from extremely dilated cysts in the cortex and medulla of severely affected kidneys to small cysts in the cortex of the less affected organs. Most small cysts arose from proximal tubules, as suggested by PAS staining and electron microscopy. Closer inspection of these cysts revealed that they appeared to be derived from an expansion of the tubular lumen. This was less obvious in the larger cysts, where gross distortions often precluded identification of the origin of the cells that lined the cysts.

Higher magnification revealed additional abnormalities in tensin$^{-/-}$ kidneys. Some cysts contained atypical casts and many glomeruli displayed enlarged Bowman's spaces. Epithelial cells surrounding each mutant glomerulus were often enlarged and cuboidal, in contrast to the flattened, barely visible simple squamous epithelium that surrounded control glomeruli. In the most severely affected regions of mutant kidneys, there were signs of focal segmental glomerular sclerosis. Focal interstitial inflammatory infiltrates were interspersed throughout the affected kidneys.

Ultrastructurally, non-cystic areas of tensin$^{-/-}$ kidneys showed typical cell–matrix junctions that were readily labeled with antibodies against other focal contact molecules. In abnormal regions, cell–matrix junctions were disrupted and tubule cells lacked polarity. Taken together, these histological findings suggest strongly that large regions of the kidneys of tensin-null mice were not functioning properly, and that fluid flow from the glomeruli to the collecting ducts was aberrant.

Comments

The loss of tensin leads to an alteration or weakening, rather than a disruption, of focal contacts in the kidney. The formation of cysts indicates that a perturbation of focal contacts affects epithelial polarity. Most other tissues appeared normal in the mutant mice, suggesting that, in most cases, tensin's diverse functions are redundant and may be compensated for by other focal contact proteins.

Acknowledgements
Elaine Fuchs
Department of Molecular Genetics and Cell Biology, University of Chicago, Chicago, IL, USA

References
1 Lo, S.H. et al. (1997) J. Cell Biol. 136, 1349–1361.
2 Davis, S. et al. (1991) Science 252, 712–715.
3 Lo, S.H. et al. (1994) J. Cell Biol. 125, 1067–1075.
4 Bockholt, S.M. et al. (1993) J. Cell Biol. 268, 14565–14567.
5 Lo, S.H. et al. (1994) BioEssays 16, 817–823.

Other names
Tissue factor, coagulation factor III

Gene symbol
Cf3

Accession number
MGI: 88381

Area of impact
Development

General description

TF is a 47 kDa membrane glycoprotein that functions as a receptor for factor VII/VIIa. It is believed to be the primary physiological initiator of blood coagulation after vascular damage. TF is also speculated to have important roles outside of coagulation, based on the potential for signal transduction through its cytoplasmic tail, and its ability to modulate the metastatic and angiogenic potential of tumor cells.

KO strain 1 construction[1]

A PGK*hprt* minigene was inserted after the TF gene promoter and replaced a 6 kb fragment of the gene, including exon 1, 2 and part of exon 3. The mutation was introduced into HPRT 129 ES cells, which were injected into C57BL/6 blastocysts. The original strain was maintained on 129/NIH Black Swiss background.

KO strain 2 construction[2]

A PGK*neo* cassette was introduced into exon 3, 76 amino acids from the N-terminus, in the first Ig-like domain. The vector was introduced into RW-4 ES cells, which were injected into C57BL/6 blastocysts. The line was maintained by crossing to C57BL/6.

KO strain 3 construction[3]

A deletion that includes the start of transcription and translation was made in the TF gene, and replaced with a PGK*neo* cassette. The vector was introduced into R1 129 ES cells, which were aggregated with CD1 morulae.

Phenotype

No $TF^{-/-}$ mice were born from crosses between $TF^{+/-}$ mice, demonstrating that TF has an essential role in embryonic development. $TF^{+/-}$ mice have half-normal TF activity, but otherwise normal hematological parameters.

$TF^{-/-}$ embryos die before E11.5. Most E9.5 and all E10.5 $TF^{-/-}$ embryos are immediately distinguishable from their $TF^{+/-}$ and $TF^{+/+}$ littermates by the extreme pallor of both the yolk sac and the embryo proper and grossly enlarged pericardial sacs and distended hearts. Pools of nucleated red cells are often visible within the yolk sac cavities and pericardial sac. A vascular plexus is present in the yolk sacs but the vessels appear empty.

The data demonstrate that TF plays a crucial role in embryonic development. The massive hemorrhaging observed in mutant embryos may be due to the failure of proper initiation of coagulation.

Comments

Inactivation of soluble coagulation factors such as fibrinogen and factor V is compatible with development to term, whereas mice with inactivation of the membrane-associated coagulation factors, TF and thrombomodulin, die at midgestation. Maternal rescue may permit mice deficient in soluble coagulation components to develop to term.

Acknowledgements
Thomas H. Bugge and Jay L. Degen
Children's Hospital Research Foundation, Cincinnati, OH, USA

References

[1] Bugge, T.H. et al. (1996) Proc. Natl Acad. Sci. USA 93, 6258–6263.
[2] Toomey, J.R. et al. (1996) Blood 88, 1583–1587.
[3] Carmeliet, P. et al. (1996) Nature 383, 73–75.

Other names
Transforming growth factor β1

Gene symbol
Tgfb1

Accession number
MGI: 98725

Area of impact
Development

General description

TGFβ is a key member of a superfamily of polypeptide growth and differentiation factors that actively participate in embryonic development, tissue and organ formation, cell growth and differentiation, wound healing and immune functions.

KO strain 1 construction[1]

A pMC1*neo* cassette was inserted into exon 6. The targeting vector was introduced into D3 129 ES cells which were injected into C57BL/6 blastocysts. The phenotype was examined on a mixed 129 × B6 background.

KO strain 2 construction[2]

A 560 bp sequence spanning part of the first exon and intron was deleted and replaced by a PGK*neo* cassette. The vector was introduced into CCE ES cells.

Phenotype

Two main phenotypes have been reported during different stages of development: an embryonic lethal phenotype and an inflammation phenotype starting at one week after birth, leading to death within 4 weeks. The predominant phenotype seen in TGFβ1-null mice is massive multifocal inflammation that leads to death, usually at or before 4 weeks of age. The infiltration of large number of inflammatory cells, primarily macrophages and leukocytes, into the lungs, heart, liver, salivary glands, pancreas and stomach causes multiorgan failure resulting in premature lethality. TGFβ1-null mice exhibit the onset of inflammation as early as post-natal day 7 coincident with enhanced and aberrant expression of both classes of MHC molecules and increased adhesion of leukocytes to vascular endothelium. Shortly after the onset of inflammation, TFGβ1-null mice exhibit autoimmune manifestations including elevated levels of antibodies to nuclear antigens and kidney deposits of immune complexes.

Comments

The lesions in salivary glands and the clinical symptoms of dry mouth and dry eyes are similar to Sjögren's syndrome in humans. TGFβ1 is the predominant isoform playing a role in immunological activation and suppression.

Acknowledgements
Ashok B. Kulkarni
NIDR, NIH, Bethesda, MD, USA

References

[1] Shull, M.M. et al. (1992) Nature 359, 693–699.
[2] Kulkarni, A.B. et al. (1993) Proc. Natl Acad. Sci. 90, 770–774.
[3] Kulkarni, A.B. et al. (1995) Am. J. Pathol. 146, 264–275.
[4] Dickson, M.C. et al. (1995) Development 121, 1845–1854.
[5] Geiser, A.G. et al. (1993) Proc. Natl Acad. Sci. USA 90, 9944–9948.
[6] Hines, K.L. et al. (1994) Proc. Natl Acad. Sci. USA 91, 5187–5191.

Other names
Transforming growth factor β3

Gene symbol
Tgfb3

Accession number
MGI: 98727

Area of impact
Development

General description

TGFβ3 is one of three mammalian TGFβ growth factors that are closely related, although each is encoded by a separate gene and located on a different chromosome. Functionally, TGFβ family members have been implicated in adhesion, proliferation, differentiation and transformation. Each isoform is thought to have a distinct biological function. A coordinated role for each TGFβ isoform in epithelial–mesenchymal interaction has been proposed. TGFβ3 is expressed in mouse embryos at several sites where such interactions are important, including the developing palatal shelves and the lung.

KO strain construction

The mouse TGFβ locus was cloned from a B6CBA/F1 genomic library. The targeting vector was constructed by inserting a PGK*neo*polyA cassette into exon 6. An HSV-*tk* cassette was included 5′ of the genomic sequences. The targeting vector was electroporated into R1 ES cells and a positive clone was used to generate chimeras. Male chimeras were mated to C57BL/6 and Black Swiss females.

Phenotype

Homozygous TGFβ3$^{-/-}$ pups failed to suckle and died within 20 hours of birth. Consistent phenotypic findings were delayed pulmonary development and defective palatogenesis. No other skeletal abnormalities were detected and the overall heart morphology was normal. Unlike other null mutants with cleft palate, TGFβ3$^{-/-}$ mice did not exhibit any other craniofacial anomalies.

Comments

TGFβ3 is essential for the normal morphogenesis of the palate and lung.

Acknowledgements
Vesa Kaartinen
Department of Pathology, Children's Hospital Los Angeles, Los Angeles, CA, USA

Reference

[1] Kaartinen, V. et al. (1995) Nature Genet. 11, 415–421.

Other names
Tyrosine 3-hydroxylase, EC 1.14.16.2

Gene symbol
Th

Accession number
MGI: 98735

Area of impact
Cardiology, neurology

General description

TH catalyzes the conversion of tyrosine to 3,4-dihydroxyphenylalanine (DOPA) in the initial and rate-limiting step in the biosynthetic pathway of catecholamine (epinephrine, norepinephrine and dopamine) neurotransmitters. TH is expressed in discrete regions in the brain including dopaminergic, noradrenergic and adrenergic neurons as well as in sympathetic post-ganglionic neurons and adrenal gland medullary cells. During vertebrate development, the gene is expressed in the primordial catecholaminergic neurons in the intermediate zone of neural tube, and in migrating neural crest cells committed to the sympathoadrenal lineage at the early embryonic stage.

KO strain 1 construction[1]

The targeting vector included 8 kb of genomic *Th* sequence consisting of the 5′ homologous region of 5.8 kb and the 3′ homologous region of 2.2 kb, in which a short fragment of 0.3 kb containing exons 7 and 8 was exchanged with a PGK*neo* cassette. For the counterselection, the HSV-*tk* and MC1-DTA gene cassettes were connected at the respective 5′ and 3′ ends of the vector. E14 ES cells were targeted and injected into C57BL/6J blastocysts. Chimeric mice were mated with C57BL/6J or MCH(ICR) strains.

KO strain 2 construction[2]

Two-thirds of the coding region including the proximal promoter and first two exons of *Th* were replaced by a neomycin-resistance cassette. A *tk* cassette was placed at the 3′ end. AB1 ES cells were targeted and microinjected into C57BL/6 blastocysts.

Phenotype

TH knockout mice died either *in utero* at a late stage of embryonic development or shortly after birth. The mortality rate of the knockout mice gradually increased past the E12.5 stage. Both TH mRNA and enzyme activity were lacking in the mutants, leading to severe depletion of dopamine, norepinephrine, and epinephrine. Gross morphology of the knockout mice was normal and development of cells showing normally catecholaminergic phenotypes was unaffected by the mutation. Expression of other catecholamine-synthesizing enzymes, aromatic L-amino acid decarboxylase (AADC), dopamine β-hydroxylase (DBH), and phenylethanolamine *N*-methyltransferase (PNMT) was also normal in the mutant embryos. The surviving mutant neonates showed bradycardia based on surface electrocardiographic analysis; the basal heart rate in the mutants was significantly reduced to approximately 70% of wild type and heterozygous values, indicating that catecholamines are required for maintaining the frequency of the spontaneous heart beat. The blood glucose concentration in the mutant neonates showed no significant difference from that of other genotypes. There were no visible effects on lung ventilation or surfactant protein synthesis in the surviving mutants after delivery. Cardiac responses were predominantly impaired in the TH knockout mice, but their metabolic and respiratory responses seemed to be normally maintained by some compensatory mechanism. The specific cause of death of the mutants was probably impaired cardiovascular function due to catecholamine depletion, which might be attributed to a defect in circulating catecholamines. The phenotypic changes in the mutants were completely resolved when a human TH transgene was introduced into mice heterozygous for the mutation, and in transgenic mice carrying multiple copies of human TH gene. The administration to pregnant females of L-DOPA, the product of the TH reaction, resulted in incomplete rescue of mutant mice *in utero*.

Comments

Disruption of the mouse *Th* gene demonstrates that catecholamines are necessary for normal development and survival of the animals during the late gestational and neonatal stages. In particular, catecholamines have an essential role in regulating cardiovascular functions during the perinatal period.

Acknowledgements
Kazuto Kobayashi
Nara Institute of Science and Technology, Nara, Japan
Toshiharu Nagatsu
Fujita Health University, Toyoake, Japan

References

[1] Kobayashi, K. et al. (1995) J. Biol. Chem. 270, 27235–27243.
[2] Zhou, Q.Y. et al. (1995) Nature 374, 640–643.
[3] Iwata, N. et al. (1992) Biochem. Biophys. Res. Commun. 182, 348–354.

Thrombin receptor

Other names
TR*tr*

Gene symbol
Cf2r

Accession number
MGI: 101802

Area of impact
Metabolism, cardiovascular

General description

tr is a seven membrane-spanning G protein-coupled receptor expressed on platelets, endothelial cells, leukocytes and mesenchymal cells. It is proteolytically activated by thrombin.

KO strain construction

A PGK*neo* gene was inserted into exon 2 of the *Cf2r* gene, thus replacing a sequence encoding the transmembrane domains 1–7. HSV-*tk* was placed at the 3′ end of the construct. RF8 ES cells were electroporated and targeted cells were injected into C57BL/6 blastocysts.

Phenotype

About 50% of $tr^{-/-}$ mice die around day 9 and 10 of embryogenesis. The other half of mutant mice survived into adulthood and appeared normal. The cause of death is not known. Platelets, but not fibroblasts, from null mice still responded to thrombin (ATP, platelet aggregation and Ca^{2+} flux). The surviving mice did not develop any bleeding diathesis and had normal bleeding times.

Comments

These results indicate that a second, tissue-specific thrombin receptor exists.

Reference

1 Connolly, A.J. et al. (1996) Nature 381, 516–519.

Thrombospondin-1

Gene symbol
Thbs1

Accession number
MGI: 98737

Area of impact
Development

General description

Thrombospondin-1 is a widely expressed extracellular matrix protein. It can act either as a pro-adhesive or anti-adhesive protein. Thrombospondin-1 modulates protease and growth factor activity, and sequesters calcium at the cell surface and within the extracellular matrix.

KO strain construction

Exons 2 and 3 and introns 2 and 3 of the *Thbs1* gene were replaced with PGK*neo* in the opposite transcriptional orientation. The construct also contained 1.2 kb and 6.1 kb of homologous sequences at the 5′ and 3′ ends, respectively. The HSV-*tk* gene was cloned into the 3′ end of the construct in pBluescript. The vector was transfected into D3 ES cells (129Sv) and targeted clones were injected into C57BL/6 recipient blastocysts. Chimeras were bred to C57BL/6 and 129Sv mice.

Phenotype

Thrombospondin-$1^{-/-}$ mice were viable, fertile and overtly normal. Some defects in the regulation of angiogenesis during wound healing and mammary gland involution were observed. Thrombospondin-$1^{-/-}$ cells showed reduced secretion of suppressors of angiogenesis. Thrombospondin-1-deficient mice displayed a mild and variable lordotic curvature of the spine that was apparent from birth. These animals also showed an increase in the number of circulating white blood cells.

Consistent with the high levels of thrombospondin-1 expression in the lung, thrombospondin-$1^{-/-}$ mice exhibited extensive abnormalities of the lungs. Histology of lungs from 4-week-old mutant mice revealed an increase in the number of neutrophils and macrophages within the lungs. The macrophages stained positively for hemosiderin, indicating that diffuse alveolar hemorrhage was occurring. In some areas, increased cellular proliferation resulted in reduction of the air space. The increased cellular proliferation was associated with the deposition of fine collagen fibrils and a thickening and ruffling of the epithelium of the airways.

Acknowledgements
Jack Lawler
Beth Israel Deaconess Medical Center, Boston, MA, USA

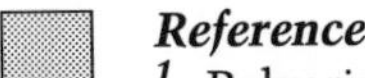

References

[1] Polverini, P.J. et al. (1995) FASEB J. 9, A272.

[2] Lawler, J. et al. (1998) J. Clin Invest. 101, 982–992.

Other names
Thymus cell antigen 1θ, CD90, D9Nds3

Gene symbol
Thy1

Accession number
MGI: 98747

Area of impact
Neurology, immunity

General description

Thy-1 glycoprotein has a single Ig-type variable domain held to the cell surface by a glycosylphosphatidylinositol (GPI) anchor. A major component of the neuronal surface in all species, it is expressed (in rodents) late in neurogenesis, the protein not appearing on axons until they have completed growing. Thy-1 also occurs at high levels on other tissues; the pattern of expression varies between species. The function of Thy-1 is unknown but experiments suggest it has a role in regulating neurite extension by neurons. It may also have an immunomodulatory role on lymphocytes.

KO strain construction

PGK*neo* was inserted at the *Bst*EII site at beginning of exon 3 of *Thy1*, disrupting the codon for third amino acid of the mature Thy-1 protein and producing constitutive gene inactivation with zero detectable protein production. A *tk* gene in the 3′ UTR was used to enhance selection of homologous recombination. 129Sv ES cells were targeted and chimeras were outbred to C57BL/6, then interbred to recover homozygotes.

Phenotype

The phenotype of Thy-$1^{-/-}$ mice was subtle. Neurologically, the brain developed and functioned normally. Enhanced GABA-ergic inhibition in the dentate gyrus led to suppression of long-term potentiation (LTP) by entorhinal afferents terminating on dentate granule cells. This suppression was complete when measured *in vivo* in anesthetized mice, but was only complete in a proportion (around half) of the animals when measured in conscious space. LTP in the CA1 region of the same mice was entirely normal, as was their spatial learning assessed in the water maze. The immunological phenotype is still under assessment, but was again relatively subtle, affecting T cell development.

Comments

These mice have now been outbred for four generations to both 129Sv and C57BL/6 lines. The same targeting vector has been used to knock out Thy-1 in C57BL/6 ES cells, and these have been outbred for three generations to C57BL/6 and 129Sv lines. The analysis of these mice is in progress.

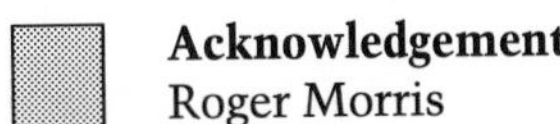

Acknowledgements
Roger Morris
Department of Experimental Pathology, UMDS Guy's Hospital, London, UK

References

[1] Nosten-Bertrand, M. et al. (1996) Nature 379, 826–829.
[2] Morris, R.J. and Nosten-Bertrand, M. (1996) Trends Neurosci. 19, 277–278.

Thyroid hormone receptor β

Other names
THRβ, c-erbA-β, Thrb1, Thrb2

Gene symbol
Thrb

Accession number
MGI: 98743

Area of impact
Hormone, transcription factors

General description

The thyroid hormone receptor (THR) α and β genes encode thyroid hormone (T3)-dependent transcription factors. THRα and THRβ are differentially expressed, suggesting that they mediate certain distinct aspects of T3 action in development and in adult homeostasis. THRβ is restricted in expression during early embryogenesis but is highly induced in several tissues, including brain, cochlea, retina, pituitary, liver and kidney during the post-natal period.

KO strain construction

The targeted deletion removed part of the exon that encodes the first zinc finger of the DNA-binding domain. It inactivated the DNA-binding and T3-binding functions of the THRβ products. Both N-terminal splice variants (THRβ-1 and THRβ-2) were inactivated. The mutation was introduced into W9.5 ES cells (derived from a 129Sv mouse) which were used to produce a mutant mouse on a mixed background of 129Sv and C57BL/6J strains.

Phenotype

Targeted inactivation of the *Thrb* gene resulted in goitre and elevated levels of thyroid hormones. Thyroid-stimulating hormone (TSH), which is released by pituitary thyrotropes and which is normally suppressed by increased levels of thyroid hormone, was present at elevated levels in Thrb$^{-/-}$ mice. These findings suggest a unique role for THRβ that cannot be substituted by THRα in the T3-dependent feedback regulation of TSH transcription. Thrb$^{-/-}$ mice provide a recessive model for the human syndrome of resistance to thyroid hormone (RTH) that exhibits a similar endocrine disorder but which is typically caused by dominant THRβ mutants that are transcriptional inhibitors. It is unknown whether THRα, THRβ or other receptors are targets for inhibition in dominant RTH; however, the analysis of Thrb$^{-/-}$ mice suggests that antagonism of THRβ-mediated pathways underlies the disorder of the pituitary-thyroid axis.

Thrb$^{-/-}$ mice also displayed defective maturation of auditory function, as evidenced by the significantly elevated thresholds required for auditory-evoked brainstem responses. Although congenital hypothyroidism results in cochlear malformations, these were absent in Thrb$^{-/-}$ mice, suggesting that

THRβ mediates only some, albeit essential, aspects of T3 action in auditory development. These results demonstrate that THRβ has an essential role that cannot be substituted by THRα in the development of hearing. Interestingly, hearing defects are generally absent in the dominant form of human RTH syndrome, indicating that in the auditory system, a dominant THRβ mutant cannot mimic the defect caused by complete loss of THRβ. This suggests the existence of tissue-specific mechanisms that modulate the activity of THRβ.

Comments

The results indicate that THRβ has certain functions *in vivo* that cannot be substituted by THRα, implying that the two THR genes serve at least some distinct functions. THR-deficient mice provide a model with which to investigate the mechanism of action of THRs *in vivo*, including the identification of the target genes that are presumably differentially regulated by THRα and THRβ.

Acknowledgements
Douglas Forrest
Department of Human Genetics, Mount Sinai School of Medicine, New York, NY, USA

References

1 Forrest, D. et al. (1996) EMBO J. 15, 3006–3015.
2 Forrest, D. et al. (1996) Nature Genet. 13, 354–357.
3 Wikstrom, L. et al. (1998) EMBO J. 17, 455–461.
4 Refetoff, S. et al. (1993) Endocrine Rev. 14, 348–399.
5 Forrest, D. (1996) J. Clin. Endocrinol. Metab. 81, 2764–2767.

Other names
Tie1

Gene symbol
Tie1

Accession number
MGI: 99906

Area of impact
Development

General description

The *Tie1* gene encodes a receptor tyrosine kinase characterized by an extracellular domain consisting of two Ig-like loops flanking three epidermal growth factor homology domains followed by three fibronectin III type repeats and transmembrane, juxtamembrane, split kinase and C-terminal domains. The ligand for Tie-1 has not been identified. *Tie1* is expressed in endothelial cells beginning at E8.0 in the dorsal aorta, heart, and in differentiating angioblasts of the head mesenchyme. It continues to be expressed in an endothelial cell-specific manner throughout gestation and persists in the endothelium of most adult tissues. Upregulation of the Tie-1 message has been observed in the endothelium of local blood vessels during ovulation and wound healing in the adult as well as in the capillaries of cutaneous and brain metastases of melanoma, suggesting a role for Tie-1 in angiogenesis. Tie-1 is also expressed in early hematopoietic progenitor cell populations.

KO strain 1 construction[1]

The homologous recombination event deletes the start of the Tie-1 protein-coding region, placing the gene for β-galactosidase under the control of the *Tie1* promoter, creating the Tie^{lcz} allele. Two *loxP* sites around the *neo* gene were included for subsequent Cre recombinase mediated excision of *neo*, thereby generating the Tie^{lczn-} allele. Heterozygous Tie^{lczn-} ES cells were retargeted with the original targeting vector to create homozygous mutant ES cell lines with the Tie^{lcz}/Tie^{lczn-} genotype.

The Tie-1 targeting vector was electroporated into R1 129ES cells which were aggregated with CD1 mice. Chimeras were outbred to CD1 and C57BL/6 strains and were also inbred with 129Sv/J mice. Qualitatively, the phenotype of homozygous mutant mice (see below) was the same for all strains, but the onset and progression of the lethal phenotype varied according to the genetic background, with the 129Sv/J-CD1 hybrid being the most severely affected.

KO strain 2 construction[2]

The *Tie1* gene was interrupted at the second amino acid in the signal sequence by insertion of a PGK*neo* cassette. The vector was introduced into CJ7 129 ES cells which were injected into C57BL/6 blastocysts. The phenotype was analyzed on a 129Sv inbred or mixed 129 × B6 background.

Phenotype

Mice homozygous for both the Tie^{lcz} and Tie^{lczn-} alleles cannot survive embryonic gestation and die *in utero* from E13.5. X-gal staining of the vasculature showed that all embryos formed a functioning circulatory system during early gestation. Homozygous mutant embryos, however, manifested small hemorrhages and/or edema throughout the body surface. Histologically the heart, major blood vessels such as the dorsal aorta, the yolk sac vasculature and placenta all appeared normal in mutant embryos, indicating that the observed phenotype was the result of a defective microvasculature. The requirement for *Tie1* in this cell type was further investigated in chimeric mice using the homozygous mutant (Tie^{lcz}/Tie^{lczn-}) ES cell line. Although cells lacking a functioning Tie-1 gene could efficiently contribute to the early embryonic vasculature (E10.5), by E15.5 and also in the adult, Tie^{lcz}/Tie^{lczn-} endothelial cells were absent from the capillary network of angiogenically vascularized tissues, such as brain and kidney, supporting the view that Tie-1 function is important for a late stage of embryonic angiogenesis.

Acknowledgement
Mira Puri
Samuel Lunenfeld Research Institute, Toronto, Ontario, Canada

References

1 Puri, M.C. et al. (1995) EMBO J. 14, 5884–5891.
2 Sato, T.N. et al. (1995) Nature 376, 70–74.
3 Partanen, J. et al. (1996) Development 122, 3013–3021.

Other names
Tissue inhibitor of metalloproteinase

Gene symbol
Timp

Accession number
MGI: 98752

Area of impact
Oncogene

General description

Timp-1 is an inhibitor of the extracellular matrix-degrading matrix metalloproteinases (MMPs). Other growth-promoting activities have been described for Timp-1, and non-matrix substrates have been described for the MMPs.

KO strain construction

A replacement vector was used which placed three stop codons in three reading frames at the *Eco*RV site within exon 3, and a PGK*neo* cassette was placed immediately downstream of the stop codons. The J1 ES cell line (129Svter) was injected into BALB/c blastocysts. Chimeric mice were inbred to 129Sv and also backcrossed to C57BL/6 and BALB/c.

Phenotype

Timp-1-deficient mice showed evidence of enhanced inflammatory responses by three different assays. The best-studied model for inflammation involves infecting the corneas of mice with *Pseudomonas aeruginosa*, then assaying numbers of bacteria by plate count at several times after infection. Bacterial attachment and an initial burst of replication occurred identically in mutant and wild-type mice. At 12–24 hours after infection, when host immune responses were mobilized, bacterial titers in mutant mice dropped to a level 500–1000-fold below that measured in wild-type mice. This elevated immune response was dependent on the complement system. When complement was depleted from mutant mice, 95–99% of the phenotype was suppressed.

Timp-1 has been implicated as a suppressor of tumor metastasis. To study the relative importance of Timp-1 expression by tumor vs. host during tumor invasion, three pairs of coisogenic, tumorigenic cells containing wild-type or mutant Timp-1 alleles were generated. These were used in experimental metastasis assays in wild-type or Timp-1-deficient mice. Timp-1 expression in tumorigenic cells could either increase or decrease tumor invasion of lungs in a tumor cell-specific manner. This suggests that depending on the tumor, Timp-1 can either suppress or potentiate metastasis. Mice deficient for Timp-1 were indistinguishable from wild-type mice in metastasis assays with all tumorigenic cells tested. Although Timp-1 is a secreted protein, and lung produces other Timps, the influence of Timp-1 on lung invasion by the tumorigenic cells tested was cell autonomous, depending only on the Timp-1 genotype of tumor and not that of host.

Acknowledgements
Paul D. Soloway
Department of Molecular and Cellular Biology, Roswell Park Cancer Institute, Buffalo, NY, USA

Reference

[1] Soloway, P.D. et al. (1996) Oncogene 13, 2307–2314.

Other names
Tailless, mtll

Gene symbol
Tlx

Accession number
MGI: 1100526

Area of impact
Development

General description

Tlx belongs to the orphan nuclear receptor subfamily. It is the mouse homolog of *Drosophila tailless* (*tll*) gene. Tlx expression is initially localized to the developing forebrain (in the prosencephalon from day 8 pc) and subsequently spreads into the dorsal midbrain. As the brain develops further, Tlx expression becomes restricted to ventricular zones where undifferentiated neurons locate. It is also detected in the developing eyes and nose. In addition, preliminary data suggest that the Tlx transcript appears during early embryogenesis, long before neurogenesis starts.

KO strain 1 construction[1]

Exons 2–5, including the DNA-binding domain, the hinge region and part of the ligand-binding domain, were deleted and replaced with a β-galactosidase/PGK*neo* cassette. β-Galactosidase was fused in-frame with Tlx in exon 2. Heterozygous mice were maintained on either pure 129Sv or mixed 129Sv × C57BL/6 genetic backgrounds.

KO strain 2 construction[2]

The targeting vector was designed such that the zinc fingers in the DNA-binding domain of the *Tlx* gene (isolated from strain 129) were replaced with a β-galactosidase-neomycin cassette. Targeted ES cells were injected into C57BL/6 blastocysts. Heterozygotes were crossed to C57BL/6 or to 129SvEv mice.

Phenotype

KO strain 1 Tlx$^{-/-}$ mice exhibited neurological abnormalities both in embryos and in adult animals. In addition, some lines of knockout mice displayed early embryonic lethality. KO strain 2 Tlx$^{-/-}$ mice survived but exhibited severe weight loss after birth and had to be hand-fed to achieve eventual weight gain. Histological abnormalities were restricted primarily to the limbic system, including olfactory bulbs, cortex, hippocampus, olfactory cortex and amygdala. There were decreased numbers of calretinin- and somatostatin-positive cells in the cerebral cortex, and abnormal and ectopic termination of mossy fibers onto

CA3 cells of Ammon's horn. The pre- and parasubiculum were severely reduced as were the medial and lateral entorhinal cortices. Behavioral abnormalities were also observed, including severe aggression in 100% of males and 50% of females. KO strain 2 $Tlx^{-/-}$ animals were also hyperexcitable and had defects in their spatial learning abilities.

Comments

Tlx is required for the differentiation of specific subpopulations of cells in the prosencephalon, giving insights into regionalization within the ventricular zone. The behavioral phenotype of the KO strain 2 mice is reminiscent of Kluver–Bucy syndrome. The phenotypes of these two $Tlx^{-/-}$ strains are substantially different, possibly due to the deletion of different exons in the targeting constructs.

Acknowledgements

A. Paula Monaghan and Günther Schütz
German Cancer Research Center, Heidelberg, Germany
Ronald M. Evans and Ming-Yi Chiang
Salk Institute, La Jolla, CA, USA

References

[1] The KO strain 1 $Tlx^{-/-}$ mouse is not published.
[2] Monaghan, A.P. et al. (1997) Nature 390, 515–517.
[3] Monaghan, A.P. et al. (1995) Development 121, 839–853.
[4] Yu, R.T. et al. (1994) Nature 370, 375–379.

Other names
Tissue non-specific alkaline phosphatase, liver/bone/kidney AP, TNSALP, Akp-2

Gene symbol
Akp2

Accession number
MGI: 87983

Area of impact
Metabolism, development

General description

Mammalian alkaline phosphatases (APs) are anchored to the cytoplasmic membrane via a phosphatidylinositol glycan moiety and catalyze dephosphorylation and transphosphorylation of a wide spectrum of substrates at alkaline pH. Four AP genes exist in humans which encode three proteins with restricted tissue expression (tissue specific placental AP, intestinal AP, and germ cell AP), and the non-restricted tissue non-specific AP (TNAP). The mouse genome contains four AP loci: embryonic AP (EAP), intestinal AP (IAP), a putative pseudogene, and TNAP. TNAP is expressed in the inner cell mass of blastocysts and in migrating primordial germ cells. It is also expressed in the developing neural tube between 8.5 and 13.5 days pc. At later stages of embryogenesis, cartilage and ossified bones predominantly express TNAP. In adult mice, TNAP is expressed in a wide variety of cell types such as osteoblasts, neutrophils, renal tubules, capillaries in brain and myoid cells in the testis.

KO strain 1 construction[1]

The targeting construct contained a 1.5 kb 5′ arm of homology covering part of intron 4, exon 5, intron 5 and most of exon 6. The *neo* gene was inserted at the 3′ end of exon 6. The 3′ 6.0 kb arm of homology contained intron 6, exon 7, intron 8, exon 8 and intron 9. The diphtheria toxin A gene was cloned in reverse orientation at the 5′ end of the construct. The construct was targeted into D3 ES cells and the targeted cells were injected into C57BL/6 blastocysts.

KO strain 2 construction[2]

Exons 3, 4, 5 and parts of exons 2 and 6 were replaced by a *lacZ*/neomycin amino-phosphotransferase (G418) resistance fusion gene (bgeobpA) ligated in-frame 20 nucleotides downstream of the initiator ATG codon in exon 2. AB1 ES cells were targeted and injected into C57BL/6 blastocysts.

Phenotype

TNAP$^{+/-}$ heterozygotes appeared healthy and survived for up to 2 years while none of the TNAP$^{-/-}$ animals survived until weaning. The average lifespan of

TNAP$^{-/-}$ mice in the 129J background was 8.8 ± 2.3 days, while in the 129J × C57BL/6J background, their lifespan was 10.6 ± 3.4 days. TNAP$^{-/-}$ pups were born with normal appearance and weight, including healthy milk lines, and were indistinguishable from either wild-type or heterozygous siblings. Most of the TNAP$^{-/-}$ mice became progressively exhausted and, at the time of death, their body weights were only 50% to 30% of those of control littermates. Severe epileptic seizures were frequently observed in TNAP$^{-/-}$ mice 1–2 days before their deaths. The seizures had various forms of presentation such as constant running in the cage, high-pitched vocalizations, biting of tongue and loss of consciousness in a supine position associated with apnea for periods of 30 seconds or more. These attacks continued for periods ranging from 30 min to a few hours and occurred daily. Body temperature was palpably decreased during these attacks and death often followed a prolonged seizure attack.

The adult tissues that normally express TNAP (such as the ribs and kidney) were completely negative in homozygous mice. The cortex of the thymus of TNAP$^{-/-}$ mice was thinner than that of littermate controls and contained many apoptotic cells. The nerve roots emerging from the spinal cord and descending within the dura in TNAP$^{-/-}$ animals were found to be consistently thinner than those from heterozygous control mice, even after correcting for their size difference. The average ratio relating the area of the nerve roots to that of the spinal cord was 9.1 ± 3.9% compared to 19.5 ± 4.7% in corresponding sections of littermate TNAP$^{+/-}$ or TNAP$^{+/+}$ controls at the L2, L3, L5 and L6 positions. The staining of 8-day-old TNAP$^{-/-}$ bones clearly showed poor mineralization in the parietal bones, scapulae, vertebral bones and ribs. The parietal bones were thinner, and the scapulae and vertebrae showed areas of hypomineralization. Evidence of spontaneous fractures was present in the fibulae. Osteoblasts with abnormal morphology were observed on the surface of trabeculae and in the periosteal region of the diaphysis of TNAP$^{-/-}$, but not TNAP$^{+/-}$, mice. At the electron microscope level, some of the mature osteoblasts in the parietal bones of TNAP$^{-/-}$ mice contained abnormal vacuoles. Although highly expressed in primordial germ cells, the TNAP mutation did not affect the development or migration of the germ cells.

Comments

TNAP is apparently not essential for the initial events leading to bone mineral deposition, but does play a role in the maintenance of this process after birth. Other phenotypic manifestations may be a consequence of the lack of TNAP in the developing neural tube between stages E8.5 and E13.5 of embryogenesis. The autonomic nervous system may be compromised in TNAP$^{-/-}$ mice. Vitamin B6 administration was able to suppress the epileptic seizures and prolong the life of these mice, although ultimately they succumbed to the disease.

Acknowledgements
Jose Luis Millan
The Burnham Institute, La Jolla, CA, USA

References

[1] Narisawa, S. et al. (1997) Dev. Dyn. 208, 432–446.
[2] MacGregor, G.R. et al. (1995) Development 121, 1487–1496.
[3] Waymire K.G. et al. (1995) Nature Genet. 11, 45–51.
[4] Hahnel, A.C. et al. (1990) Development 110, 555–564.
[5] Narisawa, S. et al. (1994) Dev. Dyn. 201, 227–235.

Other names
Tumor necrosis factor α

Gene symbol
Tnf

Accession number
MGI: 104798

Area of impact
Immunity and inflammation, cytokines

General description

TNFα is a pro-inflammatory and immunoregulatory cytokine produced by activated macrophages, T cells and other immunocompetent cells. It is essential for the maintenance of physiological immune homeostasis and is implicated in the pathogenesis of inflammatory and autoimmune disorders, such as rheumatoid arthritis, multiple sclerosis, graft-vs.-host disease and septic shock.

KO strain construction

The *Tnf*α locus was cloned from a 129Sv mouse genomic DNA library, and the replacement vector used for the targeted inactivation was constructed by replacing the first exon of the *Tnf*α gene with an MC1*neo*-polyA expression cassette. The ES cell line CCE, derived from the 129Sv mouse, was used for the gene targeting. The chimeras were mated to C57BL/6 females, and the homozygous deletion of the *Tnf*α gene was confirmed by Southern and Northern blotting.

Phenotype

TNFα knockout mice were viable and fertile, and showed no apparent gross phenotypic abnormalities, indicating that TNFα is not required for normal mouse development. The TNFα null mice readily succumbed to *Listeria monocytogenes* infection and showed reduced contact hypersensitivity responses. Furthermore, TNFα knockout mice were resistant to the systemic toxicity of LPS upon D-galactosamine sensitization, yet they remained sensitive to high doses of LPS alone. In addition TNFα-deficient mice lacked primary B lymphocyte follicles and follicular dendritic cell (FDC) networks in their spleen, peripheral and mesenteric lymph nodes and Peyer's patches, demonstrating the essential role of TNFα in the follicular organization of secondary lymphoid organs. In contrast, well-structured T cell-rich areas containing normal numbers of dendritic cells were found in all secondary lymphoid organs from TNFα knockout mice. Immunization of TNFα knockout mice with T cell-dependent antigens failed to induce formation of germinal centers in the spleen. However, despite the absence of B cell follicles, immunoglobulin class-switching could still occur; yet deregulated humoral immune responses against either T cell-dependent or T cell-independent antigens were observed.

Comments

Results obtained from TNFα knockout mice confirm the important role of TNFα in anti-bacterial host defense and LPS-induced sepsis, consistent with its role as a pro-inflammatory cytokine. Furthermore, the results establish the essential requirement of this cytokine for the structural organization of B cell follicles and follicular dendritic cell networks in all secondary lymphoid organs. In addition, the knockout mice demonstrate the importance of TNFα for the physiological maturation of the humoral immune response.

Acknowledgements
George Kollias
Department of Molecular Genetics, Hellenic Pasteur Institute, Athens, Greece

References

1 Pasparakis, M. et al. (1996) J. Exp. Med. 184, 1397–1411.
2 Pasparakis, M. et al. (1996) Cytokine Growth Factor Rev. 7, 223–229.
3 Pasparakis, M. et al. (1997) Proc. Natl Acad. Sci. USA 94, 6319–6322.

Other names
Tumor necrosis factor β, lymphotoxin alpha (LTα), LT-A

Gene symbol
Lta

Accession number
MGI: 104797

Area of impact
Immunity and inflammation

General description

TNFβ is a multifunctional cytokine produced primarily by activated lymphocytes. It is structurally related to TNFα. Both TNFα and TNFβ mediate various inflammatory and immunoregulatory responses, including tumor cytotoxicity, cell proliferation and differentiation, anti-viral responses, and the activation of numerous cellular genes and kinases. In the past, it was not clear whether TNFβ possessed any activities separate from those of TNFα, since the secreted forms of TNFα and TNFβ compete for the same two cell surface receptors. It has recently been shown that TNFβ can exist as a membrane-associated form when complexed with lymphotoxin β (LT-B). The LT-A/LT-B complex is recognized by a specific receptor, suggesting that TNFβ may have unique functions distinct from those of TNFα.

KO strain construction

The *Lta* genomic clone was derived from a C57BL/6 library. A 420 bp *Apa*I/*Kpn*I fragment of the *Lta* gene containing a portion of exon 2, all of exon 3, and a portion of exon 4 was deleted and replaced by the neomycin-resistance gene from the plasmid pMC1*neo*A$^+$. The targeting vector was electroporated into the D3 ES cell line (strain 129) and positive clones were injected into C57BL/6J blastocysts. Chimeric males were bred to C57BL/6J or CF-1BR female mice. The null mutation was confirmed by Northern blotting of RNA from spleen cells of mutant mice. Normal-sized TNFβ transcripts were absent and only a large transcript containing both TNFβ and *neo* sequences was observed. The expression of TNFα was normal. The results of a bioassay detecting both TNFβ and TNFα were consistent with an absence of TNFβ in the supernatants of ConA-stimulated TNFβ$^{-/-}$ spleen cells.

Phenotype

TNFβ$^{-/-}$ mice were viable and fertile and showed apparently normal thymic development. However, they exhibited a striking phenotype dominated by defects in the development of secondary lymphoid organs. TNFβ$^{-/-}$ mice lacked lymph nodes and Peyer's patches, and possessed spleens in which the usual architecture was disrupted. Some of the mutant mice developed abnormal lymph node-like structures, mainly in the mesenteric fat.

Abnormal clusters of lymphocytes accumulated in the periportal and perivascular regions of the liver and lungs. However, lymphocytes from TNF$\beta^{-/-}$ mice appeared phenotypically normal, expressing the expected ratios of T and B cell surface markers as well as the lymphocyte homing marker L-selectin. Bone marrow cells from TNF$\beta^{-/-}$ mice were able to reconstitute the lymphoid organs of SCID mice. When examined for humoral immune responsiveness, TNF$\beta^{-/-}$ mice were impaired in their ability to respond to different antigens, including UV-inactivated HSV-1 (KOS strain) and KLH. Cell-mediated responses appeared to be normal in TNF$\beta^{-/-}$ mice.

Comments

TNFβ appears to be critical for lymphoid organogenesis, a function not shared by TNFα. Lymphocyte expression of TNFβ is not required for the homing of these cells to existing lymph node structures.

TNF$\beta^{-/-}$ mice will provide a model for understanding the molecular interactions needed for complete lymphoid organ development, as well as the effect(s) of this mutation on the immune response in general.

Acknowledgements
Theresa Banks
Chiron Technologies, Center for Gene Therapy, San Diego, CA, USA
Michael Mucenski
Oak Ridge National Laboratory, Oak Ridge, TN, USA

References

[1] DeTogni, P. et al. (1994) Science 264, 703–707.
[2] Banks, T.A. et al. (1995) J. Immun. 155, 1685–1693.
[3] Mariathasan, S. et al. (1995) J. Inflammation 45, 72–78.
[4] Matsumoto, M. et al. (1996) Science 271, 1289–1291.
[5] Matsumoto, M. et al. (1996) Nature 328, 462–466.
[6] Rennert, P.D. et al. (1996) J. Exp. Med. 184, 1999–2006.

Other names
TNFR1, CD120a, TNF receptor α chain

Gene symbol
Tnfr1

Accession number
MGI: 98781

Area of impact
Immunity and inflammation, signal transduction

General description

The tumor necrosis factors TNFα and TNFβ affect the growth and differentiation of many cell types, and mediate inflammation and cellular immune responses. TNF is also a key mediator of septic shock induced by bacterial superantigens and LPS. TNF exerts its effects through two independent cell surface receptors, one of 55 kDa (TNFRp55 or TNFR1) and the other of 75 kDa (TNFRp75 or TNFR2). While the extracellular domains of these receptors show significant homology, the intracellular regions are distinctly different. Although TNF binds to both receptors, most of its known cellular effects have been ascribed to its interaction with TNFRp55. Engagement of TNFRp55 has been associated with activation of various second messenger systems and the induction of NF-κB activity.

KO strain 1 construction[1]

The targeting vector was designed such that a neomycin-resistance cassette was inserted into the *Tnfr1* gene at bp 535 of the coding sequence. The vector contained a 3.2 kb genomic fragment (derived from a BALB/c EMBL3 library) which included four exons of the *Tnfr1* gene. The targeting vector was electroporated into D3 ES cells (129Sv) and positive clones were injected into C57BL/6 blastocysts. Chimeric mice were bred to (C57BL/6 × DBA/2) F1 mice and heterozygous offspring were intercrossed to obtain TNFRp55$^{-/-}$ mice.

The null mutation was confirmed by examination of the binding of iodinated mouse TNFα (binds to both mouse TNF receptors) and human TNFα (binds only to mouse TNFRp55). Experiments with ConA-stimulated mutant spleen cells showed that there was no surface expression of TNFRp55, while surface expression of TNFRp75 was unaffected. Nuclear extracts of mutant cells were devoid of NF-κB-binding activity, confirming a lack of functional TNFRp55 and impaired TNF signaling.

KO strain 2 construction[2]

The targeting vector was designed such that exons 2, 3 and part of exon 4 of the mouse *Tnfr1* gene (derived from C57BL/6) were replaced by a neomycin-resistance cassette. This deletion removed coding sequences for the cysteine-rich domains I and II of the receptor which are essential for ligand binding. The short arm of homology was 868 bp and the long arm was about 5 kb. The HSV-*tk* gene was appended to the 3′ end of the genomic sequences in the opposite transcriptional orientation. The targeting vector was electroporated into E14 ES cells and positive clones were injected into C57BL/6 blastocysts. Chimeric offspring were crossed with C57BL/6 mice and the resulting heterozygotes interbred to obtain homozygous TNFRp55$^{-/-}$ mice.

The null mutation was confirmed by Northern blot and by the lack of binding iodinated human TNFα to mutant thymocytes. Binding to TNFRp75 was normal.

Phenotype

Male and female mutant mice were born at the normal frequency and were healthy and fertile up to 10 months of age. Primary and secondary lymphatic organs were phenotypically indistinguishable from the wild type: thymocyte and lymphocyte numbers and development were normal. Clonal deletion of self-reactive T cells was normal. TNFRp55$^{-/-}$ mice were resistant to doses of LPS or SEB lethal to wild-type mice. Livers in mutant animals showed only non-specific hepatotoxicity, and not the complete necrosis observed in wild-type livers. After challenge with sublethal doses of LPS, the peak concentration of systemic TNFα was slightly higher than normal, but the secretion of IL-6 was decreased 3-fold. TNFRp55$^{-/-}$ mice infected with lymphocytic choriomeningitis virus (LCMV) showed footpad swelling comparable to that in wild-type mice, although the reaction was delayed by one day. These results indicated that CTL induction and effector function were essentially normal. However, TNFRp55$^{-/-}$ mice showed greatly increased susceptibility to *Listeria monocytogenes* infection and readily succumbed to the infection after only 2–3 days. TNFRp55$^{-/-}$ mice were also unable to clear *Candida albicans* infections and died quickly[3]. TNFRp55$^{-/-}$ mice treated with dinitrofluorobenzene to induce contact hypersensitivity showed hyperresponsiveness which was normally suppressed by UVB irradiation[4].

Comments

TNFRp55 is not required for embryonic development and does not appear to play a role in the selection of the T cell repertoire. The effects of TNFRp55 deficiency in TNFRp55$^{-/-}$ mice do not become apparent until the animal is challenged with bacterial pathogenic material, after which neither TNFRp75 nor other cytokine receptors can substitute for TNFRp55 function. Engagement of TNFRp55 is a key step in the development of lethal toxic shock. TNFRp55 is indispensable for the elimination of *L. monocytogenes*, showing that TNF plays an essential role in innate immunity. TNFRp55 is also required for resistance to *C. albicans.*

Acknowledgements
Tak W. Mak
Ontario Cancer Institute and Amgen Institute, Toronto, ON, Canada

References
[1] Pfeffer, K. et al. (1993) Cell 73, 457–467.
[2] Rothe, J. et al. (1993) Nature 364, 798–802.
[3] Steinshamn, S. et al. (1996) J. Immunol. 157, 2155–2159.
[4] Kondo, S. et al. (1995) J. Immunol. 155, 3801–3805.

Other names
TNFR2, CD120b, TNF receptor β chain

Gene symbol
Tnfr2

Accession number
MGI: 98782

Area of impact
Immunity and inflammation, signal transduction

General description

The tumor necrosis factors TNFα and TNFβ affect the growth and differentiation of many cell types, and mediate inflammation and cellular immune responses. TNF is also a key mediator of septic shock induced by bacterial superantigens and LPS. TNF exerts its effects through two independent cell surface receptors, one of 55 kDa (TNFRp55 or TNFR1) and the other of 75 kDa (TNFRp75 or TNFR2). While the extracellular domains of these receptors show significant homology, the intracellular regions are distinctly different. Although TNF binds to both receptors, most of its known cellular effects have been ascribed to its interaction with TNFRp55. TNF has been shown to induce thymocyte proliferation *in vitro* via engagement of TNFRp75.

KO strain construction

The targeting vector was designed such that a neomycin-resistance cassette was inserted into exon 2 of the mouse *Tnfr2* gene derived from strain 129, disrupting the signal peptide region of TNFRp75. The genomic fragment used was 9 kb in length, and the selection cassette was positioned so that there was 6 kb of 5′ homology and 3 kb of 3′ homology. The targeting vector was electroporated into D3 C12 ES cells and positive clones were injected into C57BL/6J blastocysts. Chimeric males were mated with C57BL/6J females and heterozygous offspring were intercrossed to obtain homozygous TNFRp75$^{-/-}$ mutant mice.

The null mutation was confirmed by challenging mice with LPS, which results in the shedding of the extracellular domain of TNFRp75 into the serum in normal mice: no soluble TNFRp75 was detected in serum of mutant mice.

Phenotype

Mutant mice were born at the expected frequency, were viable and appeared overtly normal. A transient increase in serum TNF was observed in TNFRp75$^{-/-}$ mice after LPS challenge, but mutant mice did not show increased sensitivity to high doses of LPS. Unlike TNFRp55-deficient mice, TNFRp75$^{-/-}$ mice showed wild-type sensitivity to LPS plus D-galactosamine treatment. However, TNFRp75$^{-/-}$ mice were much less sensitive to high doses of TNF than wild-type mice. Injection of murine TNF under the skin of

TNFRp75$^{-/-}$ mice resulted in a marked decrease in tissue necrosis compared to wild type. No significant differences in lymphocyte numbers or subpopulations were observed, and TNFRp75$^{-/-}$ lymphocytes appeared to have normal functionality. There was no change in V_β selection in TCR. TNFRp75$^{-/-}$ mice appeared to be less severely affected than TNFRp55$^{-/-}$ mice by *Listeria monocytogenes* infection, in that they showed slightly increased mortality over the wild type. TNFRp75$^{-/-}$ mice were reduced in their ability to clear *Candida albicans* but did not show increased lethality, unlike TNFRp55$^{-/-}$ mice[2].

Comments

TNFRp75 is not required for normal T cell development or function. TNFRp75 enhances sensitivity to TNF toxicity and LPS challenge, which are mediated primarily through TNFRp55. TNFRp75 may help recruit TNF for interaction with TNFRp55, or may assume some other ancillary function that promotes transduction in TNF signaling.

References

1 Erickson, S.L. et al. (1994) Nature 372, 560–563.

2 Steinshamn, S. et al. (1996) J. Immunol. 157, 2155–2159.

Other names
Topoisomerase I, Topo-I

Gene symbol
Top1

Accession number
MGI: 98788

Area of impact
Development

General description

Eukaryotic Top-1 has a key role in many fundamental biological processes. The enzyme alters DNA topology by catalyzing transient single-stranded breakage of the phosphodiester backbone. This event allows passage of the intact strand through the break site, and subsequent religation of the nicked strand reforms the intact helix. This simple type of topological manipulation is known to be a crucial process for basic DNA and RNA metabolism in mammalian cells.

KO strain construction

For positive selection, a neomycin-resistance gene (*neo*) was inserted between a 3.8 kb *Bgl* II/*Pst* I fragment and a 1.4 kb *Bgl* II/*Hind* III fragment. This insertion eliminated exon 15, which corresponds to a region known to be essential for enzymatic activity. Negative selection markers encoding the HSV-*tk* were also used. The Top-1 targeting vector was electroporated into the ES cell line BK4 (a subclone of E14TG2a, derived from mouse strain 129Ola). A selection protocol using the Top-1 inhibitor camptothecin facilitated isolation of ES cell clones containing an inactivated allele; isolation of correctly targeted clones was enhanced 75-fold over normal selection procedures. Targeted ES cells were injected into C57BL/6 blastocysts.

Phenotype

The disrupted *Top1* allele is embryonic lethal when homozygous, and development of such embryos fails between the 4- and 16-cell stage. Both sperm and oocytes containing the inactive allele maintained viability through the fertilization point, and thus gene expression of *Top1* is not required for gamete viability. These studies demonstrate that Top-1 is essential for cell growth and division *in vivo*. The *Top1* gene was also shown to be linked to the agouti locus.

Comments

Cell lines derived from the heterozygous Top-$1^{+/-}$ animals will be useful for mutagenesis studies to investigate the mechanisms of Top-1 action. Since such cell lines contain only a single copy of the wild-type *Top1* allele, the isolation of mutants is simplified and hence, a variety of complementation studies can be attempted. Temperature-sensitive mutant cells would provide a valuable tool for discerning how the enzyme acts in replication, transcription and recombination.

Acknowledgements

Scott G. Morham
University of North Carolina at Chapel Hill, Chapel Hill, NC, USA

References

[1] Morham, S. G. et al. (1996) Mol. Cell Biol. 16, 6804–6809.
[2] Bjornsti, M. et al. (1989) Cancer Res. 49, 6318–6323.
[3] Giovanella, B. C. et al. (1989) Science 246, 1046–1048.

Other names
Tissue-type plasminogen activator

Gene symbol
Plat

Accession number
MGI: 97610

Area of impact
Metabolism, cardiovascular

General description

The plasminogen activation system is an intricate system of serine proteases, protease inhibitors and protease receptors, which convert inactive plasminogen (Plg) to the active protease plasmin (plasminogen/plasmin system of fibrinolytic system). The plasminogen/plasmin system has been implicated in blood clot dissolution, ovulation, embryogenesis, cell migration and brain function. Plasminogen transition to active plasmin is mediated by two physiological plasminogen activators, tissue-specific plasminogen activator (t-PA) and urokinase-type plasminogen activator (u-PA). t-PA is primarily responsible for removing fibrin from the vascular system and has specific and localized affinity for fibrin in blood clots. By contrast, u-PA has no affinity for fibrin. Fibrinolysis can contribute to various diseases including thrombosis, atherosclerosis, metastasis, haemangioma formation and glomerulonephritis.

KO strain construction

A *neo* cassette was inserted between bp 810 and bp 1170 of the *Plat* gene which encodes parts of the kringle-2 and protease domains (including the catalytically crucial amino acid 326 histidine). *tk* was placed at the 5′ end of the construct. Targeted D3 ES cells were injected into C57BL/6 blastocysts. t-PA$^{-/-}$u-PA$^{-/-}$ double mutants were generated by the breeding of single mutant mice.

Phenotype

t-PA$^{-/-}$ mice developed normally, had a normal lifespan, were fertile, and were macroscopically normal up to 14 months of age. t-PA$^{-/-}$ mice had reduced thrombolysis and increased occurrence of endotoxin-mediated thrombosis. t-PA$^{-/-}$u-PA$^{-/-}$ double mutant mice survived embryogenesis but displayed a high rate of spontaneous fibrin deposition in various organs (skin, lung, gonads, intestine), reduced fertility, reduced body weight, shortened lifespans (50% died between 30 and 40 weeks of age), and postnatal growth retardation. t-PA$^{-/-}$u-PA$^{-/-}$ mutant mice also developed non-healing ulcerative skin lesions, intestinal adhesions, and ischemic tissue necrosis in the intestine and uterus due to thrombosis.

Reference

[1] Carmeliet, P. et al. (1994) Nature 368, 419–424.

Other names
Thrombopoietin, c-*mpl* ligand

Gene symbol
Thpo

Accession number
MGI: 101875

Area of impact
Hematopoiesis

General description

TPO, the ligand for the proto-oncogene c-*mpl*, is the major regulator of platelet production. TPO has an N-terminal domain homologous to erythropoietin (EPO) and a C-terminal glycosylated domain unrelated to any known protein. Northern blot analysis indicates that TPO is produced primarily by the liver and the kidney. Experiments *in vitro* and *in vivo* with recombinant TPO indicate that it stimulates both megakaryocyte colony formation and megakaryocyte maturation.

KO strain construction

The targeting vector consisted of a 10 kb genomic *Xho*I/*Sac*I fragment derived from a mouse 129 library encompassing the mouse *Thpo* gene. Twenty-three amino acids of the third coding exon (in the EPO homology domain responsible for receptor binding and activation) were replaced by a *neo*-resistance cassette. The targeting vector was electroporated into ES.D3 C-12 cells, a subclone of D3 ES cells. Positive clones were microinjected into C57BL/6J blastocysts. Chimeric males were mated with C57BL/6J females and heterozygotes were interbred to obtain TPO$^{-/-}$ homozygotes. The null mutation was confirmed by Northern blotting of liver and kidney RNA.

Phenotype

TPO-deficient mice were born healthy at the expected Mendelian frequency and adult mice appeared overtly normal. TPO$^{-/-}$ mice had a greater than 85% decrease in platelet counts and megakaryocyte numbers but had normal levels of all other hematopoietic cell types. Platelet and megakaryocyte numbers in TPO$^{+/-}$ mice were intermediate between those in TPO$^{-/-}$ mice and wild-type mice. Megakaryocytes present in the bone marrow of TPO$^{-/-}$ mice were of lower ploidy than control megakaryocytes. Bone marrow from TPO$^{-/-}$ mice showed decreased numbers of megakaryocyte progenitor cells as well as reduced numbers of CFU-GM, BFU-E and CFU-MIX, indicating that TPO acts on an early common progenitor cell and that its effect is not limited to the megakaryocytic lineage. Administration of recombinant murine TPO to TPO-deficient mice led to a full recovery of platelet and progenitor cells in approximately 12 days. The delay in reaching normal levels was probably due to the low initial starting population of megakaryocyte progenitors. The platelet levels in $TPO^{-/-}$ mice eventually surpassed untreated wild-type levels. TPO treatment significantly increased the absolute number of myeloid-, erythroid- and mixed-progenitors in bone marrow and spleen.

Cytokines with megakaryocytopoietic activity such as IL-3, IL-6, IL-11 and EPO were also capable of stimulating platelet or megakaryocyte production in c-*mpl*- and TPO-deficient mice, suggesting that these cytokines do not affect megakaryocytopoiesis through the induction of endogenous TPO. Although they were not capable of restoring a normal platelet count in c-*mpl*- and TPO-deficient mice, the activity of these cytokines may be responsible for the low level of megakaryocytes and platelets still present in TPO-deficient mice.

Comments

While TPO can regulate both proliferation and differentiation of megakaryocyte progenitors, TPO is not essential for mature platelet production *in vivo*. The gene dosage effect observed in heterozygous mice suggests that the TPO gene is constitutively expressed and is not transcriptionally regulated as a function of the platelet count. The TPO level can be directly controlled by the platelets themselves, which are able to bind and remove TPO from the circulation.

Acknowledgements

Frederic de Sauvage

Genentech Inc., San Francisco, CA, USA

References

[1] de Sauvage, F. et al. (1996) J. Exp. Med. 183, 651–656.
[2] Carver-Moore, K. et al. (1996) Blood 88, 803–808.
[3] Fielder, P. et al. (1996) Blood 87, 2154–2161.
[4] de Sauvage, F. et al. (1994) Nature 369, 533–538.
[5] Kaushansky, K. (1995) Blood 86, 419–431.

Other names
Tumor necrosis factor-associated factor 2

Gene symbol
Traf2

Accession number
MGI: 101835

Area of impact
Immunity and inflammation, signal transduction

General description

TRAF2 is an intracellular signal-transducing protein recruited to the TNFRp55 and TNFRp75 receptors following TNF stimulation. TRAF2 also interacts with CD30 and CD40, two additional members of the TNF receptor superfamily involved in immune regulation. TRAF2 is expressed ubiquitously.

KO strain construction

The targeting construct replaced the entire ring finger coding region and the intervening intron of the *Traf2* gene (derived from 129Sv/J) with the PGK*neo* gene cassette placed in the reverse orientation. The targeting vector was electroporated into E14 ES cells (129Ola). Male chimeras were backcrossed to C57BL/6J females and offspring heterozygous for the mutation were intercrossed to generate TRAF2$^{-/-}$ mice.

Phenotype

TRAF2$^{-/-}$ mice appeared normal at birth but became progressively runted until death at about 10–14 days of age. Atrophy of thymus and spleen, depletion of bone marrow B cell precursors, and peripheral lymphopenia were observed. Thymocytes and other hematopoietic progenitors were highly sensitive to TNF-induced cell death and serum TNF levels were significantly elevated in TRAF2-deficient animals. Examination of TRAF2$^{-/-}$ cells revealed a severe defect in TNF-mediated JNK/SAPK activation, but only a mild alteration of NF-κB activation.

Comments

These results suggest that TRAF2-independent pathways of NF-κB activation exist, and that TRAF2 is required for an NF-κB-independent signal that protects against TNF-induced apoptosis.

Acknowledgements
Wen Chen Yeh and Tak W. Mak
Ontario Cancer Institute and Amgen Institute, Toronto, ON, Canada

Reference
[1] Yeh, W.-C. et al. (1997) Immunity 7, 715–725.

Other names
Tax responsive element-binding protein 5, Human X box-binding protein (hXBP-1)

Gene symbol
Xbp1

Accession number
MGI: 98970

Area of impact
Cardiovascular, immunity

General description

TREB5 is a b-ZIP protein which binds to a Cre-like element in both HTLV-I and MHC class II genes. The antisense RNA of either the TREB5 or c-*fos* genes downregulates expression of cell surface class II genes. TREB5 and c-Fos proteins are also known to form a stable heterodimer on class II genes. The TREB5 gene is expressed at high levels in bone, cartilage and exocrine glands during the mouse embryo stage.

KO strain construction

A *neo*-resistance gene driven by the PGK-1 promoter was inserted into the exon 2. An 8.0 kb and a 3.0 kb genomic fragment were used as the 5′ and 3′ homologous regions, respectively. The DT-A gene driven by the MC-1 promoter was used for negative selection. CCE ES cells, derived from the mouse strain 129SvEv, were used for generating chimeric mice. All chimeras were mated to A/J mice.

Phenotype

Heterozygous mutant mice did not exhibit any obvious abnormalities; however, homozygous mutant embryos died between E10.5 and E14.5. The major defect of TREB5 homozygous mutant embryos was in the heart. Cellular necrosis was found predominantly in the atrium and the truncus arteriosus with its following ventricle. The center of the myocardium consisted of eosinophilic debris and cells exhibiting karyopycnosis and karyorrhexis which were surrounded by swelling cells. In contrast, necrotic alterations were not found in the non-myocardial tissue of either the endocardial cushion or conotruncul ridge. Homozygous mutant embryos displayed reduction of ventricular wall thickness. Trabeculation in the ventricular chamber was present but decreased. Dead homozygous mutant embryos exhibited coagulation necrosis in all tissues except the heart. These alterations in heart and other tissues suggest that the first defect in homozygous mutant embryos is of the myocardium followed by other tissues undergoing necrotic alterations: the lethality is caused by anoxia as a result of cardiac dysfunction. Histological analysis of the homozygous mutant embryos at E10.5 and E11.5 also revealed abnormalities in the neural tube. The neural tubes exhibited a waviness of the neuroepithelium. These morphological abnormalities in the neural tubes were not observed in homozygous mutant embryos at E12.5.

Comments

Disruption of the TREB5 gene causes a dysfunction in heart development culminating in the embryo's death. This observation suggests that TREB5 plays an essential role in maintenance and/or growth of cardiac myocytes during cardiogenesis.

Acknowledgements
Shigeru Noguchi
Harvard Medical School, Boston, MA, USA

References

1 Yoshimura, T. et al. (1990) EMBO J. 9, 2537–2542.
2 Liou, H.C. et al. (1990) Science 247, 1581–1584.
3 Ono, S.J. et al. (1991) Proc. Natl Acad. Sci. USA 88, 4304–4308.
4 Ono, S.J. et al. (1991) Proc. Natl Acad. Sci. USA 88, 4309–4312.
5 Clauss, I.M. et al. (1993) Dev. Dyn. 197, 146–156.

Other names

Neurotropic tyrosine kinase, receptor type 1, tyrosine kinase receptor A, gp140^{trkA}, NGF receptor

Gene symbol

Ntrk1

Accession number

MGI: 97383

Area of impact

Neurology

General description

Developing neurons send their processes to target tissues and compete for survival factors called neurotropins. These neurotropins bind to tyrosine kinase receptors on neurons. There are three receptors called TrkA, B and C. The TrkA locus encodes two isoforms of gp140^{trkA}, which bind nerve growth factor (NGF), but also NT-3 and NT-4.

KO strain construction

The exons of the *Ntrk1* gene encoding part of the catalytic kinase domain (amino acids 710–760) were deleted by *neo*^{r} insertion. The vector was electroporated into D3 ES cells (129).

Phenotype

The knockout represented a deletion of the tyrosine kinase domain of the TrkA receptor. All animals homozygous for the deletion were anatomically normal at birth but thereafter showed a reduction in growth rate. They exhibited sensory deficits with respect to pain sensitivity, temperature and noxious olfactory stimuli. The majority died by the age of weaning and virtually all were dead by post-natal day 50. Males were infertile.

Gross deficits were observed in sensory and sympathetic neurons. Trigeminal and dorsal root ganglia from P0 animals showed a 70% decrease in neuron number compared to wild-type littermates, which appeared to represent predominantly large neurons. The superior cervical ganglia of P10 animals also showed large reductions in neuronal cell number.

In the epidermis of the whisker pad there was a permanent absence of all innervation at this site, which normally consists of fine-caliber unmyelinated axons which are PGP9.5 positive, and a smaller component which are CGRP positive (606.9 ns95). In the CNS there did not appear to be a significant loss of basal forebrain cholinergic neurons. There was some reduction in cholinesterase projections to the hippocampus and cerebral cortex in these mice[2].

Comments

TrkA is important in the development of sympathetic neurons and sensory neurons responsive to temperature and pain.

References

[1] Smeyne, R.J. et al. (1994) Nature 368, 246–249.
[2] Klein, R. (1994) FASEB J. 8, 738–744.

Other names
Neurotropic tyrosine kinase, receptor type 2, gp145^{trkB}, tyrosine kinase receptor B, BDNF receptor

Gene symbol
Ntrk2

Accession number
MGI: 97384

Area of impact
Neurology

General description

TrkB is a neurotropin receptor (gp145^{trkB}) that binds brain-derived neurotropic factor (BDNF) and neurotropin-4 (NT-4). TrkB is widely expressed in the CNS and PNS. The TrkB gene is alternatively spliced to yield a smaller isoform, gp95^{trkB}, that lacks the intracellular catalytic domain.

KO strain construction

neo^r was inserted into exon 2 and *tk* was spliced to the 3′ end. The vector was electroporated into D3 ES cells (129) and selected in G418 gancyclovir. Homologous recombinants were injected into C57BL/6 blastocysts. Chimeras were bred to C57BL/6 and heterozygous mice were intercrossed to produce +/+, +/− and −/− progeny (129 × B6) for analysis.

Phenotype

The knockout represented a deletion of the tyrosine kinase domain. Most animals died by post-natal day 3. This may have been due to an inability to feed. Decreased numbers of motoneurons were observed in the facial and trigeminal nuclei (40% of wild type), spinal cord (L1–L5) and nodose ganglion. Mice were somewhat insensitive to touch. Thirty per cent fewer neurons were seen in the DRG (mid-sized neurons involved in mechanoreception) and 60% fewer in the trigeminal ganglion. However, the hippocampus and cortex, regions in which TrkB is highly expressed, were apparently normal. There was also no reduction in sympathetic innervation.

Comments

The TrkB receptor for BDNF is important for the development of motoneurons.

Reference
[1] Klein, R. et al. (1993) Cell 75, 113–120.

Other names
Neurotropic tyrosine kinase, receptor type 3, gp145^{trkC}, tyrosine kinase receptor C, NT-3 receptor

Gene symbol
Ntrk3

Accession number
MGI: 97385

Area of impact
Neurology

General description

TrkC is a tyrosine receptor kinase for neurotropin-3 (NT-3). This NT-3 ligand is expressed in muscle and promotes cell survival in culture. gp145^{trkc} is widely expressed in the CNS and PNS.

KO strain construction

neo^r was inserted into exon 2 (k2) of the kinase domain. *tk* was inserted on the 5′ end. The vector was electroporated into D3 ES cells (129). Homologous recombinants were injected into C57BL/6 blastocysts. Chimeras were bred to C57BL/6. Heterozygotes were intercrossed to produce offspring for analysis.

Phenotype

The knockout represented a deletion of the tyrosine kinase domain of the receptor. Most animals (80%) died by post-natal day 20. However, the remaining 20% survived. TrkC knockout animals showed unusual and disoriented movements. TrkC is widely expressed, yet the TrkC$^{-/-}$ brain was grossly normal. There was a 10–20% reduction in the number of trigeminal neurons. TrkC/TrkB double knockouts also showed a grossly normal hippocampus and brain. Large proprioceptive neurons did not appear to be present in the dorsal root ganglia of these animals.

Comments

TrkC plays a key role in the survival of sensory neurons sensitive to proprioception and movement.

Reference

[1] Klein, R. et al. (1994) Nature 368, 249–251.

Other names
Thymocyte winged helix, Mf3, HGH-E5.1, fkh-5

Gene symbol
Mf3

Accession number
MGI: 103153

Area of impact
Development

General description

TWH is a member of the winged helix family of transcription factors. TWH is initially widely expressed in the neuroepithelium. By midgestation, TWH expression is restricted to specific groups of cells in the posterior diencephalon, midbrain, hindbrain and spinal cord. In the spinal cord, TWH is expressed in a subset of motor neurons, ventral interneurons, ependymal and glial cells.

KO strain construction

Most of the TWH-coding sequences of the gene were replaced with a *lacZ* cassette containing a nuclear localization signal. This cassette was fused to the 5′ untranslated region. The targeting construct was transfected into 129 ES cells (W9.5). Target cells were injected into C57BL/6 blastocysts, and the phenotype analyzed on a 129 × B6 hybrid background.

Phenotype

Mice homozygous for the TWH mutation were born at the expected frequency and were of normal weight and size. Most mutant pups died within 48 hours of birth, with additional increased mortality over the first 2 weeks. TWH mutant pups had a markedly reduced rate of growth; by P7, TWH mutant animals were half the weight of their normal littermates. This reduction in growth rate persisted until the pups were weaned to solid food (3–4 weeks).

TWH$^{-/-}$ mice appear to have a neuromuscular deficit. The mutants have decreased gross motor activity and reduced motor strength compared to heterozygote and wild-type mice. Some mutant mice also display a broad-based gait and an abnormal clutching reflex. No sensory deficits were detected in the mutant mice.

The pattern of spinal cord neurons is altered in the TWH mutant. The ventral interneurons and glial cells that normally express TWH are found to be more widely distributed in the mutant, and the sizes of these populations are also altered. More of these interneurons and glial cells are generated in the mutant. There are also alterations in the sizes of the motor neuron subsets which appear to be due to a defect in the generation rather than survival of these cells.

Comments

The phenotype of the TWH mutant suggests that TWH regulates the early differentiation of neural progenitors.

Acknowledgements
Suzanne C. Li
Columbia University, New York, USA

Reference

[1] Dou, C. et al. (1997) Neuron 18, 539–551.

Twist

Gene symbol
Twist

Accession number
MGI: 98872

Area of impact
Development

General description

Twist is a basic helix-loop-helix protein homologous to the *Drosophila twist* gene, which is required for normal mesoderm development in flies. Mouse *Twist* is first expressed in anterior mesoderm at E7.5, and then is expressed in somites and lateral plate mesoderm. Later expression is strong in mesenchyme cells of the head and branchial arches.

KO strain construction

The targeting vector deletes exon 1 that encodes the entire coding sequence of the twist protein. This exon was replaced by a PGK*neo* cassette. The vector was introduced in AB-1 129Sv cells, which were injected into C57BL/6 blastocysts. The phenotype was analyzed on both a C57BL/6 × 129 hybrid and a 129 inbred background.

Phenotype

All homozygous embryos were dead by E11.5. Abnormalities in neural fold closure were apparent as early as E8.5. By E9.5, exencephaly was apparent as were branchial arch defects. These anomalies appear to result from defects in head mesenchyme cells. Chimeric analysis showed that neural tube defects were associated with the genotype of the underlying mesenchyme and not the neurepithelium.

Comments

Mutations in the human *TWIST* gene have been associated with Saethre–Chotzen syndrome, which involves craniostenosis and craniofacial and limb anomalies.

Reference
[1] Chen, Z.F. et al. (1995) Genes Dev. 9, 686–699.

Other names
Ubiquitin-conjugating enzyme, E2

Gene symbol
Ubc

Accession number
MGI: 98889

Area of impact
Development

General description

The ubiquitin/proteasome pathway is a major mechanism for the selective breakdown of proteins in eukaryotic cells. Ubiquitin-dependent degradation of proteins depends on their conjugation to the "tagging" polypeptide ubiquitin. Attachment of ubiquitin to selected proteins is carried out by the sequential action of three classes of enzymes: Uba (E1), ubiquitin-activating enzymes; Ubc (E2), ubiquitin-conjugating enzymes; and Ubr (E3), ubiquitin-protein ligases. The Ubc enzymes are encoded by a multigene family.

KO strain construction

The Ubc-deficient mouse was created by insertional mutagenesis of pre-implantation embryos with the recombinant retrovirus mp10, which carries a neomycin-resistance gene. J1 ES cells were infected with mp10 and positive clones containing a single copy of the proviral genome were injected into BALB/c or C57BL/6 blastocysts. A male chimera was bred to a female 129Sv mouse to generate the transgenic Ubc-deficient mouse. The provirus was located in the first intron of the *UbcM4* gene, which is the mouse homolog of the previously described human *UbcH7* gene that is involved in the *in vitro* ubiquitination of several proteins, including the tumor suppressor protein p53.

Phenotype

When both *UbcM4* alleles were mutated, the level of steady-state mRNA was reduced by about 70%. About a third of homozygous mutant embryos died around day 11.5 of gestation. Embryos that survived this stage were growth-retarded and died perinatally. The lethal phenotype was most likely caused by impairment of placental development since this was the only organ consistently showing pathological defects. The placental labyrinth was drastically reduced in size and vascularization was disturbed.

Comments

The Ubc-deficient mouse represents the first example in mammals of a mutation in a gene involved in ubiquitin conjugation. Its recessive-lethal phenotype demonstrates that the ubiquitin system plays an essential role during mouse development.

Acknowledgements
Klaus Harbers
HPI Institute for Experimental Virology & Immunology, University of Hamburg, Hamburg, Germany

Reference

1 Harbers, K. et al. (1996) Proc. Natl Acad. Sci. USA 93, 12412–12417.

UbCKmit

Other names
Ubiquitous mitochondrial creatine kinase

Gene symbol
Ckmt1

Accession number
MGI: 99441

Area of impact
Metabolism

General description

Creatine kinase (CK) isoenzymes catalyze the reversible conversion of phosphocreatine (PCr) plus ADP into creatine plus ATP. Cytosolic muscle CK exists as dimers composed of the M (muscle) and B (brain) subunits from which three isoenzymes, MM, BM and BB, are derived. The cytosolic M and B subunits are encoded by the M-CK and B-CK genes. In addition, two mitochondrial CK isoforms (Mi-CK), called ubiquitous and sarcomeric Mi-CK (UbCKmit and ScCKmit) exist which can form octameric and dimeric structures. Mi-CKs are localized at the inner mitochondrial membrane and linked to oxidative phosphorylation.

CK replenishes ATP levels during periods of energy consumption in the heart and skeletal muscle. In addition to its role as energy buffer during muscle contraction, CK functions in energy transport. PCr can be produced by two metabolic pathways catalyzed by the muscle CK homodimers and mitochondrial CKs. PCr diffuses to the myofibrillar M bands and serves to replenish ATP through action of the MM-CK. UbCKmit and the cytosolic B-CK isoenzymes are expressed in a variety of tissues with high and alternating energy requirements, including the brain, retina, smooth muscle, uterus, placenta and spermatozoa. ScCKmit is coexpressed with M-CK in cardiac and skeletal muscle cells.

KO strain construction

Exons 7 and 8 containing an active cysteine residue were replaced by a hygromycinB-polyA resistance cassette. E14 ES cells were injected into C57BL/6 blastocysts.

Phenotype

UbCKmit$^{-/-}$ mice were viable and fertile and had no apparent abnormalities. Mobility of spermatozoa and mitochondrial pyruvate oxidation were also normal.

Reference

1 Steeghs, K. et al. (1995) Biochim. Biophys. Acta 1230, 130–138.

Other names
Urokinase-type plasminogen activator

Gene symbol
Plau

Accession number
MGI: 97611

Area of impact
Metabolism, cardiovascular

General description

The plasminogen activation system is an intricate system of serine proteases, protease inhibitors and protease receptors which convert inactive plasminogen (Plg) to the active protease plasmin (plasminogen/plasmin system of fibrinolytic system). Plasminogen transition to active plasmin is mediated by two physiological plasminogen activators, tissue-specific plasminogen activator (t-PA) and urokinase-type plasminogen activator (u-PA). The plasminogen/plasmin system has been implicated in blot clot dissolution, ovulation, embryogenesis, cell migration or brain function. Whereas t-PA is the main activator of plasmin in blot clots, u-PA binds to specific receptors on the cell surface to regulate cell-mediated proteolysis and cell invasion/migration. A role for u-PA in macrophage function, ovulation, angiogenesis, wound healing and metastasis, and invasion of tumor cells has been proposed.

KO strain construction

A *neo*-resistance cassette replaced the whole *Plau* gene except 23 amino acids at the 3′ end of the gene. *tk* was placed at the 5′ end of the construct. Targeted D3 ES cells were injected into C57BL/6 blastocysts.

Phenotype

u-PA$^{-/-}$ mice developed normally, had a normal lifespan and were fertile. These mice occasionally developed spontaneous fibrin deposits in inflamed and normal tissues (intestine, liver, skin). They also exhibited defective plasmin-induced macrophage functions, and had a higher incidence of endotoxin-triggered thrombosis. u-PA$^{-/-}$ mice developed rectal prolapses and extensive non-healing ulcerations of the eyelids, ears and the face. The u-PA mutation had no effect on thrombolysis. Macrophages had no defects in cell migration and invasion of the peritoneal cavity following thioglycollate injection.

Reference

[1] Carmeliet, P. et al. (1994) Nature 368, 419–424.

Other names
Urokinase-type plasminogen activator receptor, Plaur, CD87

Gene symbol
Plaur

Accession number
MGI: 97612

Area of impact
Hormones

General description

The urokinase plasminogen activator receptor (uPAR) is a member of the Ly6 superfamily of glycosylphosphatidylinositol-anchored membrane glycoproteins. uPAR is expressed on many diverse types of cells, including monocytes, macrophages, activated T cells, endothelial cells, smooth muscle cells, trophoblasts, and tumor cells. uPAR binds urokinase plasminogen activator (u-PA) with high affinity and potentiates u-PA-mediated plasminogen activation *in vitro*. uPAR is also implicated in u-PA : inhibitor internalization, monocyte chemotaxis and regulation of integrin function.

KO strain construction

The targeting vector included a human *hprt* minigene cassette, flanked 5′ by a 2.5 kb *Plaur* gene fragment containing parts of intron 1 and exon 2 and an HSV-*tk* cassette, and 3′ by a 1.5 kb intron 3 fragment. The vector deleted exon 3 of the *Plaur* gene. E14TG2a ES cells (derived from 129 mice) were injected into C57BL/6 blastocysts and further inbred to a C57BL/6J background.

Phenotype

Detergent-phase extracts from lung, spleen or kidney of uPAR-deficient (uPAR$^{-/-}$) mice had no detectable uPAR antigen or u-PA crosslinking activity. Activated peritoneal macrophages isolated from uPAR$^{-/-}$ mice failed to potentiate pro-u-PA-mediated cell surface plasminogen activation *in vitro*, and macrophages isolated from uPAR$^{+/-}$ mice had half-normal potentiation of plasminogen activation. Lack of uPAR led to a change in the distribution of cell-associated and cell-free u-PA. Cells from pulmonary lavages of uPAR$^{-/-}$ mice displayed a partial loss of u-PA activity, and cell-free peritoneal lavage fluid from uPAR$^{-/-}$ mice demonstrated increased u-PA activity. Conversion of pro-u-PA to active two-chain u-PA in the urogenital tract was unimpeded in uPAR$^{-/-}$ mice. Despite these biochemical findings, uPAR$^{-/-}$ mice were born with normal appearance and behavior. There was no indication of fetal loss of uPAR$^{-/-}$ mice. uPAR$^{-/-}$ mice displayed normal weight gain, survival and longevity. Female and male fertility, and reproductive success of multiparous females were normal in uPAR$^{-/-}$ mice. Standard hematological parameters were also normal. An extensive histological analysis of tissues from uPAR$^{-/-}$

mice failed to yield any signs of pathology, including fibrin-rich lesions frequently observed in liver, stomach, lung, rectum, and other organs of u-PA$^{-/-}$ mice. Healing times of incisional skin wounds and excisional corneal wounds were normal in uPAR$^{-/-}$ mice.

u-PA and tissue-type plasminogen activator (t-PA) display a strong functional overlap in plasminogen activation. *In vivo*, t-PA-mediated plasminogen activation efficiently compensates for the loss of u-PA. To determine the role of uPAR in u-PA-mediated plasminogen activation *in vivo*, mice with a combined deficiency of uPAR and t-PA were generated. uPAR$^{-/-}$ t-PA$^{-/-}$ mice, unlike t-PA$^{-/-}$ or uPAR$^{-/-}$ mice, developed hepatic and occasionally pulmonary fibrin deposits, demonstrating a role for uPAR in fibrinolysis. However, uPAR$^{-/-}$t-PA$^{-/-}$ mice did not display the pervasive multiorgan pathology, high mortality, reduced reproductive success, and impaired skin and corneal wound-healing characteristic of plasminogen-deficient or plasminogen activator-deficient mice. The surprisingly mild phenotype of uPAR$^{-/-}$t-PA$^{-/-}$ mice compared to plasminogen- or plasminogen activator-deficient mice demonstrates that u-PA function in plasminogen activation in most physiologic processes is uPAR-independent.

Comments

The studies demonstrate a role for uPAR in hepatic fibrinolysis, and provide further evidence that u-PA and t-PA act as complementary fibrinolytic factors. The largely unimpeded u-PA function in uPAR$^{-/-}$ mice challenges the paradigm that uPAR-mediated cell surface binding of u-PA is essential for efficient u-PA-mediated plasminogen activation.

Acknowledgements
Thomas H. Bugge and Jay L. Degen
Children's Hospital Research Foundation, Cincinnati, OH, USA

References
1 Bugge, T.H. et al. (1995) J. Biol. Chem. 270, 16886–16894.
2 Bugge, T.H. et al. (1996) Proc. Natl Acad. Sci. USA 93, 5899–5904.
3 Bugge, T.H. et al. (1995) Genes Dev. 9, 794–807.
4 Bugge, T.H. et al. (1996) Cell 87, 709–719.
5 Carmeliet, P. et al. (1994) Nature 368, 419–424.

Other names
Clara cell secretory protein (CCSP), CC10, CC16, blastokinin, urine protein-1

Gene symbol
Utg

Accession number
MGI: 98919

Area of impact
Immunity and inflammation, lung and kidney function

General description

Uteroglobin is a steroid-inducible secreted protein of unknown physiologic function which is expressed abundantly in Clara cells of the lung and to varying degrees within the epithelia of the genitourinary tract of mammals. Notably, mice and rats express very little uteroglobin in their genitourinary tracts. Uteroglobin has been shown to bind a variety of small hydrophobic ligands such as progestins and metabolites of polychlorinated biphenyls. Recently, a high-affinity receptor for uteroglobin which is expressed on several cell types was identified. Results from *in vitro* studies suggest that uteroglobin may participate in the regulation of the inflammatory response.

KO strain 1 construction[1]

A 5.2 kb *Hin*dIII fragment of the mouse (129) *Utg* gene, spanning approx. –2.1 kb through +3.1 kb, was truncated by the insertion of PGK*neo* into a unique *Pst*I site located at the 3′ end of exon 1. HSV-*tk* was inserted 3′ of the gene sequences for negative selection. 129Ola ES cells were targeted and germ line chimeras were bred with 129J strain females for establishment of 129J/129Ola hybrid uteroglobin$^{-/-}$ mice. The mutation has also been backcrossed for nine generations into the C57BL/6 strain. The mice were specific pathogen-free as determined by screening with a comprehensive 16 agent serologic panel.

KO strain 2 construction[2]

A *Bam*HI/*Eco*RI 3.2 kb DNA fragment containing the three exons of the 129SvJ mouse uteroglobin gene and its flanking sequences was subcloned into the corresponding site of the pPNW vector[3]. A 0.9 kb PCR-amplified fragment containing partial exon 2 and its flanking sequences with built-in *Not*I and *Xho*I sites were subcloned into the vector to generate the targeted construct pPNWUG. In pPNWUG, a 1.2 kb DNA fragment, including partial exon 2, was replaced with the PGK*neo* cassette, disrupting the uteroglobin gene. The targeting vector was electroporated into R1 ES cells and positive clones were injected into C57BL/6 blastocysts. Offspring heterozygous for the mutated allele were intercrossed to generate uteroglobin$^{-/-}$ mice.

Phenotype

KO strain 1 uteroglobin$^{-/-}$ mice exhibited mild and transient proximal tubular dysfunction in kidneys of 2-month-old mice which resolved by 6 months of age.

No perturbations were observed in glomerular morphology, fibronectin deposition or kidney fibrosis. Mice showed normal reproductive performance and growth characteristics when compared to strain-matched wild-type mice. Non-ciliated airway epithelial (Clara) cells had ultrastructural alterations to their secretory apparatus, including loss of secretory granules and >95% reduction in rER. KO strain 1 mice showed increased sensitivity to injury by inhaled oxidant pollutant (ozone and hyperoxia). Uteroglobin deficiency blocked methylsulfonyl-polychlorinated biphenyl accumulation in lung tissue and airway lining fluid, demonstrating the involvement of uteroglobin in pollutant accumulation *in vivo*.

KO strain 2 uteroglobin$^{-/-}$ mice were also viable and fertile. However, they exhibited severe renal disease that was associated with massive glomerular deposition of predominantly multimeric fibronectin (FN). The molecular mechanism that normally prevents FN deposition appears to involve high-affinity binding of uteroglobin (UG) with FN to form FN-UG heteromers that counteract FN self-aggregation, which is required for abnormal tissue deposition. Sporadic pancreatic focal necrosis and inflammation, and renal tubular fibrosis were also observed. The age of severe disease onset ranged from 4–5 weeks to about 10 months. KO strain 2 uteroglobin$^{-/-}$ mice died from end-stage kidney disease.

Comments

The phenotypic consequences of uteroglobin deficiency in KO strain 1 mutants suggest an important role for Clara cells and their secretions in protection from oxidant stress. Uteroglobin also appears to have a functional role in kidney homeostasis. The phenotype of hereditary fibronectin deposit glomerulopathy in humans is very similar to that of KO strain 2 uteroglobin$^{-/-}$ mice.

The basis for the differences in phenotype (particularly in the kidney) between KO strains 1 and 2 is unknown, but may be related to either the design of the null allele, the genetic background of the mutant mice, or pathogen status.

Acknowledgements
Barry R. Stripp
Department of Environmental Medicine, University of Rochester, Rochester, NY, USA
Anil B. Mukherjee
National Institute of Child Health and Human Development, NIH, Bethesda, MD, USA

References

1 Stripp, B.R. et al. (1996) Am. J. Physiol. 271, L656-L664.
2 Zhang, Z. et al. (1997) Science 276, 1408–1412.
3 Lei, K. et al. (1996) Nature Genet. 13, 203–209.
4 Johnston, C. et al. (1997) Am. J. Respir. Cell Mol. Biol. 17, 147–155.
5 Stripp, B.R. et al. (1994) Genomics 20, 27–35.
6 Miele, L. et al. (1987) Endocrine Rev. 8, 474–490.
7 Kundu, G.C. et al. (1996) Proc. Natl Acad. Sci. USA 93, 2915–2919.

Other names
Vascular cell adhesion molecule 1

Gene symbol
Vcam1

Accession number
MGI: 98926

Area of impact
Adhesion, development

General description

VCAM-1 is an adhesion molecule that binds α4-integrins. It is a member of the Ig gene superfamily and contains seven Ig-like domains. (Other alternatively spliced forms with fewer Ig domains are expressed at low levels.) In adults, VCAM-1 may be important in the recruitment of leukocytes to sites of inflammation. VCAM-1 expression in endothelial cells is induced by inflammatory cytokines such as IL-1 and TNFα. VCAM-1 is also expressed on some monocyte-derived and epithelial cells. In embryos, VCAM-1 is expressed in the allantois and is necessary for chorioallantoic fusion.

KO strain 1 construction[1]

A targeting construct was assembled from a 6.7 kb *Eco*RI genomic clone (derived from J1 ES cells, strain 129) containing the 5′ portion of the *Vcam1* gene. This region included the core promoter and the first four exons encoding the signal peptide and Ig-like domains 1–3. Exon 2, encoding domain 1, was disrupted by deleting the majority of this domain and a portion of the following intron with restriction enzymes *Bgl*II and *Nhe*I. At this site, *neo* was inserted in the opposite transcriptional orientation to *Vcam1*. Homologous regions of the *Vcam1* gene 5′ and 3′ of *neo* were 2.0 kb and 4.3 kb, respectively. For negative selection against random integration, HSV-*tk* was inserted at the 5′ end of the construct. The targeting vector was electroporated into J1 ES cells and positive clones were injected into C57BL/6 and BALB/c blastocysts. Chimeric males were crossed to C57BL/6, BALB/c and 129 females.

KO strain 2 construction[2]

[This strain was hypomorphic rather than a true KO.]
VCAM-1 KO strain 2 was deficient for the fourth Ig-like domain of VCAM-1 and was designated VCAM-1 D4D. In the 15.8 kb targeting construct, the 5′ portion of *Vcam1* exon 6 (encoding Ig-like domain 4) and the adjacent intron were removed and replaced with PGK*neo* in the opposite transcription orientation. Regions of *Vcam1* homology 5′ and 3′ to PGK*neo* were 7.1 kb and 1.7 kb, respectively. To select against random integration, the HSV-*tk* gene flanked by a mutated polyomavirus enhancer was placed at the 3′ end of the construct. The targeting vector was used to generate homozygous mutant mice as described for KO strain 1 mice.

KO strain 3 construction[3]

The targeting vector was designed to introduce both a deletion and a frameshift mutation into the *Vcam1* gene to prevent the production of any protein arising by fortuitous splicing. PGK*neo* was inserted into exon 2 (which contains coding sequences of Ig domain I) and was flanked by sequences downstream of exon 5 (which encodes Ig domain IV). This mutated allele was expected to abolish all α4-integrin binding. The vector was electroporated into W9.5 ES cells and positive clones were injected into C57BL/6J blastocysts. Male chimeras were bred with C57BL/6J females to produce heterozygotes, which were intercrossed to generate VCAM-$1^{-/-}$ mutants.

Phenotype

In 80% of KO strain 1 VCAM-1-null embryos, the allantois failed to fuse to the chorion at 8.5 days of gestation, resulting in abnormal placental development and embryonic death within 1–3 days. In a minority of KO strain 1 VCAM-1-null embryos (<20%), the allantois was able to fuse with the chorion, but the allantoic mesoderm was abnormally distributed over the chorionic surface, indicating a possible role for VCAM-1 in early placental formation after chorioallantoic fusion. A small number of KO strain 1 VCAM-1-null embryos survived (<3%), presumably by circumventing the placentation defects. These became viable and fertile adult mice with normal organ development (including the heart), even though they lacked VCAM-1 expression.

KO strain 2 VCAM-1 D4D mice had no expression of the seven Ig-like domain form of VCAM-1. Isoforms containing six, and to a lesser degree, three, Ig-like domains were expressed at levels 5–8% of the wild type, effectively providing a single α4-integrin-binding site. One-third of VCAM-1 D4D mice survived embryonic development and became healthy and fertile adults. VCAM-1 D4D mice exhibited abnormalities of leukocyte recruitment. Untreated VCAM-1 D4D mice showed normal levels of circulating leukocytes, including neutrophils, monocytes and lymphocytes. Cellularity of lymphoid tissues, including lymph nodes and spleen, was normal. However, there was a reduced number of peritoneal macrophages (30% of wild type). When leukocyte recruitment to sites of inflammation was investigated, reduced monocyte/macrophage recruitment and granuloma formation was observed in the β-glucan lung granuloma model. Peritonitis induced by thioglycollate was comparable to that in the wild type.

About half of KO strain 3 VCAM-$1^{-/-}$ mice exhibited the same phenotype as the embryonic lethal KO strain 1 mutants. The other half survived until E11.5–12.5 but displayed several abnormalities of the developing heart, including a reduction of the compact layer of the ventricular myocardium and intraventricular septum. These hearts lacked an epicardium and contained blood in the pericardial space.

Comments

These data demonstrate key roles for VCAM-1 in the development of both the placenta and the heart.

Acknowledgements

Myron Cybulsky

Department of Pathology, The Toronto Hospital, University of Toronto, Toronto, ON, Canada

References

1 Gunter, G.C. et al. (1995) Genes Dev. 9, 1–14.
2 Li, H. et al. (1996) FASEB J. 10, A1281.
3 Kwee, L. et al. (1995) Development 121, 489–503.
4 Yang, J.T. et al. (1995) Development 121, 549–560.
5 Terry, R.W. (1997) Transgenic Res. 6, 349–356.

Other names
Vascular endothelial growth factor

Gene symbol
Vegf

Accession number
MGI: 103178

Area of impact
Development

General description

VEGF is a potent endothelial cell-specific mitogen that is a major mediator of pathological angiogenesis. It is strongly induced by hypoxia and its upregulation during tumor growth leads to the invasion of blood vessels seen in tumor angiogenesis. It is also expressed during development in complementary patterns to its receptors, Flk-1 and Flt1, suggesting that it plays a critical role in the normal development of blood vessels during embryogenesis as well.

KO strain 1 construction[1]

The third exon of *Vegf* was replaced by a PGK*neo* cassette, causing a frameshift in the VEGF coding sequence and deleting six of eight essential cysteine residues. The vector was introduced into R1 129 ES cells. Homozygous VEGF$^{-/-}$ ES cells were generated by selection in high concentrations of G418. Phenotypic analysis of both heterozygous and homozygous loss of *Vegf* was primarily carried out by aggregation of +/− or −/− ES cells with tetraploid wild-type embryos, marked with the ubiquitous *lacZ* marker ROSA26. In such aggregates, all the embryonic tissues are derived from the ES cells, while trophoblast and primitive endoderm in derivatives are wild type.

KO strain 2 construction[2]

The coding region of exon 3 was replaced by a PGK*neo* cassette. The vector was introduced into the C12 subclone of D3 129 ES cells, which were injected into C57BL/6 blastocysts. Chimeras were mated with C57BL/6 females. No heterozygotes were found among the agouti progeny.

Phenotype

Embryos heterozygous for the loss of one *Vegf* allele die in midgestation, demonstrating haploinsufficient embryo lethality for this locus. Vascular development is severely impaired in these embryos. Lumen size of the developing major vessels is reduced as is blood island development is reduced. Examination of homozygous VEGF$^{-/-}$ embryos produced by tetraploid aggregation revealed the expected increased severity of the phenotype, with death occurring by E10.5. Blood vessel development was severely impaired.

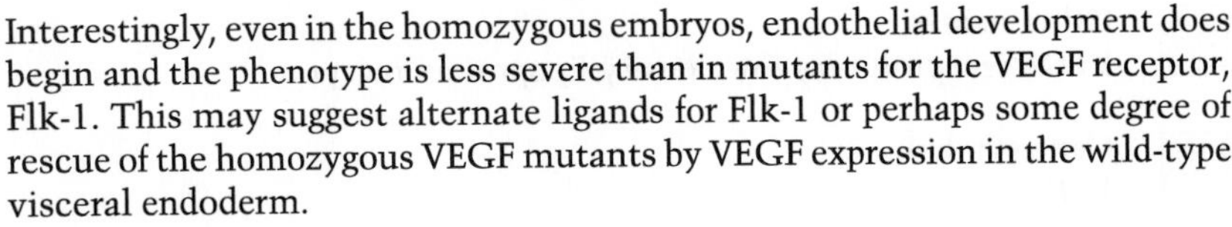

Interestingly, even in the homozygous embryos, endothelial development does begin and the phenotype is less severe than in mutants for the VEGF receptor, Flk-1. This may suggest alternate ligands for Flk-1 or perhaps some degree of rescue of the homozygous VEGF mutants by VEGF expression in the wild-type visceral endoderm.

References

[1] Carmeliet, P. et al. (1996) Nature 380, 435–439.
[2] Ferrara, N. et al. (1996) Nature 380, 439–442.

Other names
VLDL receptor, very low density lipoprotein receptor

Gene symbol
Vldlr

Accession number
MGI: 98935

Area of impact
Metabolism, cardiovascular

General description

The VLDLR is a member of the low-density lipoprotein (LDL) receptor family. VLDLR mediates the binding and uptake of apoE containing VLDL and β-VLDL, but not of apoB-100 containing LDL. The VLDLR is ubiquitously expressed with high levels in muscle and adipose tissues and low expression levels in the liver. Mutations of VLDLR in chickens cause female infertility.

KO strain construction

Residues 162–262 of exon 5 were replaced by a neomycin-resistance cassette in antisense orientation. Two copies of the HSV-*tk* gene were placed at the 5′ end. JH-1 ES cells were targeted and mutated ES cells were injected into C57BL/6 blastocysts.

Phenotype

Male and female VLDLR-null mice appeared normal and were fertile. Besides the observation that null mice were slightly smaller and leaner than their littermates, there was no significant phenotype. Lipoprotein profiles and plasma levels of cholesterol, triacylglycerol and lipoproteins were also comparable among VLDLR-null and wild-type littermates on normal, high-carbohydrate and high-fat diets.

Comments

The VLDLR is not required for the removal of VLDL from the plasma.

Reference

[1] Frykman, P.K. et al. (1995) Proc. Natl Acad. Sci. USA 92, 8453–8457.

Other names
Vitronectin

Gene symbol
Vtn

Accession number
MGI: 98940

Area of impact
Cardiovascular, adhesion

General description

VN is a 70 kDa adhesive glycoprotein that is present in the extracellular matrix and plasma. It is primarily produced in the liver, although VN expression can be observed in brain at early stages of organogenesis. VN has been identified as a ligand for $\alpha v\beta 3$ and $\alpha IIb\beta 3$ integrin receptors and has been implicated in cell attachment, spreading, and complement activation. VN binds to streptococci and staphylococci mediating bacterial adherence and opsonization for macrophages. It also seems to play a role in thrombosis and fibrinolysis and forms complexes in the plasma with the plasminogen activator inhibitor 1 (PAI-1). VN can stabilize PAI-1 in its activated form and helps to localize PAI-1 to specific regions of the extracellular matrix.

KO strain construction

A 7 kb genomic fragment containing all VN-coding sequences was replaced by a neomycin-resistance cassette. PGK-*tk* was placed at the 5′ end. D3 ES cells were targeted and mutated ES cells were injected into C57BL/6 blastocysts.

Phenotype

$VN^{-/-}$ mice had normal development, fertility and lifespans. No defects in morphology or histology were evident. Sera from these mice lacked "serum spreading factor" and PAI-1-binding activity.

Comments

The VN mutation may be functionally compensated for by other adhesive interactions.

Reference

[1] Zheng, X. et al. (1995) Proc. Natl Acad. Sci. USA 92, 12426–12430.

von Willebrand factor

Other names
vWf, FVIII related-antigen

Gene symbol
Vwf

Accession number
MGI: 98941

Area of impact
Cardiovascular

General description

von Willebrand factor (vWf) is a large multimeric glycoprotein synthesized by megakaryocytes and endothelial cells. It circulates in plasma and is found in the basement membrane. vWf is stored in granules, α-granules in platelets and Weibel–Palade bodies in endothelial cells. It is involved in primary hemostasis by mediating platelet adhesion to the subendothelium in the event of vascular injury. vWf is the carrier of coagulation factor VIII in plasma, where it protects factor VIII against proteolytic degradation.

KO strain construction

A 0.8 kb genomic DNA fragment corresponding to exons 4 and 5 of *Vwf* was deleted and replaced by a neomycin-resistance cassette in the targeting vector. An unusual recombination/insertion event led to the insertion of the *neo* in intron 5 of *Vwf* without the expected deletion of exons 4 and 5. D3 ES cells obtained from 129Sv mice were used for transfection. Targeted cells were injected into blastocysts of C57BL/6 mice.

Phenotype

Homozygous vWF-deficient mice were viable and fertile with a highly prolonged bleeding time and decreased factor VIII activity. Spontaneous bleeding events occurred in about 10% of neonates. Defective hemostasis was also observed using a vasculitis model, the Shwartzman reaction. The mice presented severe defects in thrombus formation and arteriolar shear rate.

Comments

The vWf KO and heterozygous mice provide a model for the most common human bleeding disorder, known as von Willebrand disease.

Acknowledgements
Denisa D. Wagner
Centre for Blood Research Inc., Boston, MA, USA

Reference

[1] Denis, C. et al. (1997) Blood 90, 429a.

Wnt-1

Other names
Int-1

Gene symbol
Wnt1

Accession number
MGI: 98953

Area of impact
Development

General description

Wnt1 was identified as a locus involved in virus-induced mammary carcinogenesis. The gene encodes a cysteine-rich intercellular signaling molecule related to the *Drosophila wingless* gene. There is a large family of *Wnt* genes in vertebrates, involved in many different aspects of embryonic patterning and proliferation. Expression of *Wnt1* mRNA is restricted to the adult testes and the midgestation neural tube, in a domain encompassing the midbrain/hindbrain border and in the dorsal spinal cord.

KO strain 1 construction[1]

A neo^r gene (MC1*neo*) was inserted into the *Xho*I site of the second exon. The vector was inserted as a duplication such that the mutant allele contains tandem (head-to-tail) copies of the *Wnt1* gene, both of which contain the mutated second exon. The vector was introduced in 129Ola ES cells, injected into C57BL/6 blastocysts and the phenotypes examined on a 129 × B6 hybrid background.

KO strain 2 construction[2]

An MC1*neo* cassette was inserted into a *Xho*I site in the second exon of the gene, interrupting the coding sequence at codon 57. The vector was introduced into AB-1 129 ES cells which were injected into C57BL/6 blastocysts. The phenotype was examined on a mixed 129 × B6 background.

Phenotype

All animals homozygous for the $Wnt1^{neo}$ allele show defects in midbrain and cerebellar development first recognized as a hypoproliferation of the metencephalon as early as day 8.5 of gestation. The majority of mutant individuals die perinatally, though some homozygotes have survived past weaning. Adult survivors are characterized by severe ataxia and are missing the rostral half of the cerebellum. Newborn mutant animals fall with equal frequency into two classes that can be distinguished histologically. Members of one class lack the superior and inferior colliculi as well as any detectable cerebellar tissue. Animals in the other class contain midbrain nuclei as well as some cerebellar structures. The latter are restricted to lateral regions of the newborn brain, consistent with their being the embryonic precursors to the caudal lobes of cerebellum found in the surviving *Wnt1* adults.

Comments

The spontaneous mouse mutation *swaying* is an allele of *Wnt1*. *Wnt1* and *Engrailed* interact in patterning the midbrain/hindbrain region and expression of *Engrailed* in the *Wnt1* domain can partially rescue the *Wnt1* phenotype.

Acknowledgements
Kirk Thomas
University of Utah, Salt Lake City, UT, USA

References

1 Thomas K.R. et al. (1990) Nature 346, 847–850.
2 McMahon, A.P. et al. (1990) Cell 62, 1073–1085.
3 Thomas K.R. et al. (1991) Cell 67, 969–976.
4 Danielan, P.S. and McMahon, A.P. (1996) Nature 383, 332–334

Wnt-3

Other names
Int-4

Gene symbol
Wnt3

Accession number
MGI: 98955

Area of impact
Development

General description

Wnt3 was identified as a locus involved in virus-induced mammary carcinogenesis. The gene is a member of the *wnt/wingless* family and encodes a protein presumably involved in intercellular signaling. *Wnt3* mRNA can be detected *in situ* in the midgestation neural tube.

KO strain construction

The targeting vector resulted in the insertion of neo^r (MC1*neo*) at the *Cla*I site in exon 4 of the *Wnt3* gene. The genetic background of the knockout mouse was C57BL6/129.

Phenotype

Animals homozygous for the *Wnt3* mutation died shortly after gastrulation. mRNA associated with mesodermally derived cells could not be identified in the mutant embryos.

Acknowledgements
William D. Howell
University of Utah, Salt Lake City, UT, USA

References

1 The Wnt-$3^{-/-}$ mouse is not yet published.
2 Roelink, H. et al. (1990) Proc. Natl Acad. Sci. USA 87, 4519–4523.
3 Roelink, H. and Nusse, R. (1991) Genes Dev. 5, 381–388.
4 Salinas, P.C. and Nusse R. (1992) Mech. Dev. 39, 151–160.

Gene symbol
Wnt3a

Accession number
MGI: 98956

Area of impact
Development

General description

Wnt3a is a member of the *Wnt* family of signaling molecules, related to the *Drosophila wingless* gene. It is widely expressed within the developing primitive streak and expression continues in the late streak and tailbud stages. At early somite stages, *Wnt3a* is expressed dorsally and caudally within the primitive streak, and then becomes localized to the caudal tip of the tailbud.

KO strain construction

The PGK*neo* cassette was inserted into a *Sma*I site in the third exon of *Wnt3a*, thus truncating the normal open reading frame. The vector was introduced into J7 129 ES cells, which were injected into C57BL/6 blastocysts. The mutant phenotype was examined on a mixed 129 × BL background.

Phenotype

Loss of Wnt-3a function leads to embryonic lethality between E10.5 and 12.5. Phenotypic effects were apparent by E9.5. Anterior development was normal, but, caudal to the forelimbs, the trunk was shortened, somites were disrupted or missing and the neural tube was kinked. No tailbud is formed. Further analysis revealed that in the posterior of these embryos, cells that would normally move forward from the primitive streak to form the somites instead form secondary neural tube structures. Wnt-3a thus seems to be necessary for normal proliferation and specification of paraxial mesoderm.

Comments

The spontaneous mutation *Vt, vestigial tail,* which causes loss of caudal vertebrae and tail shortening, seems to be a hypomorphic allele of *Wnt3a*.

References

[1] Takada, S. et al. (1994) Genes Dev. 8, 174–189.
[2] Yoshikawa, Y. et al. (1997) Dev. Biol. 183, 234–242.

Wnt-4

Gene symbol
Wnt4

Accession number
MGI: 98957

Area of impact
Development

General description

Wnt4 is a member of the *Wnt* family of signalling molecules, related to the *Drosophila wingless* gene. *Wnt4* is expressed in the developing kidney mesenchyme. Expression is first seen in condensed mesenchyme cells on both sides of the urethra stalk at E11.5. Expression occurs in all subsequent newly developed aggregates and their tubular derivatives. Expression occurred also in comma-shaped bodies, but was then restricted to S-shaped bodies. The dorsal spinal cord also expresses *Wnt4*.

KO strain construction

The third coding exon of the *Wnt4* gene was replaced by a PGK*neo* cassette, deleting 2.2 kb of the *Wnt4* gene. The vector was introduced into CJ7 129 ES cells, which were injected into C57BL/6 blastocysts.

Phenotype

Homozygous mice died within 3–4 hours of birth. All mutants showed small agenic kidneys, consisting of undifferentiated mesenchyme interspersed with branches of collecting duct epithelium. Death is presumed to be due to lack of kidney function. Analysis of kidney development earlier in embryogenesis revealed that initial condensation around the tips of the ureteric buds occurred normally, suggesting that induction of the mesenchyme by the ureter was not affected. By E15.5, mutant kidneys were clearly retarded. Although extensive ureteric epithelial branching was observed, mesenchyme remained undifferentiated. No comma-shaped or S-shaped bodies were observed. It appears that expression of *Wnt4* in the mesenchymal aggregates is necessary for their transition to epithelial tubules.

Reference
[1] Stark, K. et al. (1994) Nature 372, 679–683.

Gene symbol
Wnt7a

Accession number
MGI: 98961

Area of impact
Development

General description

Wnt7a is a member of the *Wnt* family of signaling molecules, related to the *Drosophila wingless* gene. *Wnt7a* is expressed in the flanking ectoderm of the trunk prior to limb bud outgrowth. As limb bud outgrowth occurs, *Wnt7a* is confined in expression to the dorsal ectoderm.

KO strain construction

A *neo*^{r} cassette was introduced into the second exon of the *Wnt7a* gene. The vector was introduced into CJ7 129 ES cells, which were injected into C57BL/6 blastocysts. The phenotype was examined on a mixed 129 × B6 background.

Phenotype

Wnt7a homozygous mutants are viable but sterile. Limb abnormalities are apparent. Ventral structures develop normally, but many dorsal structures adopt ventral fates, most strikingly the formation of ventral footpad structures on the dorsal side. Markers of ventral ectoderm are still ventrally restricted in *Wnt7a* mutants, suggesting that ventralizing activity must normally be blocked by Wnt-7a to promote dorsal fates. Antero-posterior patterning of digits is also affected, with loss of posterior elements. Loss of Wnt-7a leads to reduction in expression of *Shh* in the zone of polarizing activity (ZPA), which may lead to the antero-posterior patterning defects. This result suggests that there is cross-talk between dorso-ventral and antero-posterior patterning signals in limb development.

Reference

[1] Parr, B.A. et al. (1995) Nature 374, 350–353.

Other names
Wilms' tumor gene

Gene symbol
Wt1

Accession number
MGI: 98968

Area of impact
Development

General description

In humans, germ line mutations of the *Wt1* tumor suppressor gene are associated with both Wilms' tumors and urogenital malformations. The *Wt1* gene is expressed in the mesenchymal cells of the metanephros coincident with condensation, and continues to be expressed by epithelial cells as they differentiate into nephric tubules. *Wt1* encodes a protein with a proline/glutamine-rich N-terminal domain and four zinc fingers in the C-terminal domain which share homology with the EGR transcription factor family. Alternative splicing of the *Wt1* gene results in four transcripts. Two isoforms have been shown to bind to the EGR-1 recognition site but appear to repress transcription rather than activate it.

KO strain construction

The replacement type targeting vector resulted in the deletion of the first exon and 0.5 kb of upstream sequence of the *Wt1* gene. Genomic clones of the *Wt1* gene used in the vector were from the D3 ES cell line. PGK*neo*polyA was used for positive selection and HSV-*tk* was included for negative selection. The targeting vector was electroporated into J1 ES cells and positive clones were injected into C57BL/6 blastocysts to generate chimeras. Chimeras were crossed with C57BL/6 mice.

Phenotype

Heterozygous WT-$1^{+/-}$ mice appeared normal and had not developed tumors by age 10 months. Homozygosity for the mutation resulted in embryonic lethality. Examination of mutant embryos revealed a failure of kidney and gonad development. Specifically, at day 11 of gestation, the cells of the metanephric blastema underwent apoptosis, the ureteric bud failed to grow out from the Wolffian duct, and the inductive events that lead to formation of the metanephric kidney did not occur. In addition, the mutation caused abnormal development of the mesothelium, heart and lungs.

Comments

WT-1 has a crucial role in early urogenital development.

Acknowledgements
Jordan Kreidberg
Department of Nephrology, Children's Hospital, Boston, MA, USA

Reference

1 Kreidberg, J.A. et al. (1993) Cell 74, 679–691.

Xist

Gene symbol
Xist

Accession number
MGI: 98974

Area of impact
Development

General description

The X-linked *Xist* gene encodes a large untranslated RNA that has been implicated in mammalian dosage compensation and in spermatogenesis. It maps to the X-inactivation center and is expressed specifically from the inactive X.

KO strain construction

A vector that deletes exon 1–5 and replaces them with a PGK*neo* cassette was introduced into J1 129 ES cells. These cells were injected into BALB/c blastocysts and the phenotype examined on a 129 × BALB/c hybrid background.

Phenotype

Males carrying the mutation on their X chromosomes were healthy and fertile. Heterozygous females that inherited the mutation from their mothers were also normal and had the wild-type paternal X chromosome inactive in every cell. In contrast to maternal transmission, females that inherited the mutation from their father were severely growth-retarded and died early in embryogenesis. The wild-type maternal X chromosome was inactive in every cell of the growth-retarded embryo proper. Both X chromosomes were expressed in the mutant female trophoblast where X inactivation is normally preferential to the paternal X. However, an XO mouse with a paternally inherited *Xist* mutation was healthy and appeared normal. The lethal phenotype of the mutant females is therefore presumably due to defects in extra-embryonic tissue development when two X chromosomes are active.

Comments

Xist RNA is required for female dosage compensation but plays no role in spermatogenesis.

Acknowledgements
Rudolf Jaenisch
Whitehead Institute for Biomedical Research, Cambridge, MA, USA

References

1 Marahrens, Y. et al. (1997) Genes Dev. 11, 156–166.
2 Lee, J.T. et al. (1997) Curr. Opin. Genet. Dev. 274–280.

Other names
Xeroderma pigmentosum complementation group A correcting gene, group-A XP, XPAC

Gene symbol
Xpa

Accession number
MGI: 99135

Area of impact
DNA repair

General description

Xeroderma pigmentosum (XP) is an autosomal recessive disease characterized by skin hypersensitivity to sunlight and a dramatically (>1000-fold) increased risk of developing skin cancer following exposure to sunlight. Eight complementation groups exist in XP (XPA to XPG and an XP-variant form). The XPA gene product functions in the initial step of the nucleotide excision repair cascade, recognition of the DNA damage (DNA adducts), certainly within the global genome repair (GGR) subpathway, and probably in the transcription-coupled repair (TCR) subpathway.

KO strain 1 construction[1]

A replacement-type targeting vector was used. In an 11 kb *Bam*HI/*Bam*HI genomic fragment, a 2 kb *Apa*I fragment containing exon 3 (encoding the DNA-binding domain) and exon 4 was replaced by the PGK*neo*^{r} cassette in sense orientation. A recombinant clone of the 129Ola-derived E14 ES cell line was injected into C57BL/6 blastocysts. The chimeras were crossed with C57BL/6. A homozygous XPA$^{-/-}$ mouse strain was established by intercrossing the heterozygous F1 progeny.

KO strain 2 construction[2]

A neomycin-resistance cassette was inserted into exon 4 of the mouse *Xpa* gene in F1/1 ES cells (CBA × C57BL/6). Targeted ES cells were injected into CD1 blastocysts. Chimeric mice were bred with CD1 females.

Phenotype

XPA$^{-/-}$ mice developed normally, were healthy and fertile. Between the ages of 15 and 20 months, 15% of the mutant mice developed benign hepatocellular adenomas. The internal organs, including the CNS, showed no gross morphological abnormalities. At the age of 18 months, there were no indications of neurological degeneration. In all assays, XPA$^{+/-}$ mice and cells from these mice were indistinguishable from the wild type.

The absence of a functional *Xpa* gene had consequences both *in vitro* and *in vivo*. *In vitro*, the residual DNA repair in XPA$^{-/-}$ embryonic fibroblasts as measured by UV-induced (dose: 16 J/m2) unscheduled DNA synthesis (UDS) was less than 5% of that in fibroblasts from wild-type littermates. In addition, the strongly increased relative sensitivity of XPA$^{-/-}$ embryonic fibroblasts to genotoxic agents (determined by comparing doses at which the survival of wild-type and XPA$^{-/-}$ cells was the same) was 8.0 for UV, 3.5 for dimethylbenz[a]anthracene, 2.5 for *N*-acetylaminofluorene, 2.5 for benzo[a]pyrene, 1.8 for PhIP, and 1.1 for butylnitrosurea.

In vivo, a low daily dose of UV-B induced hyperkeratosis and erythema followed by a high incidence of skin tumors. The type of skin tumor depended on the genetic background: squamous cell carcinomas (SCCs) occurred in the XPA$^{-/-}$ F2 C57BL/6 × 129 mice, while SCCs and papillomas occurred in XPA$^{-/-}$ hairless (hr) mice (100% incidence within 20 weeks). Low doses of dimethylbenz[a]anthracene (DMBA) resulted in acute effects, including acanthosis and epidermal hyperplasia followed by an increased incidence on the skin of papillomas that never progressed to carcinomas. Oral administration of benzo[a]pyrene induced internal tumors earlier and at a higher frequency in XPA$^{-/-}$ mice compared to wild-type mice. The dominant tumor type was lymphoma, although bronchiolar or alveolar adenoma and tumors of the stomach were also observed.

Comments

Except for the absence of neurological abnormalities, the XPA$^{-/-}$ mice mimic XP-A patients very accurately. Whether XPA-deficient mice are more suitable for *in vivo* short-term carcinogenicity assays remains to be proven.

Acknowledgements
J.S. Verbeek
Department of Immunology, University Hospital Utrecht, Utrecht, The Netherlands

References

1 De Vries, A. (1995) Nature 377, 169–173.
2 Nakane, H. et al. (1995) Nature 377, 165–168.
3 Miyauchi-Hashimoto, et al. (1996) J. Invest. Dermatol. 107, 343–348.
4 Berg, R.J.W. et al. (1997) Cancer Res. 57, 581–584.
5 De Vries, A. et al. (1997) Mol. Carcinog. 19, 46–53.

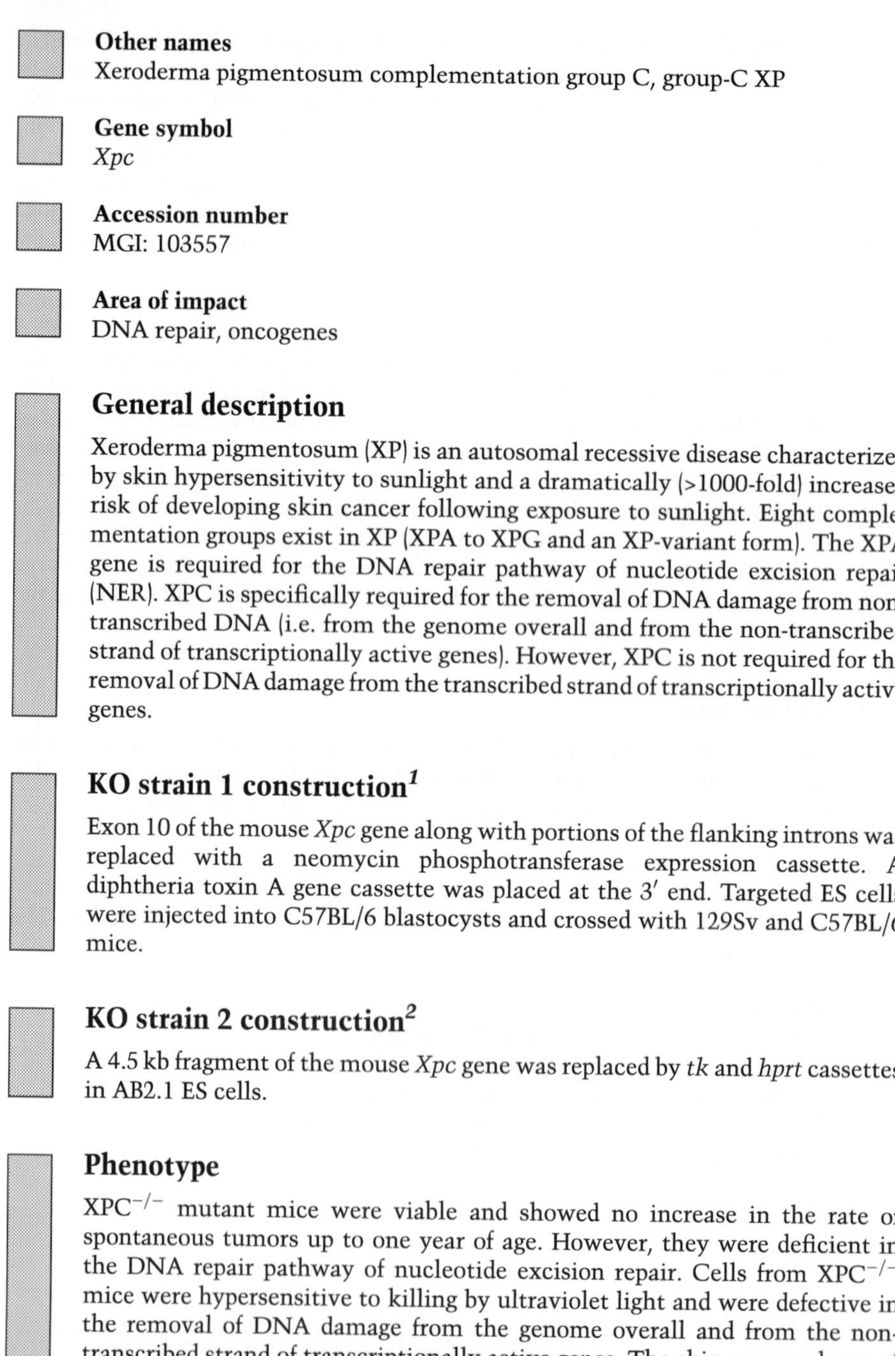

Other names
Xeroderma pigmentosum complementation group C, group-C XP

Gene symbol
Xpc

Accession number
MGI: 103557

Area of impact
DNA repair, oncogenes

General description

Xeroderma pigmentosum (XP) is an autosomal recessive disease characterized by skin hypersensitivity to sunlight and a dramatically (>1000-fold) increased risk of developing skin cancer following exposure to sunlight. Eight complementation groups exist in XP (XPA to XPG and an XP-variant form). The XPA gene is required for the DNA repair pathway of nucleotide excision repair (NER). XPC is specifically required for the removal of DNA damage from non-transcribed DNA (i.e. from the genome overall and from the non-transcribed strand of transcriptionally active genes). However, XPC is not required for the removal of DNA damage from the transcribed strand of transcriptionally active genes.

KO strain 1 construction[1]

Exon 10 of the mouse *Xpc* gene along with portions of the flanking introns was replaced with a neomycin phosphotransferase expression cassette. A diphtheria toxin A gene cassette was placed at the 3′ end. Targeted ES cells were injected into C57BL/6 blastocysts and crossed with 129Sv and C57BL/6 mice.

KO strain 2 construction[2]

A 4.5 kb fragment of the mouse *Xpc* gene was replaced by *tk* and *hprt* cassettes in AB2.1 ES cells.

Phenotype

$XPC^{-/-}$ mutant mice were viable and showed no increase in the rate of spontaneous tumors up to one year of age. However, they were deficient in the DNA repair pathway of nucleotide excision repair. Cells from $XPC^{-/-}$ mice were hypersensitive to killing by ultraviolet light and were defective in the removal of DNA damage from the genome overall and from the non-transcribed strand of transcriptionally active genes. The skin, eyes and ears of $XPC^{-/-}$ mice were hypersensitive to acute and chronic doses of ultraviolet light. $XPC^{-/-}$ mice were highly susceptible to ultraviolet light-induced skin carcinogenesis. The process of skin carcinogenesis was accelerated in

XPC/p53 double mutant mice following UV-B irradiation. XPC/p53 double mutant mice also displayed neural tube defects.

Comments

XPC mutant mice provide an important model of the human disease xeroderma pigmentosum. These mice will facilitate the study of nucleotide excision repair and the process of skin carcinogenesis.

Acknowledgements
David L. Cheo
The University of Texas Southwestern Medical Center, Dallas, TX, USA

References

[1] Cheo, D.L. et al. (1996) Curr. Biol. 6, 1691–1694.
[2] Sands, A.T. et al. (1995) Nature 377, 162–165.
[3] Cheo, D.L. et al. (1997) Mutation Res. 374, 1–9.

Gene symbol
Zap70

Accession number
MGI: 99613

Area of impact
Immunity and inflammation

General description

ZAP-70 is a non-Src protein tyrosine kinase of 70 kDa that is essential for T cell activation. ZAP-70 is highly homologous to the Syk protein tyrosine kinase. ZAP-70 is expressed in both NK cells and T cells, and plays a crucial role in intracellular signaling during T cell activation and T cell development. Following TCR stimulation, signal transduction is initiated by the ITAMs, which results in the phosphorylation of ITAM tyrosine residues. Phosphorylation of the ITAM motif in the CD3ζ subunit forms a binding site for ZAP-70. ZAP-70 itself is subsequently phosphorylated and has both positive and negative effects on antigen receptor signaling.

KO strain construction

ZAP-70-deficient mice were created through homologous recombination using a targeting vector that deleted almost the entire ZAP-70-coding region. The PGK*neo* expression cassette was used to replace an approx. 20 kb genomic DNA fragment that included all of the murine *Zap70* gene (isolated from a 129Sv library) except for 14 bp from the initiation codon. The vector contained 1.4 kb of 5′ homology and 8.0 kb of 3′ homology. The targeting vector was transfected into E14 ES cells. Positive clones were injected into C57BL/6 blastocysts to generate chimeras. Male chimeras were bred to C57BL/6 females and heterozygotes were intercrossed to obtain ZAP-70$^{-/-}$ mutants. The null mutation was confirmed by Western blot analysis of thymocyte proteins.

Phenotype

ZAP-70$^{-/-}$ mice appeared healthy and fertile under SPF conditions up until at least 10 months of age. In the spleen, splenocyte numbers were reduced by approx. 30%, no αβTCR^{+} T cells were present, and the proportion of B220^{+} cells was relatively increased. The medulla of the thymus was rudimentary in nature and the cortex was expanded. ZAP-70$^{-/-}$ mice had normal numbers of thymocytes but lacked both CD4^{+} and CD8^{+} single positive T cell populations; ZAP-70$^{-/-}$ thymocytes expressing αβTCR remained in the double positive state that is observed just prior to thymic selection. Crosslinking of the TCR with anti-CD3 did not lead to an increase in intracellular free calcium in ZAP-70$^{-/-}$ thymocytes. ZAP-70$^{-/-}$ thymocytes were resistant to deletion by peptide antigens. Reconstitution of mutant cells with human ZAP-70 was able to rescue both the phenotypic and functional defects in ZAP-70$^{-/-}$ thymocytes. ZAP-70$^{-/-}$ NK cells were normal in cytolytic and ADCC functions.

Comments

ZAP-70 is required for antigen-driven negative selection of thymocytes. TCR engagement of double positive thymocytes may activate ZAP-70 such that double positive to single positive maturation can progress. ZAP-70 is not required for B lymphocyte or for NK cell development or function.

Acknowledgements
Dennis Loh
Division of Rheumatology, Department of Medicine and Pathology, Washington University School of Medicine, St. Louis, MO, USA

Reference

1 Negishi, I. et al. (1995) Nature 376, 435–438.

Index

O